4th National Conference on Processing and Characterization of Materials

AA002837

IOP Conference Series: Materials Science and Engineering Volume 75

Rourkela, India
5 – 6 December 2014

ISBN: 978-1-5108-0081-6
ISSN: 1757-8981

Printed from e-media with permission by:

Curran Associates, Inc.
57 Morehouse Lane
Red Hook, NY 12571

Some format issues inherent in the e-media version may also appear in this print version.

Copyright© by the Institute of Physics
All rights reserved.

Printed by Curran Associates, Inc. (2015)

For permission requests, please contact the Institute of Physics
at the address below.

Institute of Physics
Dirac House, Temple Back
Bristol BS1 6BE UK

Phone: 44 1 17 929 7481
Fax: 44 1 17 920 0979

techtracking@iop.org

Additional copies of this publication are available from:

Curran Associates, Inc.
57 Morehouse Lane
Red Hook, NY 12571 USA
Phone: 845-758-0400
Fax: 845-758-2634
Email: curran@proceedings.com
Web: www.proceedings.com

TABLE OF CONTENTS

IOP Conf. Series: Materials Science and Engineering **75** (2015) 011001 doi:10.1088/1757-899X/75/1/011001

PREFACE

This volume contains selected full length technical papers amongst forty oral presentations made in the 4th National Conference on Processing and Characterization of Materials (NCPCM 2014), NIT Rourkela, Rourkela, Odisha, India, December 5 – 6, 2014.

The first conference of the NCPCM series was held at the same place in December 2011. Seeing the enthusiasm of the participants, it was decided to organize such conference in Rourkela every year. The basic idea was to establish a periodical national forum for multi–scale approaches in processing and characterization of materials in the eastern part of India. The conference NCPCM 2014 has successfully carried the tradition of previous conferences; more than fifty participants from twenty different organizations across India have registered. The conference was consisted of six technical sessions of about fifty contributory talks along with three keynote lectures. A metallography contest was also organized during the event. Out of these, thirty four best peer – reviewed contributions are published in this volume of *IOP Conference Series: Materials Science and Engineering*.

We would like to thank all the contributors, members of the organizing committee, session chairs as well as colleagues and students who helped with the preparation of the conference and, particularly, with the preparation of this volume. We convey our heartiest gratitude to the sponsors and advertisers for their contribution.

January 2015 Editorial Board, NCPCM 2014

Content from this work may be used under the terms of the Creative Commons Attribution 3.0 licence. Any further distribution of this work must maintain attribution to the author(s) and the title of the work, journal citation and DOI.
Published under licence by IOP Publishing Ltd

4th National Conference on Processing and Characterization of Materials
NCPCM 2014
5-6 December, 2014
Organized by
Department of Metallurgical and Materials Engineering
National Institute of Technology Rourkela
IIM
Indian Institute of Metals
Rourkela Chapter
Sponsors :
Council of Scientific and Industrial Research
OCL INDIA

4th National Conference on Processing and Characterization of Materials
NCPCM 2014
5-6 December, 2014
Organized by

IOP Conf. Series: Materials Science and Engineering **75** (2015) 011001 doi:10.1088/1757-899X/75/1/011001

Our Sponsors

- Council of Scientific and Industrial Research (CSIR), India
- NETZSCH Technologies India Pvt. Ltd., India
- OCL India Ltd., India
- SwamEquip, Chennai, India,
- Sarvesh Refractories Ltd., India
- Poly Refractories Pvt. Ltd., India
- Board of Research in Nuclear Sciences (BRNS), India

LIST OF COMMITTEES

Editorial Board

Prof. Anindya Basu, NIT Rourkela, India
Prof. Bankim Chandra Ray, NIT Rourkela, India
Prof. Partha Sarathi De, IIT Bhubaneswar, India
Prof. Kallol Mondal, IIT Kanpur, India
Prof. Manojit Ghosh, IIEST Shibpur, India

Organizing Committee (NIT Rourkela, India)

Patron:	Prof. Sunil Kumar Sarangi, Director
Chairman:	Prof. Subash Chandra Mishra, Head of the Department, Metallurgical and Materials Engineering
Convener:	Prof. Natraj Yedla
Co-convener(s):	Prof. S. N. Alam and Prof. S. Pal
Treasurer:	Prof. S. Pal

Technical Committee (NIT Rourkela, India)

Prof. Krishna Dutta
Prof. Debasis Chaira
Prof. Anindya Basu
Prof. Rajesh Prusty

IOP Conf. Series: Materials Science and Engineering **75** (2015) 011001 doi:10.1088/1757-899X/75/1/011001

Local Organizing Committee (NIT Rourkela, India)

Prof. A. K. Mondal

Prof. S. K. Sahoo

Prof. S. K. Karak

Mr. S. Pradhan

Mr. S. Hemberam

Mr. A. Pal

IOP Conf. Series: Materials Science and Engineering **75** (2015) 011002 doi:10.1088/1757-899X/75/1/011002

Peer review statement

All papers published in this volume of *IOP Conference Series: Materials Science and Engineering* have been peer reviewed through processes administered by the proceedings Editors. Reviews were conducted by expert referees to the professional and scientific standards expected of a proceedings journal published by IOP Publishing.

Content from this work may be used under the terms of the Creative Commons Attribution 3.0 licence. Any further distribution of this work must maintain attribution to the author(s) and the title of the work, journal citation and DOI.
Published under licence by IOP Publishing Ltd

A comparative assessment of crystallite size and lattice strain in differently cast A356 aluminium alloy

S K Mishra[1, 5], H Roy[2], A K Lohar[3], S K Samanta[3], S Tiwari[4] and K Dutta[1]
[1]Department of Metallurgical and Materials Engineering, National Institute of Technology, Rourkela –769008, India
[2]NDT & Metallurgy Group, CSIR–Central Mechanical Engineering Research Institute, Durgapur–713209, India
[3]Near Net Shape Manufacturing Technology Group, CSIR-Central Mechanical Engineering Research Institute, Durgapur–713209, India
[4]Materials Technology Department, Mahindra & Mahindra, Mahindra Research Valley, Chennai–603204, India

E-mail: srimantnitrkl@gmail.com

Abstract. In this investigation, A356 aluminium alloy has been prepared by different routes viz. gravity casting, rheo pressure die casting (RPDC) and RPDC with T6 heat treatment. X-ray diffraction studies of these samples have been done in the scanning range of 20 – 90°. X-ray peak broadening analysis has been used to estimate the crystallite size and lattice stain, in all the samples. The sample prepared by RPDC with T6 treatment has comparatively smaller crystallite size and lesser lattice strain than gravity cast and RPDC samples.

1. Introduction

A356 aluminium alloys (Al–7Si–Mg) are finding increased use in variety of structural, automotive, aerospace, and engineering applications due to its excellent cast-ability, corrosion resistance, mechanical properties and especially high strength to weight ratio [1-3]. However, the new technology for fabrication of this alloy and the methods of enhancing their properties are the subject of many investigations. Particle size and crystal morphology play important roles in various applications of this alloy, which have driven researchers to focus on the fabrication techniques, type of modification done and heat treatments etc. Analysis of X-ray diffraction peaks has become a very powerful tool for characterization of microstructure of materials either in bulk or in powder form. For the calculation of the crystallite size and the lattice strain form the broadening of X-ray peaks, the Williamson–Hall method was most commonly used.

No material has a perfect crystal structure due to their finite size causing a deviation from perfect crystallinity which leads to broadening of the X-ray diffraction peaks [4]. Broadening of x-ray diffraction peaks is easily apparent in patterns obtained with a diffractometer, and this information can be directly quantified. However, it is important to realize that broadening of diffraction peaks arises mainly due to three factors viz. instrumental effects, crystallite size and lattice strain. The broadening

[5] **To whom any correspondence should be addressed.**

Content from this work may be used under the terms of the Creative Commons Attribution 3.0 licence. Any further distribution of this work must maintain attribution to the author(s) and the title of the work, journal citation and DOI.
Published under licence by IOP Publishing Ltd

is evaluated by measuring the angular width "B", in radians, at intensity equal to half the maximum intensity (FWHM). Subtracting the instrumental effect from the obtained peak broadening, two main properties; crystallite size and lattice strain are extracted from peak width analysis. Crystallite size and lattice strain affect the Bragg peak by increasing the peak width and intensity and shift the 2θ peak position accordingly [4]. The crystallite size varies as $1/cos\theta$ and stain varies as $tan\theta$ from the peak width. The size and strain effects on peak broadening are known from the above difference of 2θ [5]. W-H analysis is an integral breadth method. Size-induced and strain-induced broadenings are known by considering the peak width as a function of 2θ [6].
In this study, the crystallite size and lattice strain are estimated by using a well known Williamson and Hall method for A356 aluminium alloy fabricated by gravity cast and rheo pressure die cast technique. Also rheo pressure die cast samples with T6 heat treatment was taken into consideration in this analysis.

2. Experimental details

2.1. Material and its fabrication technique
The material selected for this investigation is A356 aluminium alloy. The alloy is prepared by two different casting techniques i.e. gravity casting and rheo pressure die casting (RPDC). The dimension of each cast was 15×15×150 mm^3. The method of gravity casting is similar to popularly used techniques available for it; a method is described by Arunkumar et al. [7]. Rheo pressure die casting is one of the semi solid casting techniques. In this process, aluminium alloy is melted in a melting furnace which is then allowed to pass through a cooling slope to obtain semi-solid slurry (solids fraction varies from 12-20%) of the alloy. This semi-solid slurry is collected in a holding furnace which stores the slurry iso-thermally (slurry temperature at the exit of the cooling slope). Then, some quantity of slurry is taken from holding furnace by the ladling unit and poured in the sleeve of the pressure die casting machine (350T - model DC 350J-MS). Temperature of about 300°C was maintained in mould halves and sleeve through circulating hot oil to avoid premature solidification. By applying a pressure the molten alloy is forced into the mould halves and allowed for solidification. A part of the RPDC samples are subjected to T6 heat treatment process. In this, the samples are solution treated at 540°C for 30 minute followed by a water quench (25°C); then artificially aged at 180°C for 6 hrs followed by water quenching.

2.2. Microstructure and X-ray diffraction
The optical micrographs of mechanically polished samples are examined using ZEISS optical microscope at different magnification levels. To estimate the crystallite size, lattice strain etc. X-ray diffraction studies of the sample are done using an X-Ray Diffraction system (Make/Model: RIGAKU JAPAN/ULTIMA-IV) with scanning rate of 5° per minute and step size of 0.05 in the scanning range of 20-90°. Cu Kα radiation is used for this.

3. Results and discussion

3.1. Microstructural analysis
The optical microstructures of the samples of gravity cast, RPDC and RPDC with T6 treatment are illustrated in figure 1(a) to 1(c) respectively. In all the images, primary alpha aluminium is present which is surrounded by eutectic silicon. In case of gravity cast sample, typical elongated dendritic shape of primary alpha phases are distributed randomly throughout the sample, whereas in case of RPDC sample, nearly equi-axed primary alpha phase was observed which are distributed uniformly throughout the matrix. Such variation in the microstructure by varying casting technique was also reported by Tahamtan et al. [8]. The microstructure of the RPDC with T6 is indeed different from RPDC as the eutectic silicon is getting rounded of due to T6 treatment. As there is an irregular shape

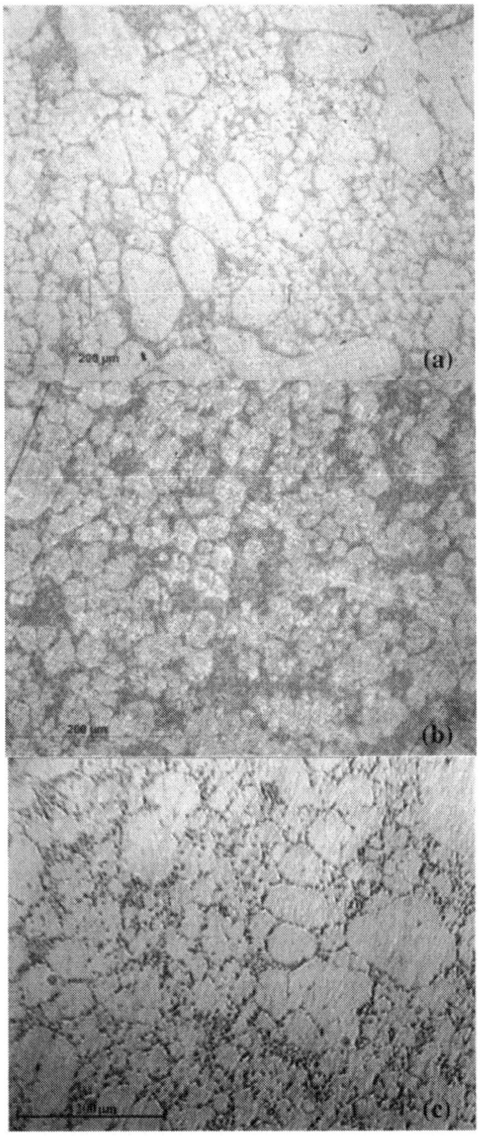

Figure 1. Optical micrographs of A356 aluminium alloy: (a) gravity cast (b) RPDC and (c) RPDC + T6 treatment.

of primary alpha phase is present in the gravity cast sample, the aspect ratio of the phase was calculated and presented in Table 1 for all the samples. It can be noted from the table that aspect ratio of primary alpha phase decreases from gravity cast to rheo cast sample. After heat treatment of RPDC samples, no much variation in aspect ratio was found.

3.2. Estimation of crystallite size and lattice strain

In order to determine crystallite size and lattice strain of the casts, X-ray diffraction profile analyses was done for all the casts. Intensity vs. 2θ plots of the investigated samples are presented in figure 2. This figure illustrates all the detectable peaks with their respective hkl values for the investigated samples. All these detectable peaks could be indexed as cubic structure found in the standard reference data (for Al: JCPDS 01–1180 and for Si: JCPDS 03–0544). Similar peaks for Al and Si at approximately same 2θ values are also reported by Chen and Alpas [9]. One can note from these figures that there is little shifting of peaks in RPDC sample towards lower 2θ value from the gravity cast sample. In contrast, the T6 heat treated RPDC sample shows peak shifting towards higher 2θ value as those of the RPDC sample. The obtained peaks of Mg were of very low intensities; thus not presented in the given X–ray diffraction plots for all three conditions; significantly low wt.% of Mg in all the casts may be the reason for not getting these peaks.

Table 1. Quantitative metallographic results of investigated alloy

Sample	Aspect ratio
A356 Al alloy: Gravity cast	1.38 ± 0.50
A356 Al alloy: RPDC	0.98 ± 0.28
A356 Al alloy: RPDC + T6	1.08 ± 0.18

The crystallite size and lattice strain can be calculated from the broadening of X-ray diffraction peaks by considering the full width at half maximum (FWHM) of the all individual peaks. If the observed X-ray peak has broadening of width B_o and the width due to instrumental effect is B_i, then the remaining broadening of the peak due to crystallite size and lattice strain is B_r, which can be expressed considering Gaussian profile as [6]:

$$B_r^2 = B_O^2 - B_i^2 \tag{1}$$

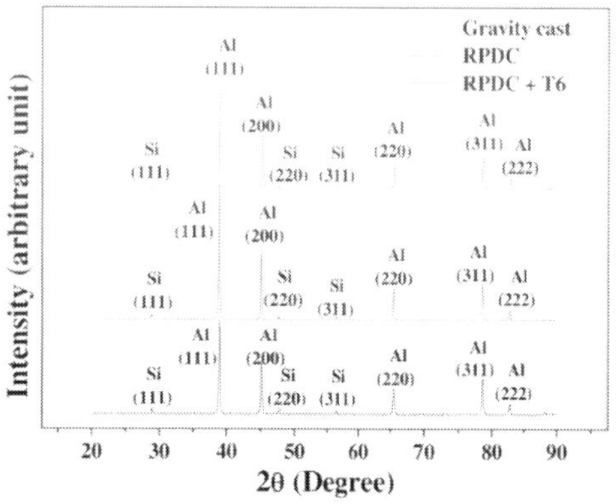

Figure 2: XRD spectrum of the investigated A356 aluminium alloy.

Many methods are available in the literature to estimate crystallite size and lattice strain. Out of those methods, Williamson and Hall [10] is one that is most commonly used. According to this, the particle size and strain contribution to the line broadening are independent to each other and both have a Cauchy-like profile; the observed line breadth is simply the sum of the breadth due to crystallite size and lattice strain [11].

$$B_{hkl} = B_{crystallite} + B_{strain} \tag{2}$$

The well known Scherrer equation [12] for determination of crystallite size from broadening of X-ray diffraction peaks is expressed as $B_{crystalline} = (k\lambda/L\cos\theta)$ where λ is the wavelength of the X-rays used, θ is the Bragg angle, L is the average crystallite size measured in a direction perpendicular to the surface of the specimen, and k is a constant (k = 0.94 for small cubic crystal). The broadening due to lattice strain in the material can be represented by the relationship $B_{strain} = 4\eta\tan\theta$, where η is the strain in the material. Hence we can write that

$$B_{hkl} = \frac{k\lambda}{L\cos\theta} + 4\eta\tan\theta \tag{3}$$

Rearranging equation (3) we get

$$B_{hkl}\cos\theta = \frac{k\lambda}{L} + 4\eta\sin\theta \tag{4}$$

Williamson and Hall plots for different samples are shown in figure 3 (a) to (c). A linear fit of the scattered results is taken into consideration. From this linear fit, the crystallite size was estimated from the y-intercept, and the strain, from the slope of the fit.

The crystallite size and lattice stain is calculated by comparing the linear fit equations with equation (6) for three different samples are presented in Table 2. One can note from this table that the lattice strain developed in A356 alloy prepared by RPDC technique is more than the gravity cast technique. This is attributed to the application of pressure in RPDC samples which may cause some lattice structure variation; resulting increased lattice strain in the material. Lattice strain is a measure of the distribution of lattice constants, such as lattice dislocations, which arise from crystal imperfections [4]. In order to minimize this strain, the rheo cast samples are subjected to T6 heat treatment. The heat treated samples has comparatively smaller crystallite size and lesser lattice strain than gravity cast and rheo cast samples.

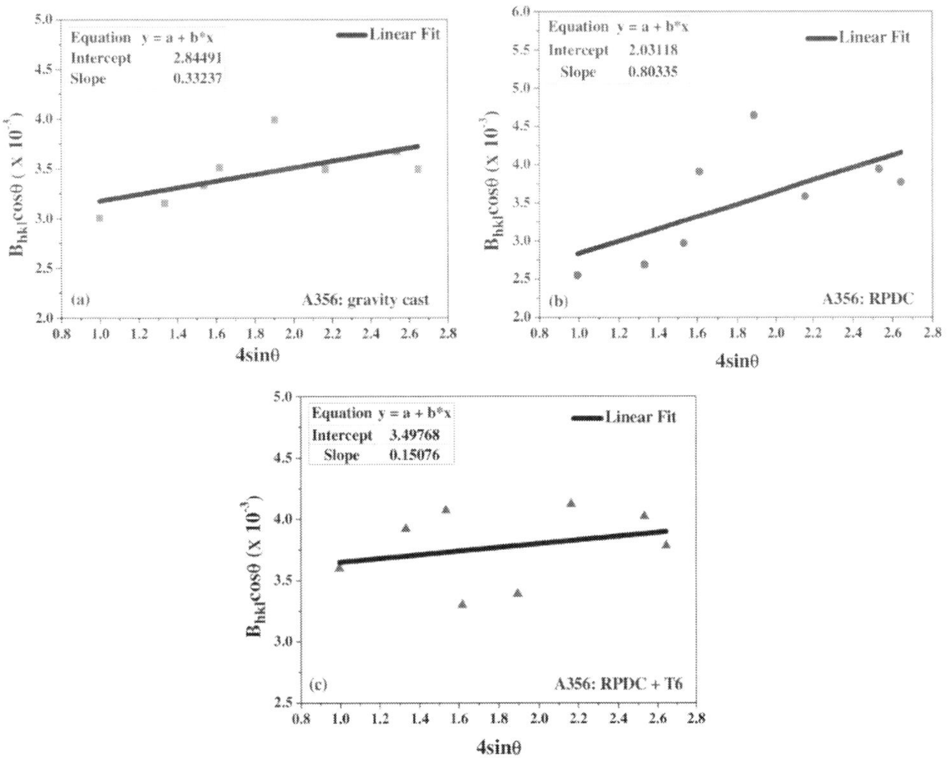

Figure 3. Williamson and Hall plots for A356 aluminium alloy: (a) gravity cast, (b) rheo pressure die cast and (c) rheo pressure die cast with T6 heat treatment.

Table 2: Crystallite size and lattice strain of different samples of A356 aluminium alloy.

Samples	Williamson and Hall method	
	Crystallite size (L) in nm	Lattice strain (X 10^{-3})
A356: Gravity cast	50.88	0.33
A356: RPDC	71.27	0.80
A356: RPDC + T6	41.38	0.15

4. Conclusions

Microstructure and X-ray diffraction have been studied for the samples of A356 aluminium alloy fabricated by gravity cast, rheo cast and rheo cast with T6 treatment. The following conclusions can be resolved from this investigation:

1. The micrographs show that, in gravity cast samples, the primary alpha phase have elongated shapes with some dendrite shapes found randomly. The microstructure of RPDC with T6 is indeed different from RPDC; as the eutectic silicon is getting rounded of due to T6 treatment. However, the aspect ratio of the primary alpha in RPDC and RPDC with T6 don't have significant variation.

2. The broadening due to crystallite size and lattice strain was analyzed by Williamson and Hall method. Both crystallite size and lattice strain developed in the sample prepared by RPDC are more compared to the sample prepared by gravity cast. However, with application of T6 heat treatment after RPDC, samples have comparatively smaller crystallite size and lesser lattice strain than gravity cast and RPDC samples.

References

[1] Sharma S R, Ma Z Y and Mishra R S 2004 *Scripta Materialica* **51** 237–241.
[2] Li J, Xie J, Wang W and Wei S 2007 *Materials Science Forum* **561-565** 147–150.
[3] Ammar H R, Samuel A M and Samuel F H 2008 *Materials Science and Engineering A* **473** 65–75.
[4] Zak A K, Abd. Majid W H, Abrishami M E and Yousefi R 2011 *Solid State Sciences* **13** 251–256.
[5] Prabhu Y T, Rao K V, Sai Kumar V S and Kumari B S 2014 *World Journal of Nano Science and Engineering* **4** 21–28.
[6] Suranarayana C and Norton M G 1998 *X-ray Diffraction: A Practical Approach* Springer Science+Business Media New York.
[7] Arunkumar S, Sreenivas Rao K V and Prasanna Kumar T S 2008, *International Journal of Heat and Mass Transfer* **51** 2676–2685.
[8] Tahamtan S, Boostani A F and Nazemi H 2009 *Journal of Alloys and Compounds* **468** 107–114.
[9] Chen H and Alpas A T 1996 *Wear* **192** 186–198.
[10] Williamson G K and Hall W H 1953 *Acta Metall1urgica* **1** 22–31.
[11] Mote V D, Purushotham Y and Dole B N 2012 *Journal of Theoretical and Applied Physics* **6:6** 1–8.
[12] Cullity B D 1956, Elements of X-ray Diffraction. Addison-Wesley Publishing Company Inc. California.

IOP Conf. Series: Materials Science and Engineering **75** (2015) 012002 doi:10.1088/1757-899X/75/1/012002

A comparative study of the mechanical performance of Glass and Glass/Carbon hybrid polymer composites at different temperature environments

M J Shukla*, D S Kumar, K K Mahato, D K Rathore, R K Prusty and B C Ray

Composite Materials Group, Department of Metallurgical and Materials Engineering, National Institute of Technology, Rourkela-769008

*Email: shuklameet25@gmail.com

Abstract. Glass Fiber Reinforced Polymer (GFRP) composites have been widely accepted as high strength, low weight structural material as compared to their metallic counterparts. Some specific advanced high performance applications such as aerospace components still require superior specific strength and specific modulus. Carbon Fiber Reinforced Polymer (CFRP) composites exhibit superior specific strength and modulus but have a lower failure strain and high cost. Hence, the combination of both glass and carbon fiber in polymer composite may yield optimized mechanical properties. Further the in-service environment has a significant role on the mechanical performance of this class of materials. Present study aims to investigate the mechanical property of GFRP and Glass/Carbon (G/C hybrid) composites at room temperature, in-situ and ex-situ temperature conditions. In-situ testing at +70°C and +100°C results in significant loss in inter-laminar shear strength (ILSS) for both the composites as compared to room temperature. The ILSS was nearly equal for both the composite systems tested in-situ at +100°C and effect of fiber hybridisation was completely diminished there. At low temperature ex-situ conditioning significant reduction in ILSS was observed for both the systems. Further at -60°C G/C hybrid exhibited 32.4 % higher ILSS than GFRP. Hence this makes G/C hybrid a better choice of material in low temperature environmental applications.

1. Introduction

Fiber reinforced composites are one of the most commonly used structural materials in today's world. The most promising properties include high strength to weight ratio, high stiffness, low density, high corrosion, high endurance limit etc. Composites find various applications in aerospace industry, automobile industry, sports and goods, marine equipment etc. In these applications, during service life composites are exposed to various environmental conditions. Some harsh environmental conditions increase the vulnerability of degradation in mechanical properties. The properties of the fiber reinforced composite are governed by the individual property and behaviour of matrix, fibers and fiber/matrix interface in a particular environment. Due to their wide use in structural application the composites are commonly exposed to high temperature. During high temperature conditioning noticeable physical and chemical changes occur in the polymer network. Disintegration of short range ordering and thermo-oxidative degradation are observed at elevated temperature. It is inferred that these changes influence the performance of composite at high temperature. Due to difference in thermal expansion co-efficient of the matrix and reinforcement, thermal stresses get induced in the composite which can result in formation of micro-cracks at the interface. The interface hence becomes more susceptible to hostile reactions at elevated temperature environment [1]. The common failure modes at high temperature include fiber fracture, matrix cracking and fiber/matrix debonding. On exposing the unidirectional graphite composite at elevated temperature it was observed that the ILSS decreased by 30 % [2] Investigation on CFRP composite at -196°C for 555 hrs showed 20 % degradation in tensile strength compared to that at room temperature [3]. Due to increasing application of Fiber Reinforced Polymer (FRP) composites in aerospace vehicles for cryogenic fuel tanks, it is necessary to understand their behaviour at cryogenic temperature. The most commonly recorded damage mechanisms at cryogenic temperature are localised surface degradation,

Content from this work may be used under the terms of the Creative Commons Attribution 3.0 licence. Any further distribution of this work must maintain attribution to the author(s) and the title of the work, journal citation and DOI.
Published under licence by IOP Publishing Ltd

delamination and generation of micro-cracks in polymer matrix and/or at the fiber/matrix interface [4] [5]. These failure mechanisms start degrading the properties of the composite at microscopic level which further extends to a noticeable proportion. A recent development in the fiber reinforced composite is the hybrid composite. This composite can have two or more types of reinforcements or matrix hybridisation with some nanoparticle. Hybrids are conceptually fabricated to take advantage of all the constituent reinforcements. As glass fibers have low specific strength and specific modulus, some volume fraction of carbon fibers (having high specific strength and modulus) should be incorporated in GFRP composite to enhance the mechanical properties and result in a better composite compared to GFRP. But due to low strain to failure and high cost of carbon fibers a balance should be maintained between the cost and performance of the resulting composite. Mechanical properties of E-glass fiber, carbon fiber and epoxy are shown in table 1.

Table 1. Mechanical Properties of E-glass fiber, carbon fiber and epoxy.

Material	Tensile strength (MPa)	Young's Modulus (GPa)	Density (g/cm3)	Thermal Exp. Co-efficient (mm/°C)
E-glass	2000	80	2.58	5.4
Carbon	2900	525	1.85	2
Epoxy	85	3.5	1.2	1.5-10

The present investigation deals with comparing the mechanical property of high and low temperature conditioned GFRP and hybrid with the unconditioned samples at room temperature. Also comparison is made between the performance of GFRP and glass/carbon hybrid.

2. Materials and experimental methods

2.1 Materials

Present investigation includes fabrication of two types of composite systems i.e. Glass fiber reinforced polymer composite (GFRP) and Glass/carbon fiber reinforced hybrid composite (G/C hybrid) Diglycidyl Ether of Bisphenol A (DGEBA) based epoxy resin was used as matrix and triethylene tetra amine (TETA) was used as hardener. Glass fibers and carbon fibers used in the study were manufactured by Saint Gobain and Toray industries, Japan, respectively. Epoxy resin and hardener were procured from Atul Industries Ltd, Gujarat under the trade name of Lapox L-12 and K-6 respectively. The epoxy to hardener ratio was taken as 10:1 as per the supplier's prescribed standard. GFRP composite laminate was fabricated with 14 layers of woven glass fibers fabric (40 vol. % of reinforcement) by hand layup method. Similarly Glass/carbon (G/C) hybrid contained 5 layers of woven fabric glass fiber (20 vol. %) and 4 layers of woven fabric carbon fiber (21 vol. %) with alternate sequence, keeping glass fiber at both ends by hand layup method. The laminates were then cured in a compression moulding press at 60°C temperature for 20 minutes at pressure of 1 MPa. After removing the laminates, they were dried in an oven at 50 °C for 3 hrs to remove moisture and other volatile entities. Samples were cut using diamond cutter as per the required dimensions as shown in figure 1(a) and 1(b). Samples were then post cured in an oven at 140°C for 6 hrs.

2.2 Experimental methods

2.2.1 Short Beam Shear (SBS) test. The test was performed with 3 point bending fixture in Instron 5967 UTM (shown in figure 1(c)) using ASTM D2344-84 standard specimens to evaluate the apparent ILSS. The specimens were tested at room temperature condition, high temperature and low temperature. In-situ high temperature testing of samples was done in the environmental chamber of Instron 5967 at +70°C and +100°C with the equilibration time of 10 minutes. Ex-situ high temperature conditioning of samples was done in an oven at +70°C and +100°C for 60 hrs. The samples were allowed to cool in an oven till the temperature comes down to room temperature and then removed out for testing. Ex-situ low temperature conditioning of samples was done at 0°C and -60°C in an ultra-low chamber for 48 hrs. The specimens were removed, brought to room temperature and then tested. Some samples were also conditioned in a liquid nitrogen bath for 8hrs and 24 hrs. The samples were drained and tested immediately. All testing was conducted with loading rate of 1 mm/min.

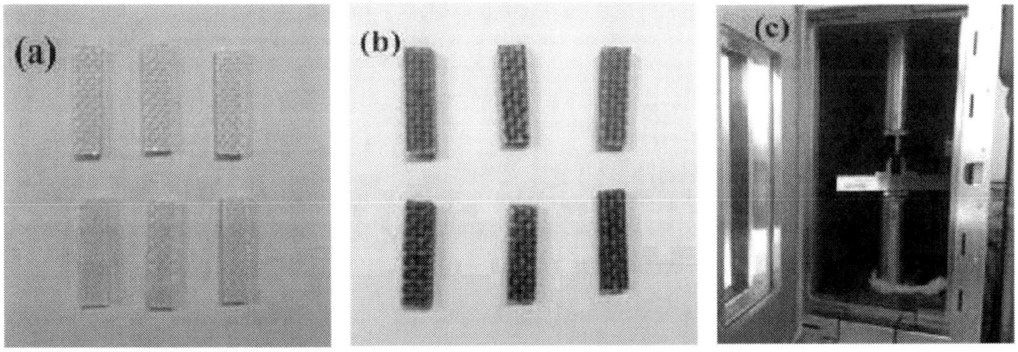

Figure 1. (a) GFRP standard specimens, (b) G/C hybrid specimens and (c) Environmental chamber of Instron 5967 UTM with 3 point bend fixture.

2.2.2 Scanning Electron Microscopy analysis. The fractured surfaces of the tested samples were further analysed for determining the failure mechanisms using JEOL-JSM 6480 LVSEM at 20KV.

3. Results and discussions

3.1 In-situ high temperature conditioning

The samples were tested at +70°C and +100°C with the equilibration time of 10 minutes. The ILSS obtained was compared with that at room temperature.

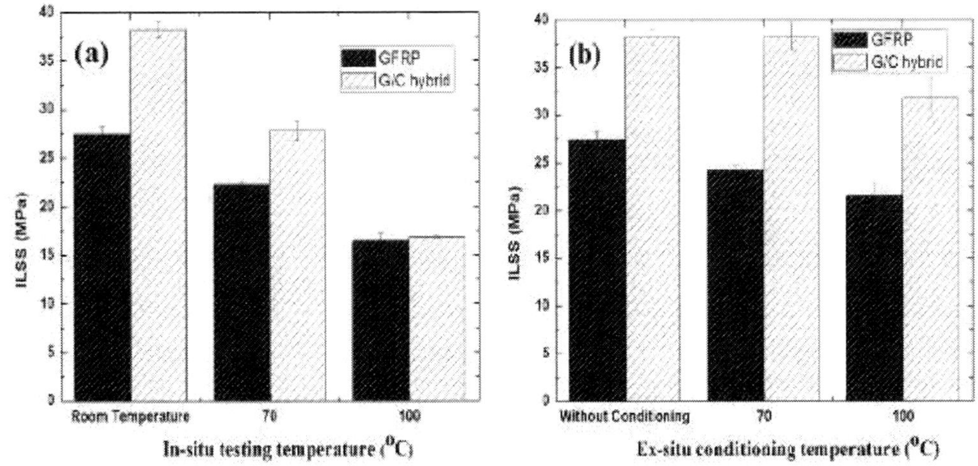

Figure 2. ILSS of GFRP and G/C hybrid samples (a) tested at +70°C and +100°C with 10 min equilibration time and (b) conditioned at +70°C and +100 °C for 60 hrs, then tested at room temperature.

It is observed from figure 2(a) that the ILSS of GFRP was decreased at +70°C and +100°C by 18.9 % and 40 % respectively as compared to that at room temperature. Similarly for glass/carbon hybrid the decrease was 27 % and 56 % at +70°C and +100°C respectively. At elevated temperatures thermal stresses are induced due to differential thermal expansion of fiber and matrix. These stresses get released on the cost of micro-cracks in the matrix and at the fiber/matrix interface. At low loading rate as used here, these micro cracks get time for coalescence and turn into potentially enough crack to cause reduction in mechanical properties of the composite. It is further observed that ILSS of hybrid is 28.2 % higher over GFRP at room temperature and there is not much difference in ILSS of both the systems at +100°C. The high ILSS in the hybrid samples w.r.t GFRP may be drawn from the presence of carbon fibers which inherently exhibit a higher strength and better interfacial bonding with epoxy. At temperature close to the glass transition of the polymer (Tg is around +120 °C for this polymer), i.e. at +100 °C the mechanical response obtained is mostly matrix dominated. At this stage, the state of the matrix is not capable for effective load transfer. Hence, irrespective of the fiber system the ILSS obtained was nearly same.

3.2 Ex-situ high temperature conditioning

The samples were conditioned at +70°C and +100°C for 60 hrs. ILSS of the conditioned samples was compared with that of samples tested at room temperature. It is observed from figure 2(b) that the ILSS of GFRP decreased at +70°C and +100°C by 11.8 % and 21.5 % respectively as compared to that at room temperature. Similarly for glass/carbon hybrid the decrease was 0.05 % and 17 % at +70°C and +100°C respectively. The quality of interface is poor in GFRP due to weak glass/epoxy adhesion as compared to strong carbon/epoxy adhesion in hybrid. At high temperature the glass/epoxy interface further degrades at a faster rate and acts as stress raisers. These stress raisers ultimately transform into potential micro cracks after certain exposure of time.

3.3 Ex-situ low temperature conditioning

The samples were conditioned at 0°C and -60°C for 48 hrs. ILSS of conditioned samples was compared with that of ILSS at room temperature.

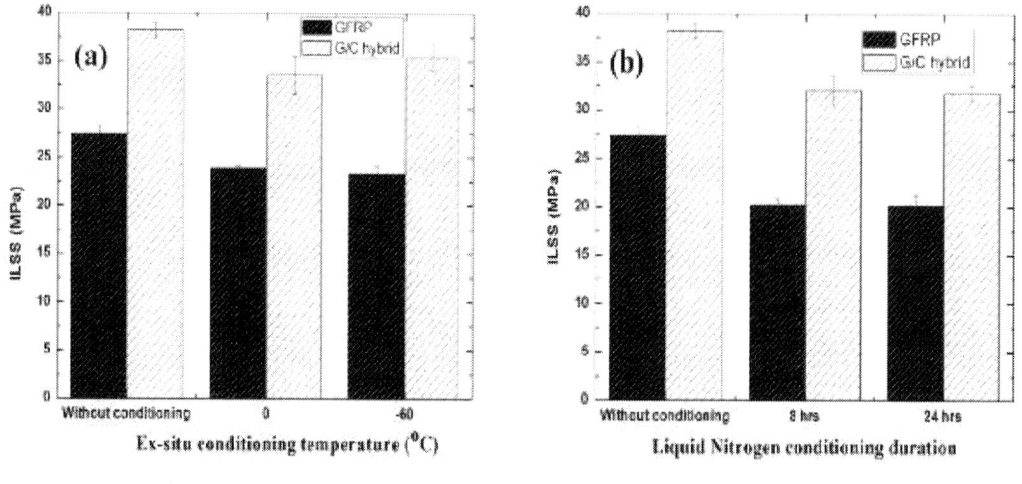

Figure 3. ILSS of GFRP and G/C hybrid samples conditioned (a) at 0°C and -60°C for 48 hrs and (b) in liquid nitrogen for 8 hrs and 24 hrs.

It can be seen from figure 3(a) that in case of GFRP there is reduction in ILSS at 0°C and -60°C by 13 % and 15 % as compared to that at room temperature. Similarly in case of hybrid the decrease at 0°C and -60°C was 12.4 % and 7.5 % with respect to ILSS at room temperature. Under low temperature exposure, polymer matrix becomes too hard and brittle to yield during load transfer from matrix to fibers. Hence this gives rise to formation of micro-cracks at the interface (being a weak region). Thus ILSS decreases. At -60°C the ILSS of hybrid is 34.2 % more than that of GFRP. Hence it can be noted that at cryogenic temperature the performance of hybrid is better than GFRP.

3.4 Liquid Nitrogen conditioning

The samples were kept in liquid nitrogen i.e. at -196°C for 8 hrs and 24 hrs. ILSS of the conditioned samples was compared with that of samples tested at room temperature. It can be seen from figure 3(b) that in case of GFRP there is reduction in ILSS for 8 hrs and 24 hrs by 26.2 % and 26.5 % as compared to that at room temperature. Similarly in case of hybrid the decrease for 8 hrs and 24 hrs was 16.2 % and 17 % with respect to ILSS at room temperature. It should be noticed that for both GFRP and hybrid samples there is a very significant reduction in ILSS for 8 hrs in comparison to ILSS at room temperature, which is then getting saturated (not much difference in ILSS for 8hrs and 24 hrs). The reason may be again attributed to micro crack density formed due to thermal stresses. When the samples were exposed to liquid nitrogen, micro cracks were immediately formed in the material. This micro crack density then gets saturated and no further decrement in ILSS takes place with further exposure time.

3.5 Fractography analysis

After 3 point bend test, fractured surfaces of samples were analysed under Scanning Electron Microscope (SEM) to study various failure mechanisms. Figure 4(a) shows the fracture surface of GFRP tested at room temperature. Initiation of failure could be fiber/matrix debonding at the interface, which upon further loading results in detachment of fibers from the matrix. Extensive fiber imprints on polymer matrix and attachment of matrix ligands on fibers were observed. Figure 4(b) shows fracture surface of G/C hybrid composite. The surface shown here represents matrix region near bundle of carbon fibers. Cohesive failure of matrix is clearly evident and can be attributed for higher interfacial strength of carbon fibers with epoxy resin. Figure 4(c) shows fracture surface of G/C hybrid composite tested at +70°C with 10 minutes equilibration time. Bare fibers with little polymer adherent on their surfaces is visible. This indicates that at high temperature failure is mainly due to loss of integrity of composite because of separation of matrix from fiber surface. Figure 4(d) shows fracture surface of liquid nitrogen conditioned GFRP sample. At this temperature, larger debonded interfaces in terms of delaminated surfaces were observed.

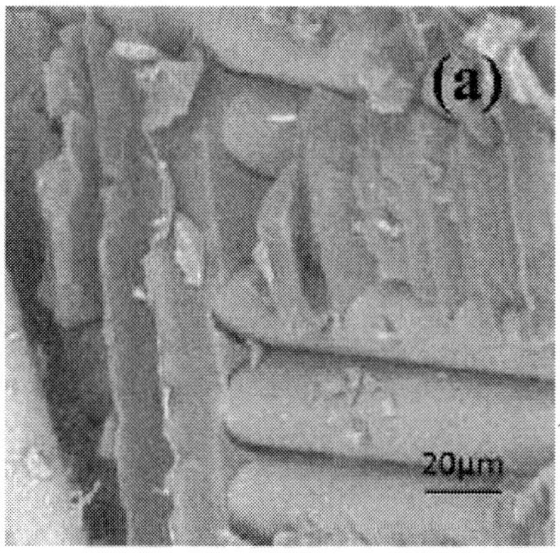

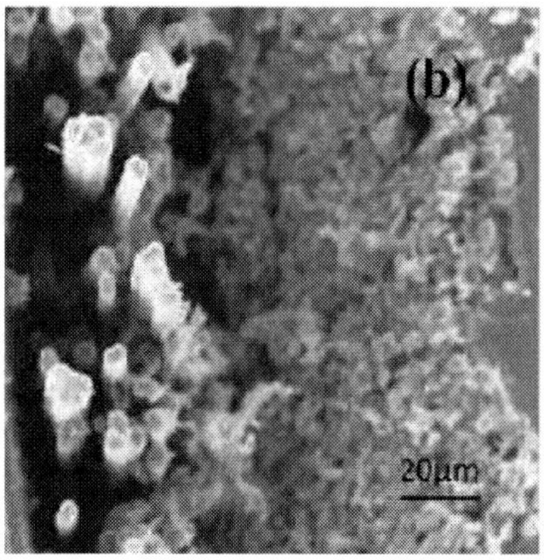

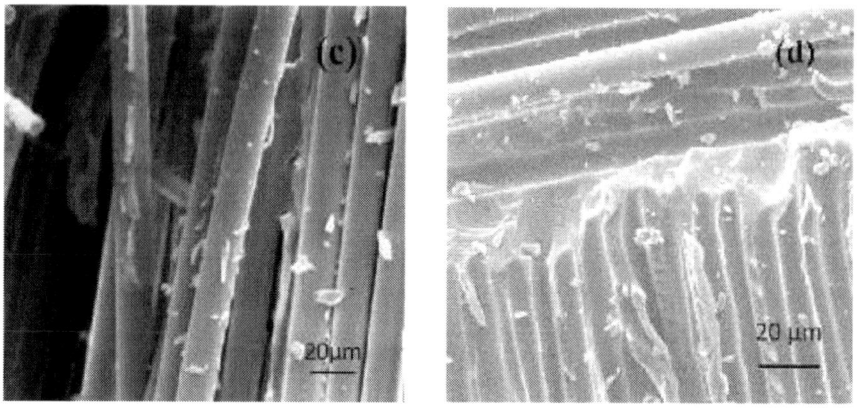

Figure 4. Scanning electron micrographs of fracture surfaces of (a) GFRP tested at room temperature, (b) G/C hybrid composite tested at room temperature, (c) G/C hybrid composite tested for in-situ at +70°C with 10 min equilibration time and (d) Liquid nitrogen conditioned GFRP for 24 hrs.

4. Conclusions

The influence of room temperature, above room temperature and below room temperature on interfacial mechanical property was studied for GFRP and G/C hybrid. The following conclusions can be drawn from the present investigation.

- At room temperature condition though the ILSS of G/C hybrid is more than that of GFRP by 28.2 %, yet the rate of degradation in ILSS for G/C hybrid is more than that of GFRP at high temperature. At +100°C the ILSS of both GFRP and G/C hybrid is found to be nearly equal due to negligible contribution from fiber at elevated temperature and the interfacial property of both the composite materials is governed by the matrix.
- While in-situ conditioning due to attainment of soft glassy state in matrix, stress transmittance through interface becomes less efficient, hence low ILSS is obtained. In case of ex-situ testing although spending more time at high temperature, the material regains its original properties upon cooling to room temperature. Loss in ILSS may now be attributed to the irreversible changes like formation, coagulation and growth of micro cracks at high temperature due to differential thermal changes between fiber and matrix.
- Decrease in ILSS is more in case of low temperature conditioning of GFRP than G/C hybrid. The ILSS of hybrid is 34.2 % (in contrast to 28.2 % without conditioning) more than that of GFRP when conditioned at -60°C for 48hrs.
- Liquid nitrogen conditioning for 8 hrs showed decrease in ILSS of GFRP and G/C hybrid by 26.2 % and 16.2 % respectively. Further conditioning for 24 hrs showed negligible change in ILSS due to saturation of density of micro-cracks with long exposure time. Hence it can be noted that at low temperature the performance of hybrid is better than GFRP.

References

[1] Ray B C 2004 Thermal shock on interfacial adhesion of thermally conditioned glass fiber/epoxy composites *Mater. Lett.* **58** 2175–7

[2] Sethi S, Rathore D K and Ray B C 2015 Effects of temperature and loading speed on interface-dominated strength in fibre/polymer composites: An evaluation for in-situ environment *Mater. Des.* **65** 617–26

[3] Bechel V T, Camping J D and Kim R Y 2005 Cryogenic/elevated temperature cycling induced leakage paths in PMCs *Compos. Part B Eng.* **36** 171–82

[4] Gong M, Wang X F and Zhao J H 2007 Experimental study on mechanical behavior of laminates at low temperature *Cryogenics* **47** 1–7

[5] Mahato K K, Shukla M J, Kumar D S and Ray B C 2014 In- service Performance of Fiber Reinforced Polymer Composite in Different Environmental Conditions: A Review *J. Adv. Res. Manuf. Mater. Sci. Metall. Eng.* **1** 55–88

Adhesive Wear Behavior of Heat Treated Spheroidal Graphite Cast Iron

M. SALIM, S.SHAMA[1], Y.H.MOZUMDER, R.K.BEHERA, SINDHOORA L.P, A.PATTANAIK, S.C.MISHRA, S.SEN
Dept. of Metallurgical and Materials Engineering, NIT Rourkela, India-769008

[1]Corresponding author e-mail: shama.shaik2501@gmail.com

Abstract. Spheroidal graphite cast iron is one of the most favorable materials in cast iron family due to its wide range of mechanical and tribological properties. In the current research priority is given towards the investigation of wear system response of spheroidal graphite cast iron subjected to various heat treatment conditions. Pearlitic/ferritic and upper bainitic matrix was obtained through normalizing and austempering treatment for an austenitizing temperature of 1000°C. Dry sliding wear test was performed at 10N, 20N, 30N for a sliding distance of 7.54m. With increase in load weight loss was observed for as-cast specimen whereas normalized specimen showed very less weight loss with increasing load. On the other hand the austempered specimen observed to lose weight when operated at 10N, whereas at 20N and 30N drastic gain in weight was observed.

Keywords: SG cast iron, heat treatment, dry sliding wear, microstructure, hardness

I. Introduction

Spheroidal graphite cast iron has gained its reputation in structural, automotive, agriculture and many other industries due to its good mechanical and tribological properties. Although it posses good mechanical properties in as-cast condition for any desired application, they can be improved through proper heat treatment application. Austempered ductile iron (ADI) has excellent mechanical properties such as high strength [1,2] toughness and good wear resistance [3,4] in addition to excellent fatigue strength [5,6] and good fracture toughness [7,8].

Tribological investigation carried out in past are mainly focused on austempered ductile iron subjected to various austempering time and temperature. Boutorabi et.al [9] performed dry sliding wear test on SG iron austempered at different temperature and time. The increase in austempering temperature and time leads to decomposition of carbon into the austenite surrounding and eliminates the formation of martensite and carbide which decreases the hardness and increase the wear rate. Haseeb et.al [10] studied the wear system response of austempered and quenched and tempered ductile iron with same hardness level. The bainitic ferrite matrix was more resistant to dry sliding wear at higher loads and the hardness below the wear surface was also increased. The wear rate in both the specimen was also found to be increasing with increase in sliding distance. It was also reported in their result that one of the major constituent of the austempered iron, i.e., bainitic ferrite, which is less prone to thermal instability than martensite, might undergo strain hardening during wear resulting in an increased hardness. Tempering treatment on ductile iron with boron increased the wear resistance but with increasing boriding time wear rate of boro-tempered ductile iron decreased [11]. Not only alloying element or heat treatment but also graphite nodule size and distribution affect the wear

Content from this work may be used under the terms of the Creative Commons Attribution 3.0 licence. Any further distribution of this work must maintain attribution to the author(s) and the title of the work, journal citation and DOI.
Published under licence by IOP Publishing Ltd

properties of SG iron. According to a study by Sugishita and Fujiyoshi [12] and Zimba et.al [13] presence of large size graphite nodules reduces the wear rate by acting as lubricating agent.
The past results reported are mainly leaned towards austempered ductile iron and the studies were focused on the effect of austempering time and temperature on wear behavior. Hence the present investigation is concentrated to study the wear behavior of as-cast pearlitic/ferritic ductile iron compared with higher amounts of pearlite in normalized and upper bainitic austempered ductile iron subjected to dry sliding condition.

II. Experimental details

2.1: Specimen Preparation

In order to investigate the structure-property relationship, ductile iron test blocks with different alloying elements were brought from LandT Kansbahal, India. The chemical composition test block by weight percentage is presented in Table 1. In order to carry out the experiment specimens of 25×10×5 mm^3 were machined from the test block and then austenitized at 1000°C for 90 minutes followed by air cooling to room temperature for normalizing heat treatment. On the other hand specimens after austenitization were quenched in KNO_3 + $NaNO_3$ (1:1 ratio) maintained at 500°C and kept there for 240 minutes in order to get complete transformation of bainitic matrix, followed by air cooling to room temperature for austempering heat treatment process. After heat treatment oxide layer from each of the specimen was removed by conventional filing and emery paper polishing method.

Table 1: Chemical composition of specimens in wt. %

Elements	Wt. %
C	3.45
Si	2.07
Mn	0.15
S	0.008
P	0.024
Cr	0.02
Ni	0.15
Mg	0.043
Fe	Balance

2.2: Vickers hardness and wear testing

Vickers hardness was measured by applying a load of 20Kg and dwell time being 10seconds on each heat treated and as-cast specimen. Ducom TR-208-M1 Ball-on-Plate (diamond tip) type wear tester was used in order to investigate the wear system response of the as-cast heat treated specimens. Test was conducted at 10N, 20N, 30N loads for a sliding distance of 7.54m at linear velocity of 0.063m/s. The weight loss for corresponding specimens was measured with the help of electronic balance of 0.1mg accuracy, prior to the weight measurement specimens was cleaned ultrasonically with acetone before and after the wear took place. Ball on plate type wear monitor [14-17] with specimen as flat plate and a spherical tipped diamond cone of 120° angle 0.4mm tip diameter was used in order to investigate the wear system response of the as-cast heat treated specimens. The mechanism of Ball on plate wear test is very much similar to that of Pin on disc wear tester. However the minor difference is that in pin-on-disc specimen is in the form of cylindrical pin, held stationary in specimen holder and disc is the counter body which rotates against the pin, whereas in case of Ball on plate wear test instead of pin specimen is a flat one rotates against a counter ball which is fixed. Also, in pin on disc machine only the disc rotates whereas in Ball on plate mechanism both the specimen and indenter rotate at same relative speed. The Ducom TR-208-M1 Ball on plate type wear monitor along with schematic diagram of pin on disc wear monitor is presented in Fig. 1 (a) and Fig. 1 (b) respectively.

Fig. 1: (a) Ducom TR-208-M1 Ball on plate type wear monitor

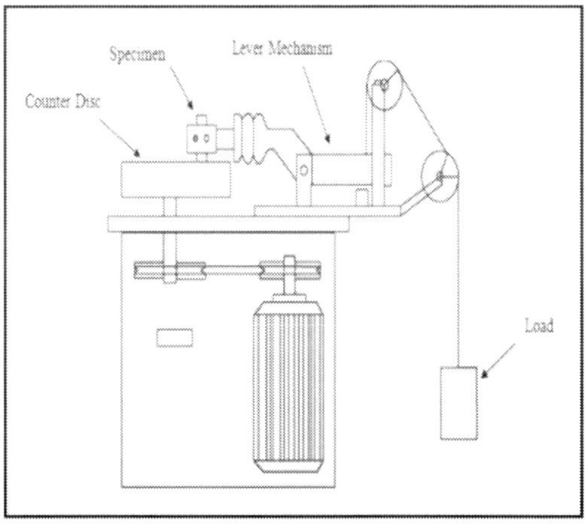

Fig. 1: (b) Schematic diagram of pin on disc wear test machine [18]

2.3: Microstructural and wear morphology investigation

In order to correlate the wear response with matrix structure standard metallographic technique was followed for microstructural investigation. Specimens were first polished with belt polisher followed by 1/0, 2/0, 3/0, 4/0 grades of emery paper and finally cloth polishing was done with alumina slurry followed by diamond polishing. Metallographic images were taken with the help of computer integrated optical microscope at 100X magnification. The worn surfaces of each specimen under various loading conditions were observed under optical microscope at 200X.

III. Results and Discussion

3.1: Morphological Study

The microstructures of respective specimens were shown in Fig. 2. The as-cast matrix Fig. 2 (a) was bull's eye ferritic/pearlitic one with graphite nodules surrounded by ferrite. Normalized specimen Fig. 2 (b) had the same ferritic/pearlitic matrix consisting of more amount of pearlite. It was observed that pearlite nucleates around the graphite nodules. On the other hand austempered specimen Fig. 2 (c) showed upper bainitic matrix since austempered at 500°C [14], with graphite nodules embedded within the matrix.

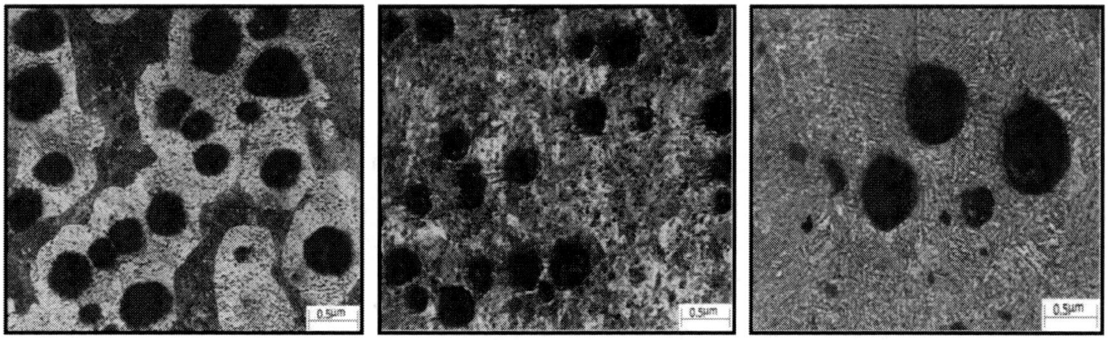

Fig. 2: (a) As-cast Fig. 2: (b) Normalized Fig. 2: (c) Austempered

Fig. 2: Microstructure of respective specimens

3.2: Hardness and Wear System Response

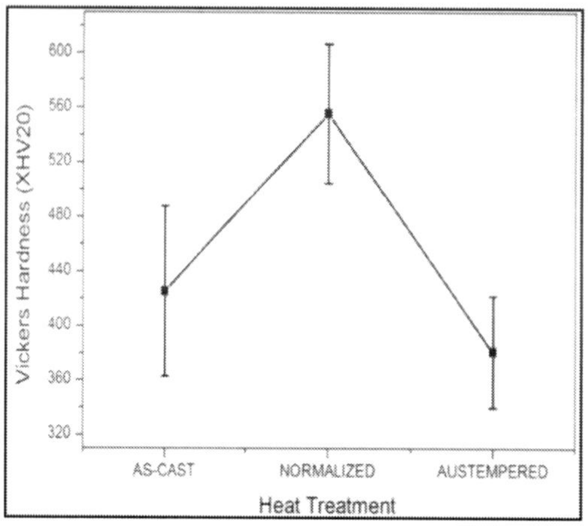

Fig. 3 Vickers Hardness of respective specimens

The Vickers hardness plot was presented in Fig. 3. It was found that normalized specimen having pearlitic/ferritic matrix had highest and austempered specimen with upper bainitic matrix had the lowest hardness value among all the three specimens. Although the as-cast specimen is having pearlitic/ferritic matrix like normalized specimen, higher pearlite content increased the hardness of the later [15]. On the other hand the lowest hardness value of austempered specimen was due to the coarse upper bainitic matrix [13]. Fig. 4 depicts the variation of weight loss/gain with variation of applied load for respective specimens. Continuous weight loss was observed for as-cast specimen when load was increased from 10N to 20N. But when load was increased to 30N a weight gain was observed. However in case of austempered specimens weight gain was observed for 20N and 30N, and there was no weight loss or gain was observed for normalized specimen at 20N. When compared with respect to hardness of respective specimens the weight loss at 10N is lowest for the hardest specimen due to higher amount of pearlite in the matrix. On the other hand the as-cast ferritic/pearlitic matrix with moderate hardness value showed significantly high loss in weight as compared to the austempered which weight loss is slightly higher than normalized. At 20N load upper bainitic matrix with lowest hardness value showed a marginal gain in weight whereas normalized specimen had neither lose nor gained any weight but as-cast specimen had significantly lose some weight. Further at 30N again weight gain was observed for both austempered and as-cast specimen whereas normalized specimen showed weight loss. However the gain in weight for austempered specimen was very less.

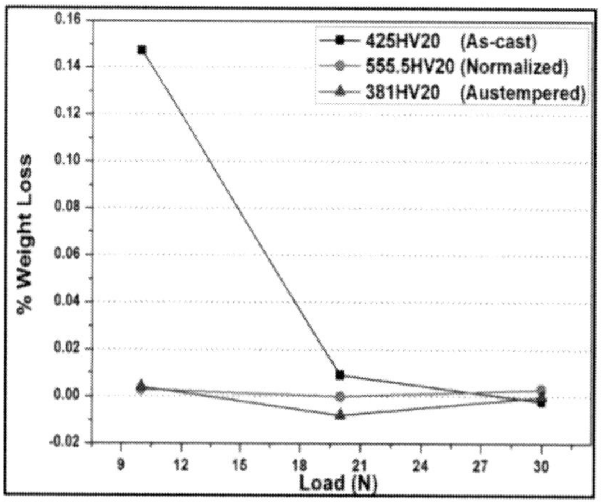

Fig. 4: Load vs. Weight loss for different hardness level

3.3: Wear morphology

The worn surface of as-cast and heat treated specimens taken under optical microscope at 200X were shown in Fig. 5. The wear direction was indicated by black arrow in every respective case. In each of the as-cast and heat treated specimen the principal mechanism was observed to adhesive wear for 10N and 20N loading condition Fig. 5(a) - (f). Shallow pits formed due to delamination [9, 12] were observed in each specimen and get deposited over the graphite nodules Fig. 5(a). Correlation between hardness and wear properties was clearly evident from the micrographs i.e., higher hardness can yield better wear resistance as can be seen in Fig. 5(a)-(c) for 10N, Fig. 5(d)-(f) for 20N and Fig. 5(g)-(i) for 30N. On further analysis of worn surfaces under NOVA NANOSEM 450, field emission scanning electron microscope, the EDAX analysis showed that austempered specimen when subjected to wear under 20N and 30N load (Fig. 6 (a) and Fig. 6(b)), oxides were observed at different areas which caused the gain in weight after the wear test. Also the as-cast specimen operated under 30N (Fig. 6(c)) load was observed to have oxides at the wear debris leading to gain in weight after the test. The red rectangles in the respective figures show the area where presence of oxide was detected.

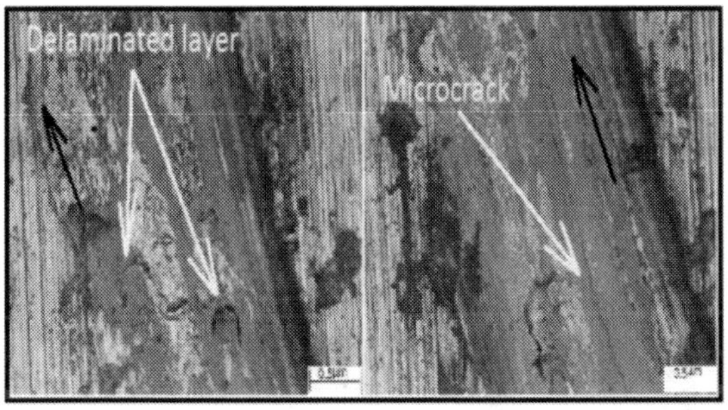

(a) As-cast 10N

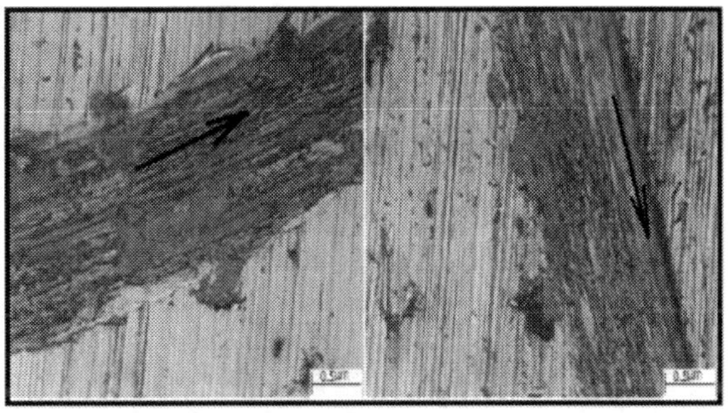

(b) Normalized 10N

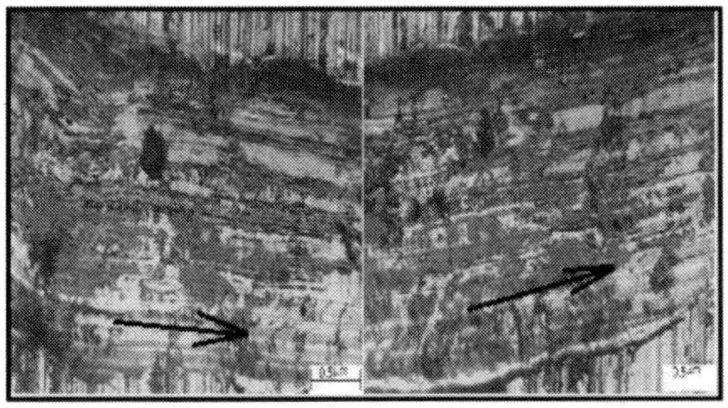

(c) Austempered 10N

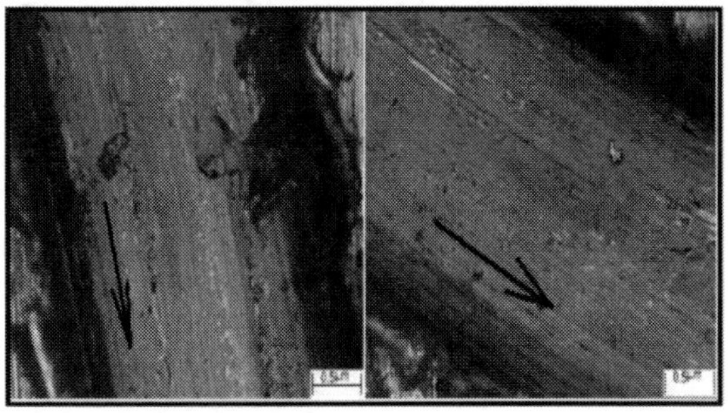

(d) As-cast 20N

(e) Normalized 20N

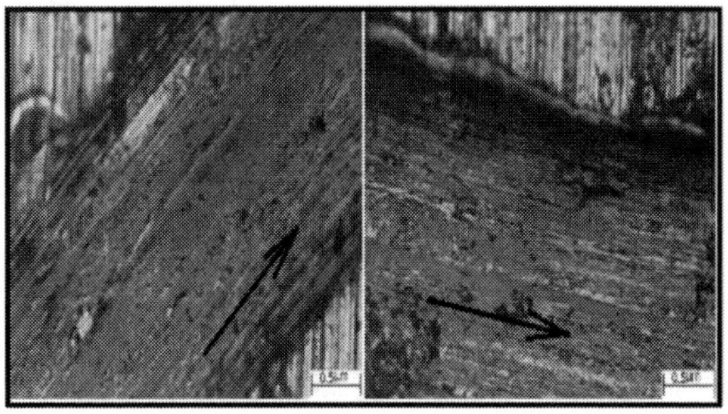

(f) Austempered 20N

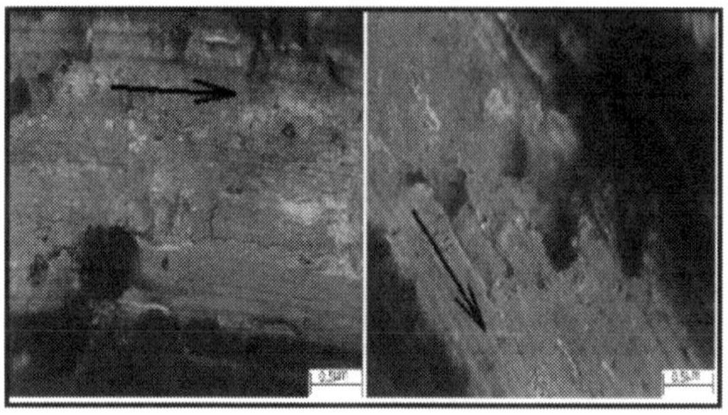

(g) As-cast 30N

(h) Normalized 30N

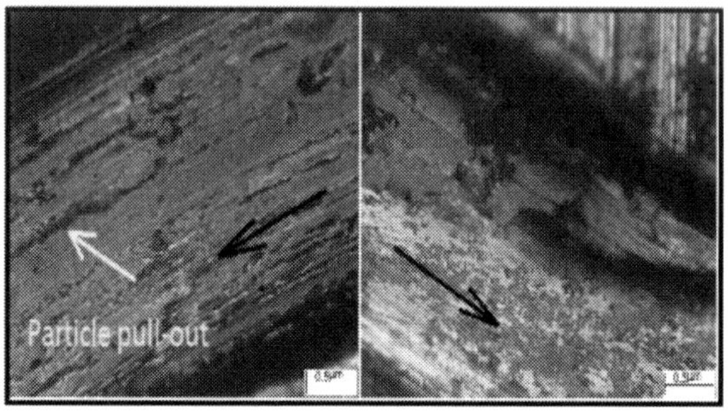

(i) Austempered 30N

Fig. 5: Worn surface of respective specimen at varying load condition

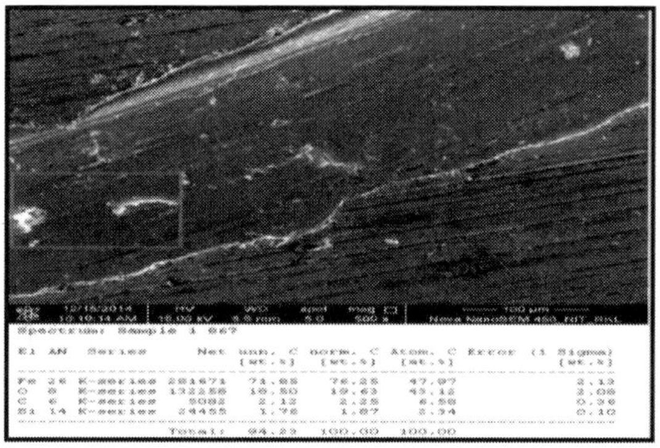

(a)

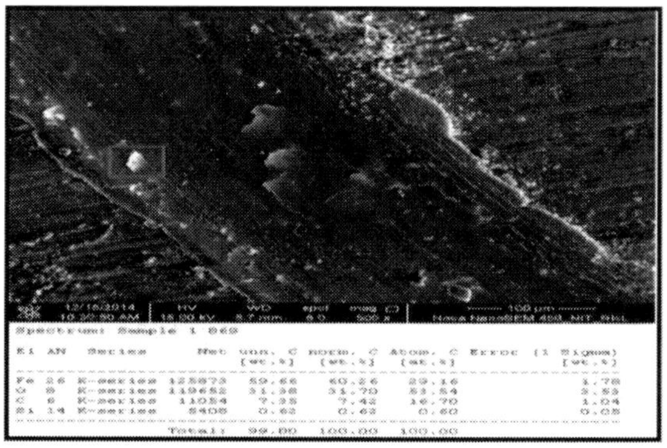

(b)

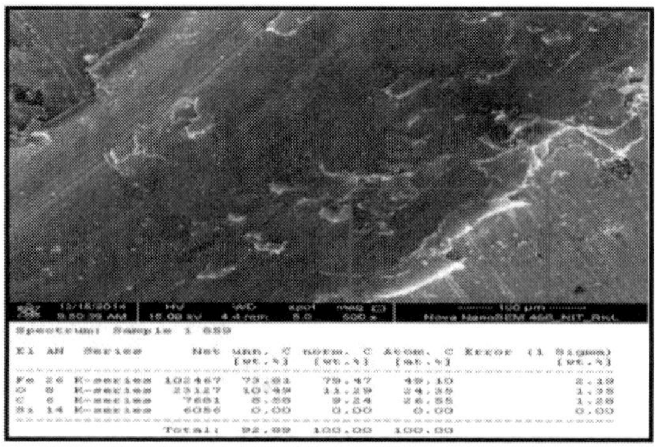

(c)

Fig. 6: FESEM and EDAX analysis of specimen observed to have oxides. (a) Austempered at 20N load, (b) Austempered at 30N load, (c) As-cast at 30N load

Also with increase in applied load the width and depth of wear track was increased. Micro cracks were also observed along the wear direction in each specimen and crack length was observed to be decreasing with increase in applied load. The features of worn surface in case of 30N load was very distinct than those for 10N and 20N. In as-cast condition Fig. 5(g), the wear surface observed to have ploughing marks at some area and nodules were started to disrupt from their original shape, suggesting adhesive wear [16]. However in case of normalized specimen Fig. 5(h) graphite shape disruption was observed along with delaminated layer over nodules. On the other hand austempered specimen Fig. 5(i) showed particle pull out from the worn surface along with breaking of wear surface.

IV. Conclusion

The wear behavior of as-cast pearlitic/ferritic ductile iron was compared with upper bainitic austempered and pearlitic /ferritic normalized specimens. From investigation the following conclusions can be drawn:

1. Austempering at higher temperature leads to transformation of pearlitic/ferritic matrix into coarse upper bainitic matrix and decreases the hardness.
2. Specimen with higher amount of pearlite was harder and appeared to be more resistant to wear.
3. A considerable gain in weight was observed for austempered specimen at 20N load whereas the normalized specimen does not show any weight loss or weight gain.
4. Wear surface morphology of as-cast and heat treated specimens for 10N and 20N does not show any significant difference. Whereas for 30N loading condition ploughing marks was observed for as-cast specimen and austempered specimen showed particle pull phenomena along with breaking of wear continuity.

V. Acknowledgement

The authors deeply acknowledge the financial support provided by Board of Research in Nuclear Science, India in order to carry out this investigation. The authors also express their gratitude towards the help of LandT Kansbahal, India for providing test blocks for this investigation.

VI. References

[1]. Gudlanch RB, Janowak JF 1983 *AFS Trans.* **94** 377
[2]. Harding RA, Gilbert GNJ 1986 *Brit Foundryman* **79** 489
[3]. Harding RA 1993 *The Foundryman* 97
[4]. Yang J, Putatunda SK 2005 *Mater. Sci. Eng. A* **406** 217
[5]. Putatunda SK, Bartosiewicz L, Krause AR, Singh I, Alberts FA 1993 Mater. *Charact.* **30** 221
[5]. Harding RA 1985 *Mater. Des.* 177
[6]. Hughes ICH 1985 *Mater. Des.* **6** 124
[7]. Putatunda S.K 2001 *Mater. Sci. Eng. A* **315** 70
[8]. S. M. A. Boutorabi, J. M. Young and V. Kondic, M. Salehi 1993 *Wear* **165** 19
[9]. A.S.M.A. Haseeb, Md. Aminul Islam, Md. Mohar Ali Bepari 2000 *Wear* **244** 15
[10]. Yusuf Kayali, Yılmaz Yalcin, Sukru Taktak 2011 *Mater. Des.* **32** 4295
[11]. Junji Sugishita and Shonoshin Fujiyoshi 1982 *Wear* **77** 181
[12]. J. Zimba, M. Samandi, D. Yu, T. Chandra, E. Navara, D.J. Simbi 2004 *Mater. Des.* **25** 431
[13]. M. Nili Ahmadabadi, H.M. Ghasemi, M. Osia 1999 *Wear* **231** 293
[14]. A. Roy Choudhury, Tamer Ezz, Satyajit Chatterjee, Lin Li 2008 *Surf. and Coat. Technol.* **202** 2817
[15]. L. Steiner, V.Bouvier, U.May, N.Huber 2010 *Tribol. Int.* **43** 1410
[16]. A. Lyubicheva, I.Soldatenkov 2012 *Tribol. Int.* **55** 77
[17]. M. Alvarez-Vera, A.Juarez-Hernandez, C.E.Gonzalez-Rivera, R.D.Mercado-Solis, M.A.L. Hernandez-Rodriguez 2013 *Wear* **301** 243
[18]. Sandeep Kumar, Sant Ram Chauhan 2012 *Int. J. Compos. Mater.* **2(5)** 101

[19]. R.A. Gonzaga 2013 *Mater. Sci. Eng. A* **567** 1
[20]. M.J. Perez, M.M. Cisneros, H.F. Lopez 2006 *Wear* **260** 879

Air jet erosion test on plasma sprayed surface by varying erodent impingement pressure and impingement angle

Ajit Behera[1]*, Asit Behera[2], S C Mishra[3], S Pani[4], P Parida[5]
[1*]Department of Metallurgical & Materials Engineering
Indian Institute of Technology, Kharagpur-721302, India
[2]School of Mechanical Engineering,
KIIT University, Bhubaneswar-751024, Odisha, India
[3]Department of Metallurgical & Materials Engineering
National Institute of Technology, Rourkela-769008, Odisha, India
[4]Department of Mechanical Engineering,
Indian Institute of Technology, Kharagpur-721302, India
[5]Department of Chemistry,
National Institute of Technology, Rourkela-769008, Odisha, India

*Corresponding Author, email id: ajit.behera88@gmail.com

Abstract

Fly-ash premixed with quartz and illmenite powder in different weight proportions are thermal sprayed on mild steel and copper substrates at various input power levels of the plasma torch ranging from 11 kW to 21 kW DC. The erosion test has done using Air Jet erosion test Reg (As per ASTM G76) with silica erodent typically 150-250 µm in size. Multiple tests were performed at increasing the time duration from 60 sec to 180 sec with increasing pressure (from 1 bar to 2.5 bar) and angle (60° & 90°). This study reveals that the impact velocity and impact angle are two most significant parameters among various factors influencing the wear rate of these coatings. The mechanisms and microstructural changes that arise during erosion wear are studied by using SEM. It is found that, when erodent are impacting the fresh un-eroded surface, material removal occurs by the continuous evolution of craters on the surface. Upper layer splats are removed out after 60 sec and second layer splat erosion starts. Based on these observations Physical models are developed. Some graphs plotted between mass loss-rate versus time period/impact Pressure/impact Angle gives good correlation with surface features observed.

Keyword: *Erosion wear, Plasma spray, Mass-loss, Surface Properties.*

1. Introduction

Solid particle erosion is to be expected whenever hard particles are entrained in a gas or liquid medium impinging on a solid at any significant velocity. Degradation of material by erosion increases operating costs, reduces energy efficiency, increases ideal time, reduces functional life, basically enhance pitting corrosion and decreases productivity across a broad range of applications [1-2]. Many investigations have been studied to increase the surface properties by deposition technique to reduce wear [3-7]. Now-a-days plasma sprayed coatings find broad applications not only in research and development area, but also in the industrial workplaces ranging from textile mills to even medical applications. In the automotive industries of many advanced countries, plasma-sprayed coatings are used to improve the resistance to erosion, abrasion and corrosion of machine components and structural parts [8]. This spray technology has the advantage of being able to process various low-grade minerals to obtain value-added products and to deposit ceramics, metals, and even a combination of these, generating near-homogenous composite coatings with the desired microstructure on a range of substrates [9-11].

Challenges related to thermal spray process concerning the improvement of spray systems and spraying of new materials. These aspects require a better understanding of the involved phenomena, i.e., physical, chemical, thermodynamics, etc. to improve the efficiency of plasma spray coating operation [12-15].

Content from this work may be used under the terms of the Creative Commons Attribution 3.0 licence. Any further distribution of this work must maintain attribution to the author(s) and the title of the work, journal citation and DOI.
Published under licence by IOP Publishing Ltd

Plasma sprayed fly-ash composite coatings have been widely used in the coatings industry for its superior erosive, sliding and abrasive wear properties. This paper describes about fly-ash+quartz+illmenite composite coating composition. The coatings produced by this thermal spray processes exhibit a broad range of coating hardness, porosity and microstructural features like splat size and volume fraction of different phases. In this study, we have evaluated the coating microstructures of various fly-ash+quartz+illmenite coatings produced from different spraying power level. The objective of our investigation is to explore the erosion wear mechanism of fly-ash+quartz+illmenite coatings in great detail, and determine how these mechanisms are influenced by the coating microstructure.

2. Experimental Procedure

Fly-ash+quartz+illmenite homogeneous mixture having 60:20:20 weight ratios were Plasma sprayed on Mild steel and copper substrates, having dimensions 1 inch diameter and 3 mm thickness. Non-transferred arc mode atmospheric plasma spraying operation was done using a 40 KW plasma spray system at the Laser & Plasma Technology Division, BARC, Mumbai. The major subsystems of plasma set up include the spray torch, powder supply, powder feeder, plasma gas supply, control console, cooling water and spray booth. The plasma input power was varied from 11 to 21 KW by controlling the gas flow rate and arc current. The powder deposited at spraying angle 90^0. Operating parameters used for coating deposition are given in table 1. Erosion wear test of these coatings are investigated by using Air Jet erosion test Reg (As per ASTM G76). For erosion study, only a small area of the sample was exposed to particle erosion, while the rest of the sample was wrapped in polyester tape, which allows impact of particle only in the exposed area. The tests were stopped after a short interval of time to calculate the mass loss. Multiple tests were performed at increasing time from 60 sec to 180 sec with increasing pressure from 1 bar to 2.5 bar and constant stand-of-distance of 65 mm. It is known that hard ceramic coating failure occurs at 90° angle [16]. So for high erosion two impact angle 60° & 90° were taken. The morphology of substrate (before worn and after worn) was studied using JOEL T-330 scanning electron microscope. Solid particle erosion test parameters are given in Table-2.

Table 1. Operating parameters during coating deposition of fly-ash+ quartz+ illmenite.

Operating parameters	values
Plasma arc current (amp)	260-500
Arc voltage (volt)	40-44
Torch input power (kW)	11,15,18,21
Plasma gas (argon) flow rate (IPM)	28
Secondary gas (N2) flow rate (IPM)	3
Carrier gas (Ar) flow rate (IPM)	12
Powder feed rate (gm/min)	15
Torch to base distance (TBD) (mm)	100

Table 2. Testing parameter during erosion.

parameters	values
Testing Temperature	25 °C
Impingement Angle	60° & 90°
Erodent impingement pressure	1.0, 1.5, 2.0, 2.5 bar
Test duration	60 sec, 120 sec & 180 sec

3. Results and Discussion

3.1 Investigation on the surface wears during erosion of fly-ash+quartz+illmenite coatings.

After wear test, all the samples have been examined under SEM. Figure 1(a), (b), (c) & (d), show the surface morphology of worn surfaces, when the erodent is impacted at an angle 60° and 90° respectively with varying pressure level of 1 bar and 2 bar. At 60° angle of impact, there are many angular grooves occurs. At 90° angle of impact, sharp vertical groves are seen which is due to dominating effect of perpendicular component of the force of impingement of the erodent. In general, it can be stated that when hard and multifaceted erodent i.e. SiC is used, deeper groves are produced and plastically flow regions are found and also limit the surface crack propagation. It observed that by increasing in erodent impact pressure, there is increasing in removal of coating mass causing exposure of substrate material. By solid particle impingement, there is a formation of grooves called crater. Here crater indicates that the particle indenting the region is more rounded, as an angular particle would have cut deeper into the surface, creating a more prominent crater-lip [17]. SEM micrograph confirms that the damage mechanism is related to single or multiple splat removal through crack propagation along the splat boundaries of single splat or a cluster of splats. This wear mechanism based on piled up of material with gradual loss of mass after subsequent impacts by forming a crater. Initial piled up material is heavily strained and some small cracks are observed & spread along splats boundaries. This is the reason of fragmentation of splat during solid particle erosion.

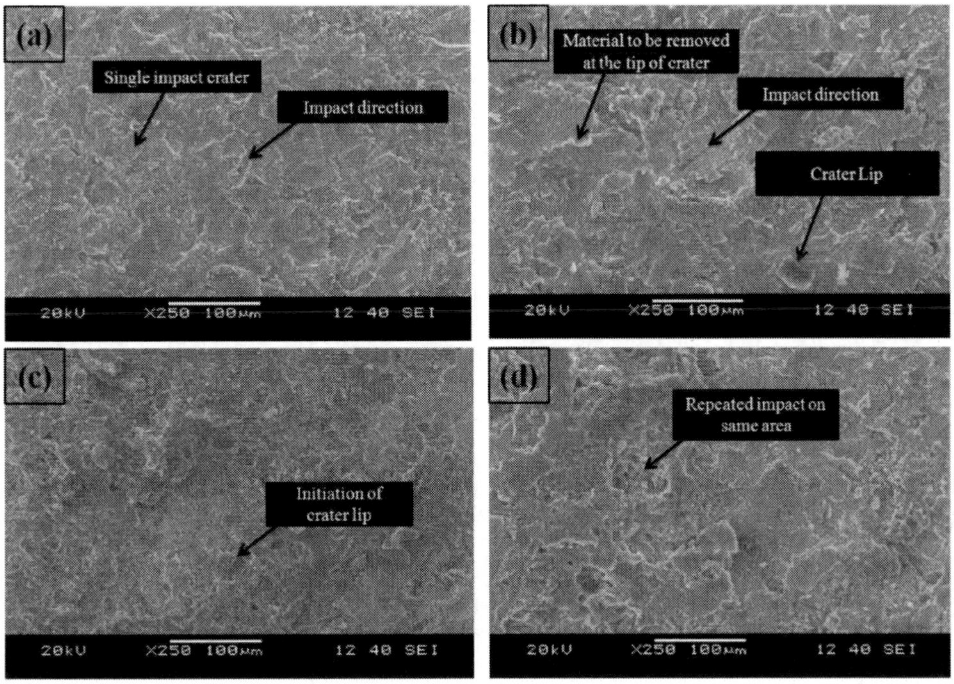

Figure 1. SEM micrographs of eroded surfaces of coatings deposited at 18 kW, eroded at angle of impact (a) 60° & 1 bar; (b) 60° & 2 bar; (c) 90° & 1 bar; (d) 90° & 2 bar using SiC as the erodent.

3.2 XRD Analysis

Figure 2 shows the XRD analysis of (a) raw material which is to be coated, and (b)-(e) shows the coating at a different power level. Here it is observed that by increase in deposition power level, there is increase in different phase formation. These oxide formation is due to the higher energy transformation in the presence of atmospheric air. These different oxide phases act as the ceramic coating and can hinder the corrosion rate.

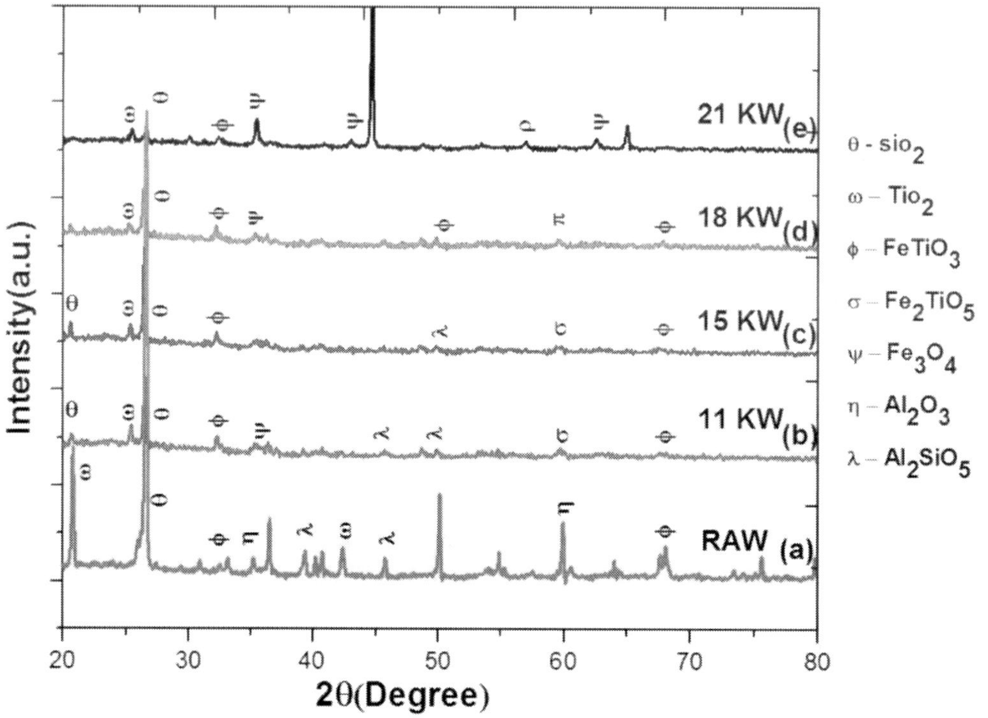

Figure 2. X-ray diffraction pattern showing different phases present in different sample (Different sample power level are indicated in right side of the pattern).

3.3 Hardness measurement

By micro-hardness testing, It is found that an average hardness is within a range of 532.77 Hv to 659.75 Hv and erosion rate in the range of 60.1×10^{-4} to 283.74×10^{-4}. This variation arises due to distribution of composite coating and due to transformation of different phases in the time of specimen preparation (basically due to different power supply for coating). Additionally, the presence of titanium, iron, and silicon compounds and oxides in the coating phase greatly increased the hardness. These composite phases were more effective in improving wear resistance. Wear behavior of coated composite layers also depends on different powder feed rates and input energy conditions [18]. It is found that, the sample prepared by 18 kW power level has a greater hardness and less erosion rate for all test conditions. Figure-3 shows the graph between erosion rate and micro-hardness of plasma sprayed fly-ash+quartz+illmenite coating. The figure indicates erosion rate decreases as increasing in coating hardness. But in some cases, erosion rate becomes a little higher as there is an increase in hardness as compared to others. This is due the presence of larger amount of void present in the coating, semi-molten phases etc.

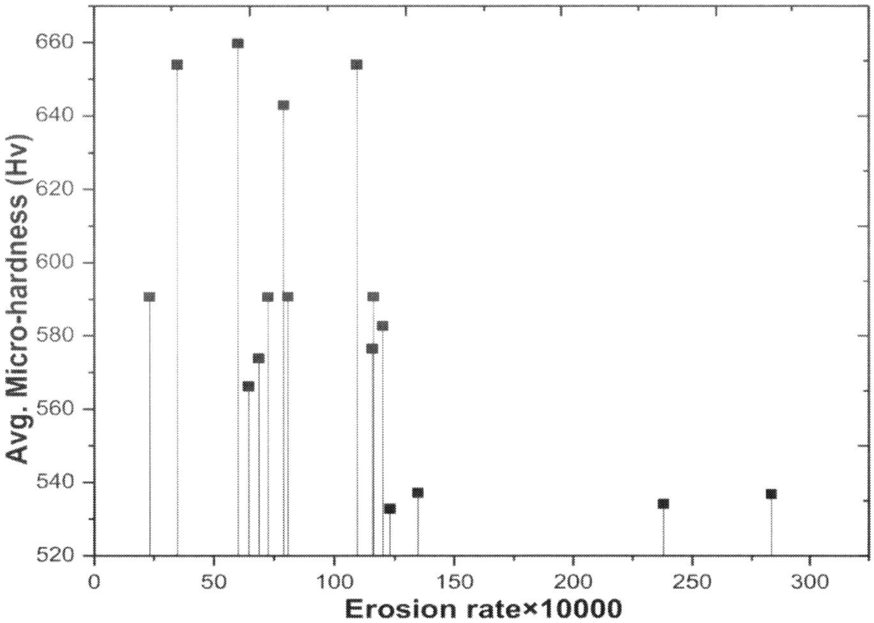

Figure 3. Erosion rate vs micro-hardness of plasma sprayed fly-ash+quartz+illmenite coating (Erosion at 90° angle of incident).

3.4 Graphical presentation of coating mass with respect to erosion time

Figure-4 shows that mass of coating sample versus time in order to identify the erosion rate. Here time duration taken from 60 sec to 180 sec. Between 60 sec to 120 sec, when the particles impact on the fresh uneroded coating surface, each erodent causes material removal by forming prominent lips and craters [19]. Here it is found that the mass loss increases with increase in time. Mass loss rate can be obtained by the average mass loss at a particular time divided by the time duration. In between 120 sec and 180 sec, in some case, there is a decrease in mass loss. This is due to: (a) strain-againg as sticking of the surface materials [20] or (b) the efficiency of erodent retarded due to the rough uneven surface which absorbs the particle impact energy and hinder for mass loss [21] or (c) due to brittle to ductile transition behavior [22]. The variation of mass loss with time, in case of the coating eroded by SiC for 18 kW are illustrated in Fig 4(a) & (b) and for 21 kW illustrated in Fig 4(c) & (d). From the figures, the mass of the coating material decreases with the time period. The initial mass loss is high and it follows a steady state after a certain time of exposure. Finnie [23] has explained that the drastic drop of erosion rates is due to transition of type of fracture mechanism i.e. from brittle to ductile behavior. From these figures, it is concluded that the initial wear rate is high irrespective of the angle of impact. With increasing exposure time the rate of wear starts decreasing and in the transient regime, a steady state is attended [24]. It is also found that, maximum erosion occurs at a normal impact angle. This behavior is observed in case of brittle materials where the erosion rate continuously increases with increasing impact angle and attains a maximum at normal (90°) impact, in addition, under brittle erosion conditions the magnitude of erosion rate is determined only by the normal component of impact velocity [25].

IOP Conf. Series: Materials Science and Engineering **75** (2015) 012004 doi:10.1088/1757-899X/75/1/012004

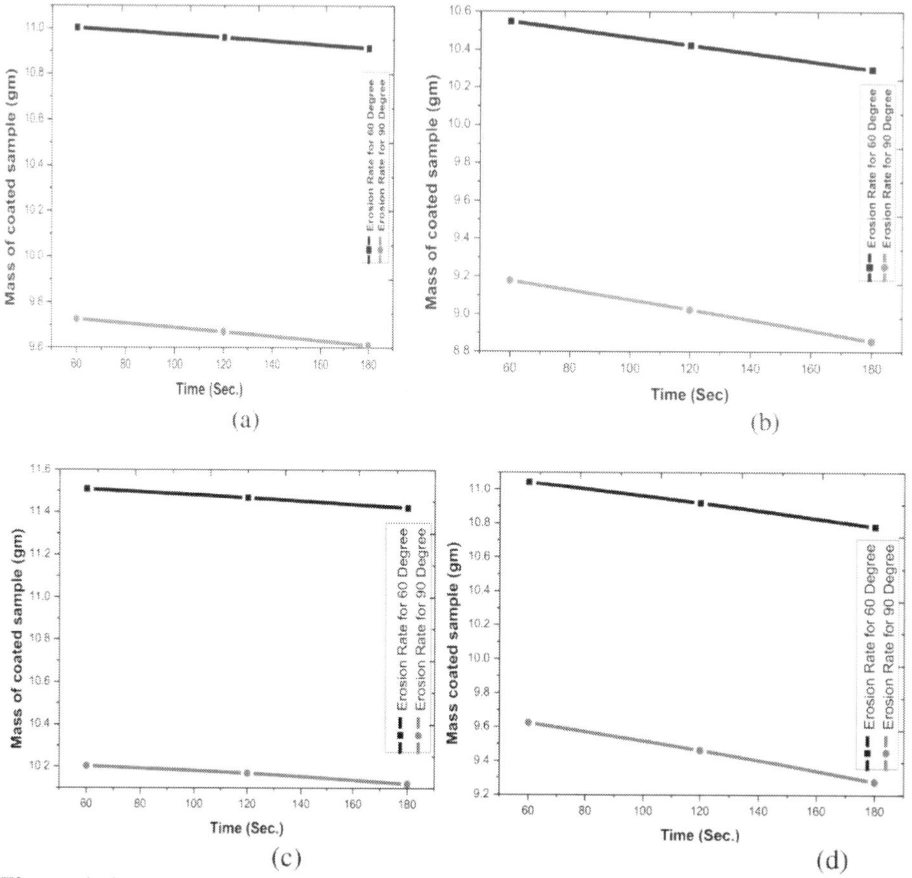

Figure 4. Coating mass vs time for coating substrate at different power level and erodent impact pressure (a) for 18 kW and 1 bar, (b) for 18kW and 2 bar; (c)for 21kW and 1 bar; (d) for 21kW and 2 bar.

4. Conclusions

Due to phase transformations and semi-molten powder deposition during plasma spraying, changes the coating characteristic such as erosion rate and hardness. The flyash+quartz+illmenite coating gives much harder than substrate metals for which it can be recommended for tribological applications. It is clear that erosion wear strongly influence by the size of erodent, impact velocity, impact angle standoff distance and hardness, etc. Maximum erosion took place at a normal angle of impact. Here it is found that, this type of erosion mechanism initiates through plastic deformation.

For recommending plasma spraying for a specific application, study of erosion wear behavior is the main requirement. It is concluded that the composite coating materials are more effective for improvement of wear resistance and flyash+quartz+illmenite can be considered as a potential coating material suitable for various tribological applications.

5. References

[1] Cook B A, Peters J S, Harringa J L and Russell A M 2011 Enhanced wear resistance in AlMgB14-TiB_2 composites, *Wear* **271** 640-646.

[2] Chattopadhyay R 2001 *Surface Wear: Analysis, Treatment, Prevention*, vols. xi-xii, ASM International, Materials Park, OH, USA, 267-290.

[3] Ulutan M, Yildirim M M, Buytoz S, and Elik O C 2011 Microstructure and Wear Behavior of TIG Surface-Alloyed AISI 4140 Steel *Tribology Transactions* **54** 67-79, DOI: 10.1080/10402004.2010.519859.

[4] Sahu S P, Satapathy A, Mishra D, Patnaik A and Sreekumar K P 2010 Tribo-Performance Analysis of Fly Ash-Aluminum Coatings Using Experimental Design and ANN *Tribology Transactions* **53** 533-542, DOI: 10.1080/10402000903491317.

[5] Toparli M, Celik E, Birlik I, Dokumaci E, Ak-Azem N F 2009 Tribological Properties of Electric Arc–Sprayed CuSn Coating for Bearing Elements *Tribology Transactions* **52** 389-394, DOI: 10.1080/10402000802593122.

[6] Zbek Y Y O, Durman M and Akbulut H 2009 Wear Behavior of AISI 8620 Steel Modified by a Pulse-Plasma Technique *Tribology Transactions* **52** 213-222, DOI: 10.1080/10402000802369721.

[7] Mishra S C, Sahu A, Das R, Satapathy A and Sen S 2008 Microstructure, Adhesion and Erosion wear of plasma sprayed Alumina-Titania composite coatings *J. Reinf. Plast. Compos.* **00** 1-11.

[8] Louda P, Tumova S and Rozek Z 2006 *Application of nanotechnology at automotive industry*, In International Conference Vacuum and Plasma Surface Engineering, Liberec, Technical University of Liberec. ISBN 80-7372-129-5.

[9] Mishra S C, Das S, Satapathy A, Sarkar S, Ananthapadmanabhan P V A and Sreekumar K P 2009 Investigation on Composite Coating of Low Grade Minerals *J. Reinf. Plast. Compos.* 24 3061-3067.

[10] Satapathy A, Sahu S P and Mishra D 2009 Development of protective coatings using fly ash, premixed with metal powder on aluminium substrates *Waste Manage. Res.* 1-7.

[11] Behera A and Mishra S C 2012 Dependence of Adhesion Strength of Plasma Spray on Coating Surface Properties J. Materials and metallurgical and engg. **2** 23-30.

[12] Grunling H W and Mannsmann W 1993 Plasma sprayed thermal barrier coatings for industrial gas turbines: morphology, processing and properties, Journal De Physique IV, Colloque C7 *supplkment au Journal de Physique*. **3** 903-912.

[13] Smith M F, Hall A C, Fleetwood J D and Meyer P 2011 *Coatings* **1** 117-132; doi:10.3390/coatings1020117.

[14] Tsui Y C, Doyle C and Clyne T W 1998 Plasma sprayed hydroxyapatite coatings on titanium substrates Part 1: Mechanical properties and residual stress levels *Biomaterials* **19** 2015-2029.

[15] Kulkarni A A, Goland A, Herman H, Allen A J, Ilavsky J, Long G G and Carlo F D 2005 Advanced Microstructural Characterization of Plasma-Sprayed Zirconia Coatings Over Extended Length Scales *J. Therm. Spray Technol.* **14(2)** 239-250.

[16] Branco J R T, Gansert R, Sampath S, Berndt C C and Herman H 2004 Solid Particle Erosion of Plasma Sprayed Ceramic Coatings *Materials Research* **7** 147-153.

[17] Rajahramn S S, Harvey T J, Walker J C, Wang S C and Wood R J K 2012 Investigation of erosion-corrosion mechanisms of UNS S31603 using FIB and TEM *Tribology International* **46** 161-173.

[18] Yılmaz S O 2005 Wear behavior of gas tungsten arc deposited FeCrC, FeCrSi, and WCo coatings on AISI 1018 steel *Surf. Coat. Technol.* **194** 175-183.

[19] Howard R L and Ball A 1995 The solid particle and cavitation erosion of titanium aluminide intermetallic alloys *Wear* **186-187** 123-128.

[20] Ramesh C S, Keshavamurthy R, Channabasappa B H and Pramod S 2009 Influence of heat treatment on slurry erosive wear resistance of Al6061 alloy *Materials and Design* **30** 3713-3722.

[21] Davis J R *Surface Engineering for corrosion and wear resistance* ASM International, ISBN: 0-87170-700-4.

[22] Barkoula N M, Gremmels J and Karger-Kocsis J 2001 Dependence of solid particle erosion on the cross-link density in an epoxy resin modified by hygrothermally decomposed polyurethane *Wear* **247** 100-108.

[23] Mishra S C, Praharaj S and Satpathy A 2009 Evaluation of Erosion Wear of A Ceramic Coating With Taguchi Approach, *J. Manufacturing Engineering* **4** 241-246.

[24] Sparks A J and Hutchings I M, 1991 Transitions in Erosive Wear behavior of Glass Ceramic *Wear* **149** 99-110.

[25] Behera A 2012 M.Tech Thesis, *Processing and characterization of plasma spray coatings of industrial waste and low grade ore mineral on metal substrates* submitted In Department of Metallurgical & Materials Engineering, National Institute of Technology, Rourkela, India.

[26] Maozhong Y, Baiyun H and Jiawen H 2002 Erosion wear behaviour and model of abradable seal coating *Wear* **252** 9-15.

Characterisation of Sol-Gel Synthesis of Phase Pure $CaTiO_3$ Nano Powders after Drying

P K Mallik[1*], G Biswal[2], S.C Patnaik[1], S K Senapati[2]
[1]Department of Metallurgical and Materials Engineering, IGIT Sarang, Odisha India
[2]Department of Mechanical Engineering, IGIT Sarang, Odisha India

E-mail: prafulla.mallik@gmail.com

Abstract: According to a few recent studies, calcium titanate (CT) is a material that is similar to hydroxyapatite in biological properties. However, calcium titanate is not currently being used in the biomedical applications as to hydroxyapatite. The objective is to prepare nano calcium titanate powders from the equimolar solution of calcium oxide, ethanol and Titanium (IV) isopropoxide via sol-gel synthesis. The phase analysis and morphology of powder particles were studied by X-ray diffraction (XRD), while the composition and size of powder particles were determined by Transmission electron microscope (TEM) attached with energy dispersive x-ray spectrometer (EDS). As results, XRD confirm the presence of phase pure crystalline $CaTiO_3$ after drying at 100^0C for 24 hours, while TEM analysis confirms about 13 nm sizes of $CaTiO_3$ particles and some agglomerated particle of 20-30 nm. Moreover, EDS analysis indicates that the approximately stoichiometric Ca/Ti ratio 1:1 was obtained in the $CaTiO_3$ powders. Finally, it can be concluded that described sol-gel synthesis could be novel method for the production of nano $CaTiO_3$ particles at lower temperature compared to any other methods of production.

1. Introduction

Currently, researchers in chemistry, physics, biochemistry, and material engineering are interested in new class of nanosized materials for electronic and biomedical applications. In this regard, calcium titanate ($CaTiO_3$) was reported as good substrate for apatite growth in vitro with the potential to enhance osseointegration and osteoconductivity.[1-2] Furthermore, some researchers suggested that $CaTiO_3$ possess large dielectric constant (~190) and high dielectric loss (~3.5).[3-4] The beneficial effect of addition of $CaTiO_3$ to HA encourages the development of electrically active biomaterial for bone replacement.[5-6]Apart from this, $CaTiO_3$ is widely used as synthetic rock for immobilization of high level nuclear wastes.[7] Several special methods are known for the synthesis of $CaTiO_3$ nanomaterials such as sol–gel [8], coprecipitation [9], organic–inorganic solution [10], and combustion [11] etc. $CaTiO_3$ is one of the materials that were initially synthesized via the mechanochemical route from the varying mixture of $CaCO_3$, CaO and TiO_2.The major drawbacks of this processes are contamination and energy consumption. However, the study on sol gel synthesis effect on fine particles and uniform distribution of particles have been created much interest among researchers because of its several advantages to downstream processes like reducing the sintering time and temperature, reducing the phase transformation temperature, decreasing the thermal decomposition temperature, and increasing the particle reactivity.[12-14] In addition, small sizes of nanoparticles can affect the sintering or miscibility properties of the material relatively at lower temperatures.[15] On the whole, in nanoparticles high numbers of atoms are located on the surface of the grains or interface boundaries which lead to an increase in surface activity between materials and host bodies.[16]

Content from this work may be used under the terms of the Creative Commons Attribution 3.0 licence. Any further distribution of this work must maintain attribution to the author(s) and the title of the work, journal citation and DOI.
Published under licence by IOP Publishing Ltd

The objective of the present work is to synthesis of phase pure $CaTiO_3$ using sol-gel process in which calcium oxide (CaO) and Titanium (IV) isopropoxide ($Ti(OC_3H_7)_4$ as starting materials, ethanol as the dispersed medium, nitric acid (HNO_3) as catalyst for the reaction and distilled water as used for the hydrolysis reaction. During this process, Titanium (IV) isopropoxide and CaO were hydrolysis into Ti $(OH)_4$ and $Ca(OH)_2$ gels. As result, with their combined reaction of gels were decomposed into precipitates $CaTiO_3$.The precipitate was dried at 100°C for 24 hours to obtain crystalline $CaTiO_3$ powders. X-ray diffraction (XRD) and Transmission Electron Microscopy attached with energy dispersive X-ray spectroscopy (TEM-EDS) analysis were carried out to study the microstructural and morphological behavior of dried $CaTiO_3$ sol-gel powders.

2. Experimental procedure

2.1. *Materials and methods*

The precursor materials used for the synthesis of $CaTiO_3$ were calcium oxide (CaO) and Titanium (IV) isopropoxide ($Ti(OC_3H_7)_4$, ethanol and nitric acid (HNO_3) and distilled water. Initially, CaO was dispersed in ethanol. The dispersed medium was kept on a hot plate and the suspension was stirred by an electromagnetic stirrer. Following this, an equivalent amount (keeping the Ca/Ti ratio same as $CaTiO_3$) of Titanium Isopropoxide solution (0.17M) was added drop wise in the dispersed CaO medium. The total solution was stirred at 80°C for 3-4 hrs to allow the reaction to take place towards completion. Then, distilled water and concentrated HNO_3 was added drop wise till the pH of the solution increases to 10. For the production of 25g $CaTiO_3$, the amount of CaO, ethanol, distilled water, and HNO_3 solution were required for according to the stoichiometric molar ratio of ($Ti(OC_3H_7)_4$:C_2H_5OH:H_2O:HNO_3::1:20:4:0.8). Subsequently, the solution was kept at room temperature for one day to precipitate the reaction product, which was further dried at 100°C for one day in a drying oven. After drying, the powder lump was crushed by using an agate mortar to make the material in powder form.

2.2. *Powders characterisation*

The calcined powders were analysis through the different characterisation technique to study the composition, microstructures and morphology. X-ray diffraction (XRD) analysis was performed on a Diffractometer (SHIMADZU, XRD700) using CuK_α radiation for all analyses at 40 kV and 30mA in order to identify the phases of the powders. Crystal size of the calcined powders was calculated with Scherer formula from data of FWHM of XRD spectrum. The XRD patterns were recorded in the 2θ range = 30–80° using a step size of 0.02° and a counting time of 5s per step. The microstructural and compositional characteristics of powder particles were investigated by TEM (Transmission Electron Microscopy, GEOL GM1400) attached with energy dispersive X-ray spectroscopy (EDS, OXFORD).

3. Results and discussions

3.1. *Phase analysis*

From the figure 1 show XRD peaks of the crystalline $CaTiO_3$ sol-gel powders after drying at 100°C for 24 hours, which indicates the characteristic peaks corresponding to crystalline $CaTiO_3$ are accurately matched to the three major peaks from JCPDS card number 75-2099. There were no traces of CaO and TiO_2 in powder or in calcined $CaTiO_3$ samples. The interesting result is that the crystalline of $CaTiO_3$ gel powders obtained after drying at 100°C for 24 hours. A similar result has been reported for sol-gel derived $BaTiO_3$ powders prepared from the barium alkoxid [17]. Such result indicates that the degree of crystallization of $CaTiO_3$ gel powders is dependent on the mechanisms of hydrolysis and polycondensation reactions of the mixture rather than the calcination temperature of drying powders [18].

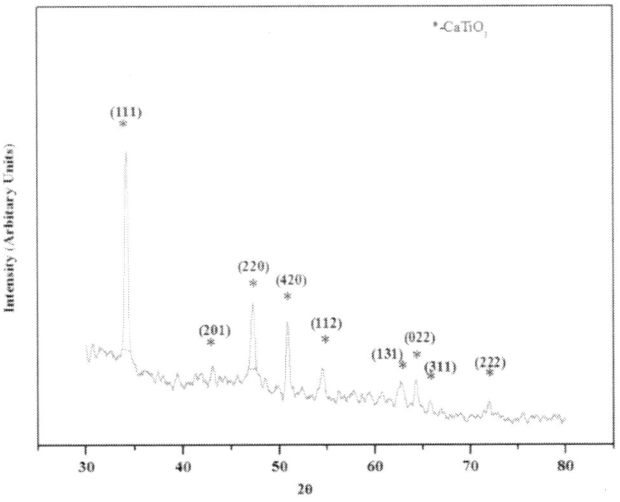

Figure 1. XRD spectrum of calcined $CaTiO_3$ sol-gel powders after drying at 100°C for 24 hours.

The above results clearly indicate that the synthesis of phase pure crystalline $CaTiO_3$ powder after drying can be possible using the equimolar mixture of CaO, ethanol and Titanium (IV) isopropoxide.

3.2. Morphological Analysis

From the figure 2 and figure 3 show the bright field image of TEM micrograph of $CaTiO_3$ particles is in the nanometer range. It can be observed that the average particle size of $CaTiO_3$ nanoparticles is approximately13 nm. But there are some agglomerated particles of size about 20 nm to 30 nm, as shown in figure 3. The agglomeration of $CaTiO_3$ nano particles is due to the effect of high surface energies between the monolithic $CaTiO_3$ crystals which is evident from the bright field TEM image, as shown in figure 3. Moreover, the results of compositional analysis using EDS attached to TEM show that all the elements are belong to the Ca, Ti and O spectrum, as shown in table 1. This indicates that the presence of impurities was not detected in any more. This suggests that the purity of sol-gel derived $CaTiO_3$ powders is better compared to powders prepared by solid state reaction or mechanochemical route of synthesis.

From the above results, XRD analysis indicates that the pure crystalline $CaTiO_3$ powder can be prepared from the above molar mixture of the starting materials without any traces of CaO and TiO_2, which is not possible to achieve through the conventional powder preparation methods such as mechanical chemical synthesis. From bright field TEM micrographs, it was observed that the approximate microcrystals size of $CaTiO_3$ with about 13 nm were obtained after drying at 100°C for 24 hours from the mixture of ethanol, CaO and Titanium (IV) isopropoxide for about 3-4 hours. As smaller the particles due to strong surface energy, these microcrystals may grow in size and their average size in reaches about 20 nm to 30 nm, as shown in figure 3. EDS analysis clearly reveals the stoichiometric ratio of $CaTiO_3$ powders, which is merely achieved through the other processing methods. Therefore, it is suggested that crystalline nano $CaTiO_3$ can be produced by described sol-gel process after drying exploit better reproducible functional and the structure properties of ceramics for the technological applications.

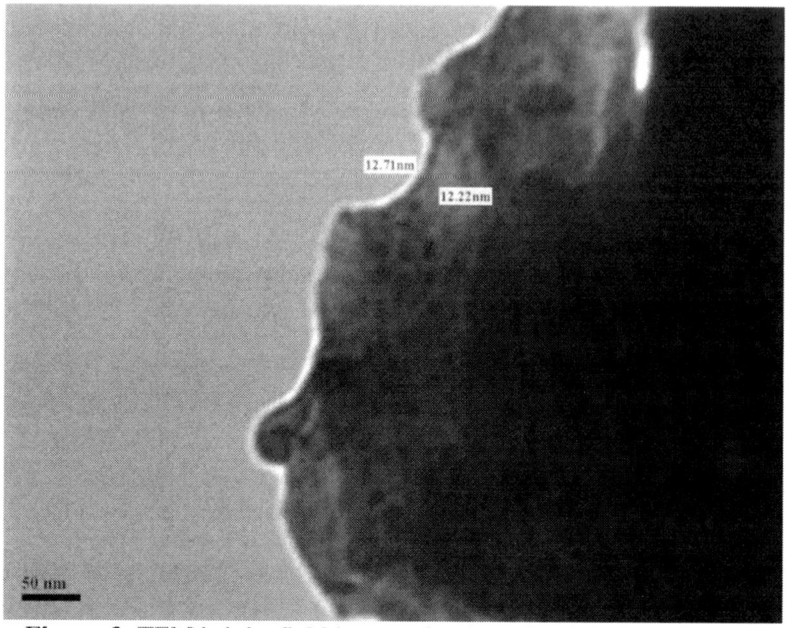

Figure 2. TEM bright-field images for the sol-gel derived $CaTiO_3$ powders after drying at 100°C for 24 hours.

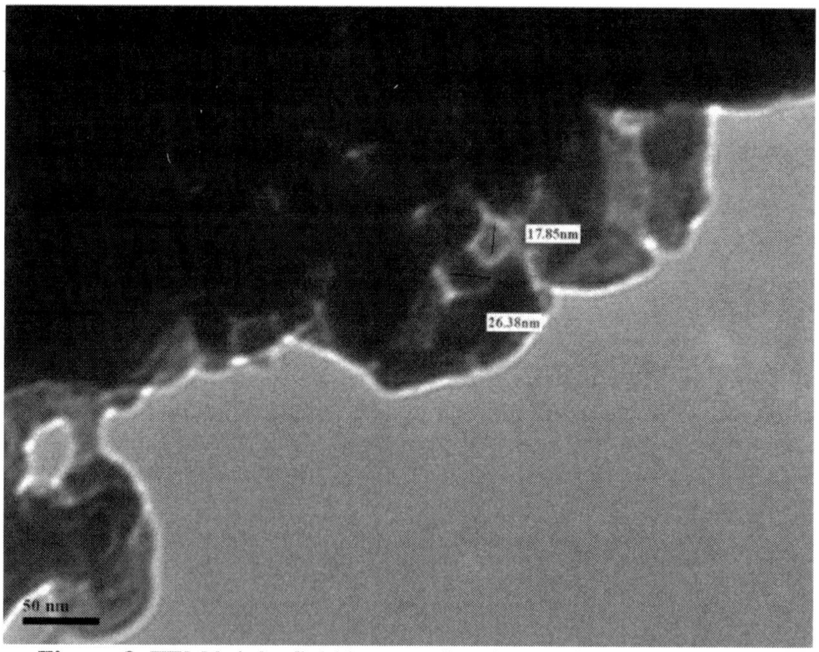

Figure 3. TEM bright-field images showing agglomerated the sol-gel $CaTiO_3$ powders after drying at 100°C for 24 hours.

Table 1: Quantitative EDS spectrum of $CaTiO_3$ sol-gel powders.

Element	Weight%	Atomic%
O K	5.27	13.27
Ca K	42.98	43.21
Ti K	51.75	43.52
Total	100	100

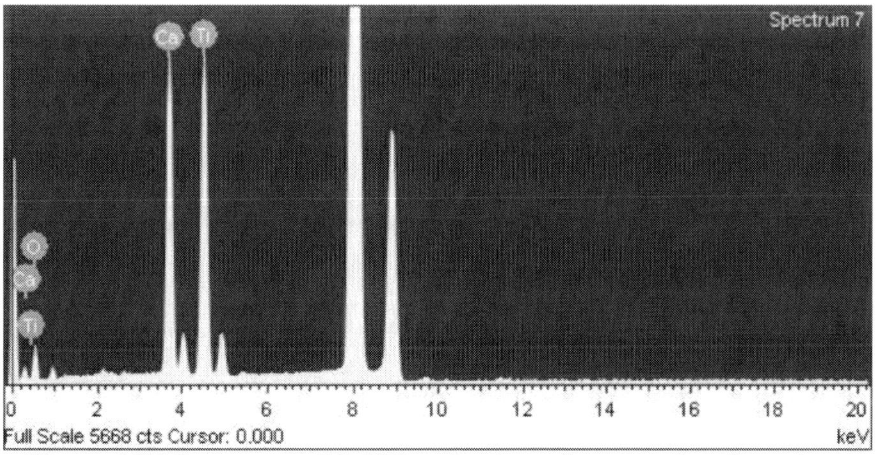

Figure 4.EDS spectrum of $CaTiO_3$ sol-gel powders was taken from above TEM bright field images.

4. Conclusion

Based on the experimental results and analysis, the following conclusions have been presented.

1. From XRD analysis confirms presence of pure crystalline $CaTiO_3$ phases after drying at 100°C for 24 hours, which cannot be synthesized from grinding the mixtures of CaO-TiO_2 powders and it also shows the absences of any trace of CaO and TiO_2 powders.
2. From TEM observation, the approximate particle size of the powders is about 13 nm and also it shows the agglomeration of nano particles of size around 30 nm.
3. EDS analysis indicates composition and Ca/Ti ratio of $CaTiO_3$ powders are equal to stoichiometric ratio 1:1.
4. Finally, it can be conclude that phase pure crystalline calcium titanate ($CaTiO_3$) nanoparticles with stoichiometric ratio can be prepared through sol-gel process after drying at 100°C for 24 hours from the equimolar ratio CaO and Titanium (IV) isopropoxide as starting materials.

Acknowledgments

Authors would like to thank to CPET Bhubaneswar for their kind help for XRD and TEM analysis.

References

[1] Thomas W, Ergun J, Doremus C, Lanford H R and William A 2003 *J. Biomed. Mater. Res. Part. A* **67** 975.
[2] Dubey A K, Tripathi G and Basu B 2010 *J. Biomed. Mater. Res. Part. B: Appl. Biom.* **95** 320.
[3] Yang Z Z, Yamada H and Miller G R 1986 *J. Mater. Sci.* **21** 405.

[4] Dubey A K, Mallik P K, Kundu S and Basu B 2013 *J.of Eur. Cer.Soc.* **33** 3445.
[5] Mallik P K and Basu B 2013 *J.of Biomed. Mat. Res. Part. A* **00**000.
[6] Thriviraman G, Mallik P K and Basu B 2013 *Biomaterials* **34** 7073.
[7] Shrivastava O P, Shrivastava R 2002 *Prog Cryst Growth Charact Mater.* **45** 103.
[8] Holliday S, Stanishevsky A 2004 *Surface & Coatings Tech.***188**741.
[9] Gopalakrishnamurthy H S, Rao M S, Kutty T R N1975 *Thermochim. Acta.* **13** 183.
[10] Lee S J, Kim Y C, Hwang J H 2004 *J. Ceram. Process. Res.***5** 223.
[11] Patil K C, Aruna S T and Ekambaram S 1997 *Current Opinion in Solid State & Mat Sci.* **2**156.
[12] Palaniandy S, and Jamil N H 2009 *J. of Alloys and Compounds* **476** 894.
[13] Mi G, Murakami Y,Shindo D, and Saito F1999 *Powder Technology* **104** 75.
[14] Brankovi G,Vukoti V, Brankovi´Z, Varela J A 2007 *J. of the Eur. Cer. Soci.***27** 729.
[15] Koch C C 2003 *Rev. Adv in Mat. Sci.* **5** 91.
[16] Bal´aˇz P 2008 *Mechanochemistry in Nanoscience and Minerals Engineering* (Springer-Verlag).
[17] Mardiynmi K S, Dolloff R T, Smith II J S 1969 *J. Am. Ceram. Soci.* **52** 523.
[18] Pfaff G1994 *Chem.Mater.***6** 58.

IOP Conf. Series: Materials Science and Engineering **75** (2015) 012006 doi:10.1088/1757-899X/75/1/012006

Designing of Sub-entry Nozzle for Casting Defect-free Steel

Anupal Sen, B Prasad, J K Sahu and J N Tiwari

OCL India Ltd., Rajgangpur, Odisha, India

E-mail: sen.anupal@ocl.in

Abstract. Production of defect-free steel is a continuous demand for each & every steel maker. The design of refractories used in steel making process has a vital role on this. Manufacturing of cleaner steel is mostly attributed to the continuous casting process of the steel plant. Defects in steel mainly come from the non-metallic inclusions present in steel that are incorporated during various steel making process. During continuous casting, the main focus is to remove these inclusions to cast defect free steel. The flow pattern of molten steel in the mould plays a vital role in the removal of these impurities. The turbulence of the flowing steel in the mould should be such that more and more impurities will be floated at the meniscus of the mould and they will be captured by the casting powder. But, this turbulence should not be higher enough to capture the particles of casting powder inside the molten steel. So, in order to optimize the turbulent intensity and to get uniform distribution of temperature and velocity in the mould, the design of Sub-entry nozzle (SEN) is very crucial. SEN is used to guide the flow of steel from tundish to mould during continuous casting. The bore configuration and outlet design of SEN along with its immersed depth in the mould determine the flow pattern of steel inside the mould, which is the key factor to produce cleaner steel. This paper focuses on the different designing aspects of SEN to get an optimized and uniform flow of steel in the mould. The designing parameters include bore diameter, bore configuration, port shape, port angle, port dimension, number of ports and immersed depth of SEN inside the mould. Design optimization is done to improve the steel quality by removing the impurities.

1. Introduction

The present global economic crisis has compelled the steel makers to produce cleaner steel at lower cost. Many robust processes & sophisticated equipments are used to produce defect-free steel but it involves a lot of cost. This has opened up a challenge for the refractory makers to design refractories in such a way that defect-free steel can be casted without the use of high-cost sophisticated equipments.

Submerged nozzle or Sub-entry nozzle (SEN) is used for shrouding the flow of steel from tundish to mould during continuous casting of steel. It prevents re-oxidation and temperature drop of steel by prohibiting the exposure of steel to atmosphere. They are carbon bonded refractories. Generally, the body part of SEN is made of alumina-carbon and the slag band of SEN i.e., the portion of SEN that remains immersed into the pool of mould powder, is made of zirconia-carbon due to excellent corrosion resistance of zirconia against the nascent fluorides in mould powder.

Mould powder is given on the molten steel surface in the mould for lubrication & insulation of mould as well to protect steel from re-oxidation & freezing. It also absorbs the non-metallic inclusions from steel and makes cleaner steel. A part of SEN remains immersed inside this mould powder. During casting, the mould powder melts and flows in between mould and steel. For proper functioning of mould powder, temperature at the steel-mould powder interface should be optimum. Temperature

Content from this work may be used under the terms of the Creative Commons Attribution 3.0 licence. Any further distribution of this work must maintain attribution to the author(s) and the title of the work, journal citation and DOI.
Published under licence by IOP Publishing Ltd

IOP Conf. Series: Materials Science and Engineering **75** (2015) 012006 doi:10.1088/1757-899X/75/1/012006

distribution depends on the turbulent intensity of flowing steel in the mould and distribution of turbulent intensity, in turn, depends on the design of SEN and velocity of steel coming out of SEN bore. If the turbulent intensity is more at the steel-mould powder interface, there is chance of mould powder infiltration in the steel which will cause defects. On the other hand, if the turbulent intensity is less at the mould surface, temperature will also be low. This will result into non-melting of mould powder and subsequent incorporation of inclusions & defects in steel. Moreover, the non-metallic inclusions in steel, coming from the refractories or other sources during secondary steel melting, do not get absorbed by the mould powder if it is non-melted. This again results into defects in steel. So, the design of SEN plays a very important role in producing defect-free steel.

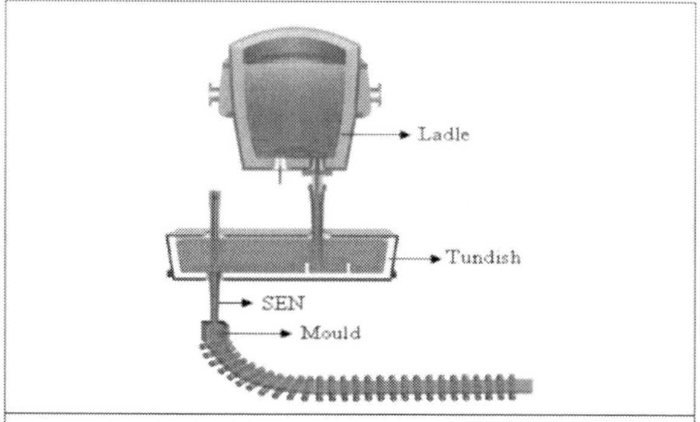

Figure 1. Use of SEN in continuous casting of steel

2. Designing of SEN by varying number of ports

In bloom or billet casters, single bore SENs are mostly used. However in bigger blooms (more than 250 mm square or round), SENs having 4 side ports should be used. In single bore SENs, since the flow of steel is at the downward direction hence turbulent intensity at the steel-mould powder interface is lesser. Temperature is also low at steel-mould powder interface resulting into more chance of mould powder freezing which may cause centre segregation defects in steel. Figure 2 shows the mathematical modeling (Ansys Fluent CFD) of temperature distribution of single port and 4-port SEN in 300 mm square mould. Modelling is done following 2D symmetric, realizable k-ε model. 2 ports are shown here instead of 4 ports as this is 2D view. It is observed that temperature at meniscus is low in case of single port SEN which may cause mould powder freezing.

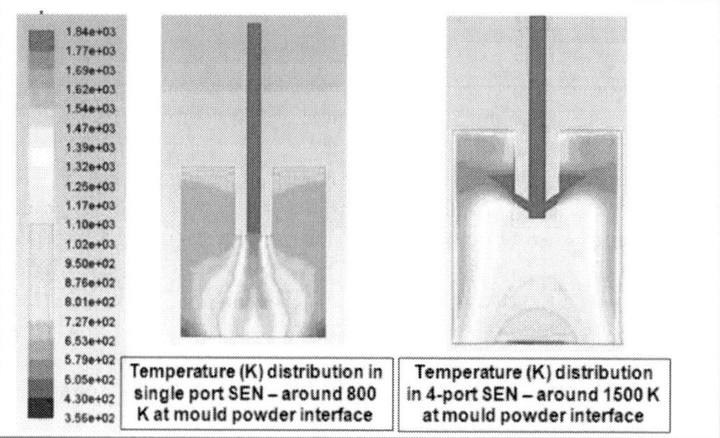

Figure 2. Temperature distribution of single port & 4-port SEN

Figure 3 shows the velocity distribution, where it is found that the optimum velocity (0.2-0.4 m/s) is observed in 4-port SEN, but in case of single port SEN it is low. In figure 4, turbulence intensity distribution is shown. Turbulence Intensity for the 4-port SEN at the steel-mould powder interface is less than 0.05 N.m. Turbulence intensity adjacent to the walls are also very low and uniformly distributed throughout the mould. This condition is optimum for inclusion absorption at the meniscus.

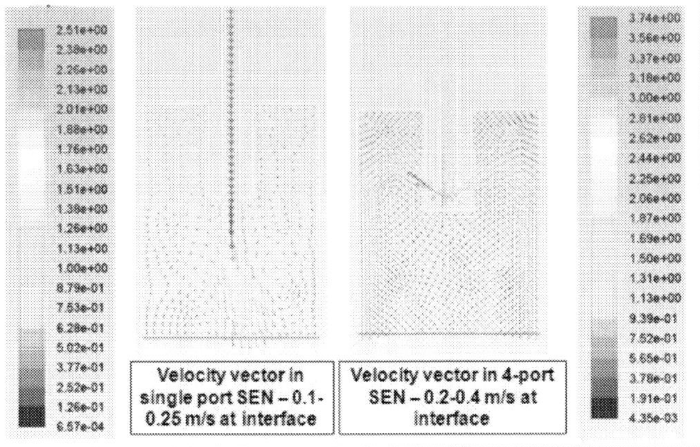

Figure 3. Velocity distribution of single port & 4-port SEN

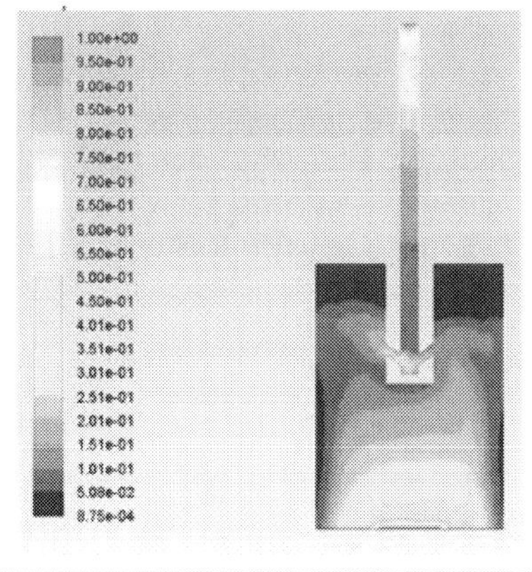

Figure 4. Turbulent Intensity distribution of 4-port SEN

3. Varying Immersed depth of SEN inside mould

Mathematical modelling is done in a mould section of 250 mm square and 500 mm height with different immersed depth of 180 mm, 150 mm and 120 mm. The temperature distribution is shown in figure 5. For 180 mm immersed depth, temperature is very low (around 500 K) at steel-mould powder interface adjacent to mould wall. This will result into non-melting of mould powder and substantial entrapment of inclusion in liquid steel. For 150 mm immersed depth, temperature is slightly more than previous case, but still lower (around 650 K) than required at steel-mould powder interface. Hence, there will be some entrapment of inclusion in liquid steel. For 150 mm immersed depth, temperature is around 800 K at steel-mould powder interface, which is closer to the required temperature of 1000 K. Hence, entrapment of inclusion in liquid steel will be less.

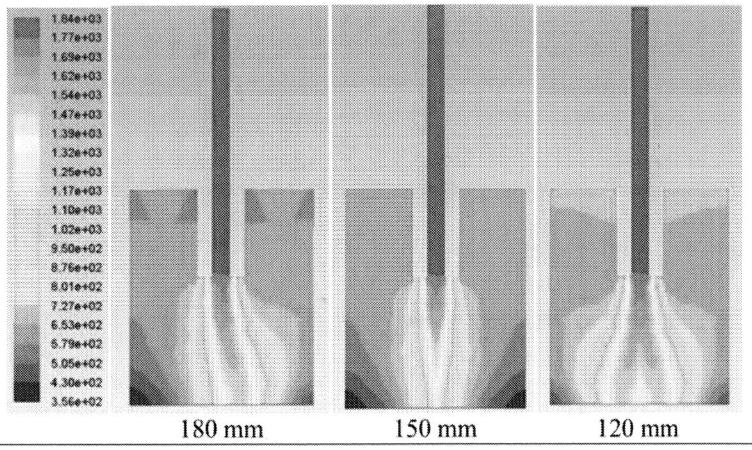

Figure 5. Temperature contour with different immersed depths

4. Effect of port angle on steel flow

In slab caster, the port angle of SEN plays a vital role on the flow pattern of steel in mould. In most slab casters, 2-port SENs are used. Molten steel coming out of the port has a double roll flow pattern in the mould. If flow intensity is more than optimum at the upper roll, then meniscus turbulence is more resulting into entrapment of mould powder inclusion in steel. On the other hand, if flow intensity is more than optimum in the lower roll, then chance of inclusion floatation is less and chance of mould powder freezing is more. Port angle is one of the parameters that determine the optimum distribution of flow of steel in mould.

Figure 6 shows that the jet angle or the angle of incoming steel with the mould surface changes with the port angle. For 10 deg port angle, the jet angle is 31 deg, for 15 deg port angle it is 33 deg and for 20 deg port angle it is 34 deg. If the jet hits the top half of the mould, then flow in upper roll will be more. Similarly, when the jet hits the bottom half of the mould, the flow in lower roll is more. For optimum flow pattern in which the linear velocity at the mould meniscus remains in between 0.2 to 0.4 m/s and turbulent intensity is below 0.05 N.m, jet angle should be such that the incoming steel hits 0.4-0.6 of the total mould height.

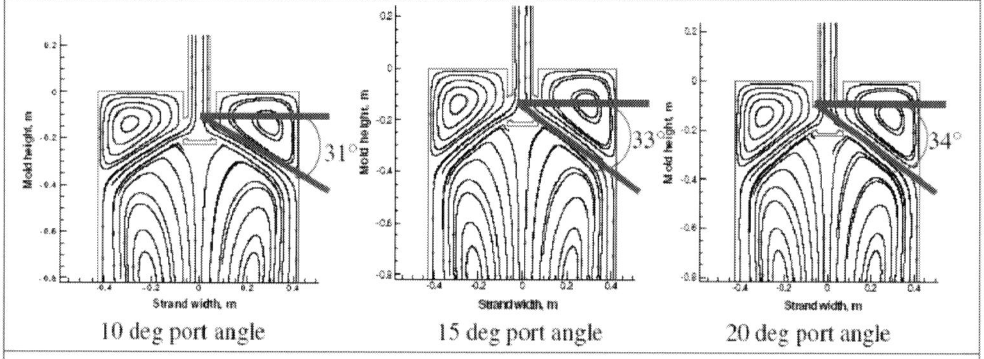

Figure 6. Variation of jet angle with port angle

If the jet angle is very high and the incoming steel hits the mould wall beyond 0.6 of mould height, then there may be chance of break-out. For lesser section width, steel jet hits at lower part of the mould even with lower jet angles. Hence, port angle will be lower for lesser mould sections. In figure 7, it is seen that with same port angle & jet angle, the steel jet hits at lower part of mould when the casting section is bigger.

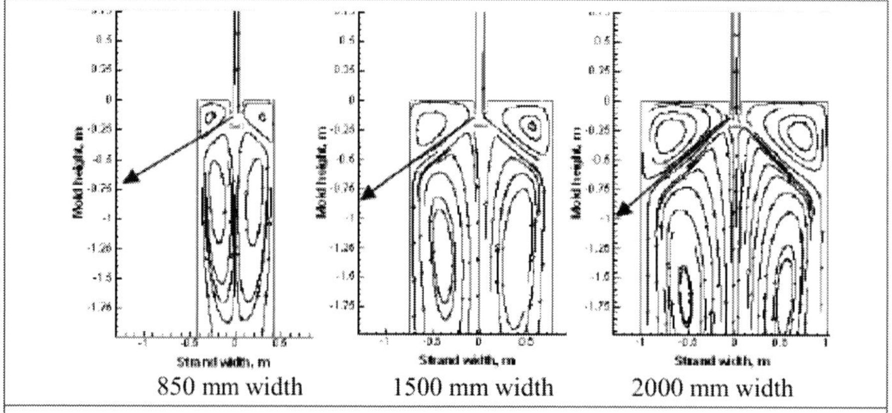

Figure 7. Variation of jet trajectory with mould section

5. SEN with different bore diameters

In figure 8, the turbulent intensity contour has been shown for two different bore diameters 80 mm and 90 mm. It is found that with higher bore diameter the velocity of steel coming out of the port is less resulting into less turbulence at the meniscus. So, if high turbulence and mould powder entrapment is a problem, then bore diameter can be increased for optimized flow.

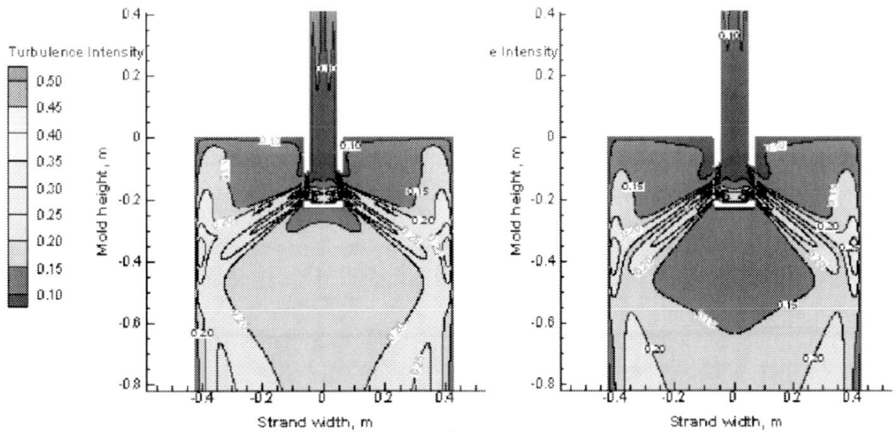

80 mm bore diameter 90 mm bore diameter

Figure 8. Turbulent intensity contour for different bore diameter

6. SEN with different port shapes

In figure 9, the turbulent intensity contour is shown for rectangular shaped port and oval shaped port. In rectangular port, turbulent intensity is less than 0.1 N.m at mould powder-steel interface. Hence, chance of mould powder freezing is there which will increase NMI in steel. In oval port, Turbulence intensity is more than 0.15 N.m. which is optimum. Hence oval or round shaped port is always preferred over the angular ports for better flow patterns.

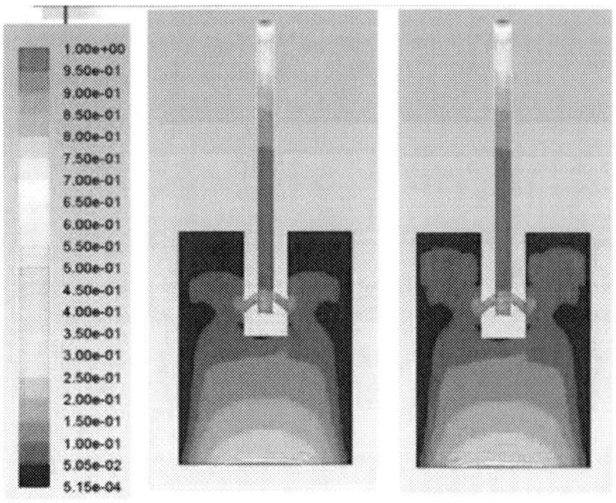

Rectangular port Oval port

Figure 9. Turbulent intensity contour for different port shapes

IOP Conf. Series: Materials Science and Engineering **75** (2015) 012006 doi:10.1088/1757-899X/75/1/012006

7. Conclusion

Defect-free steel can be produced by optimizing the flow properties of steel in the mould. Optimum temperature distribution, velocity distribution and turbulent intensity inside the mould results into floatation & absorption of impurity at the mould surface by casting powder. Higher turbulent intensity at the mould meniscus may also result into entrapment of mould powder in steel. On the other hand, lower turbulent intensity can cause freezing of mould powder which may result into centre segregation defects in steel.

There are different design parameters of SEN that can be varied to get optimum flow pattern of molten steel inside the mould. The designing aspects of SEN include bore diameter, bore configuration, port shape, port angle, port dimension, number of ports and immersed depth of SEN inside the mould. With the optimized design of SEN, defect free steel can be produced through uniform and optimized flow pattern of steel in the mould.

References

[1] Kuroda T, Mizobe A and Kurisu J 2011 *UNITECR* 1-B2-16

[2] Tianfei M, Guoqi L, Wengang Y and Jianbin Y 2009 Structure optimization of submerged entry nozzle for super wide slab continuous casting by numerical simulation *UNITECR*

[3] Huixiang Y, Junming Z and Wanjun W 2002 Numerical simulation of fluid flow in continuous slab-casting mold *Journal of University of Science and Technology Beijing* **24**(5) 492-496

[4] Fanjun M, W and Gang L 2006 Numerical value simulation of hot metal flowing in slab concaster mould *Journal of Iron and Steel Research* **1** 10-14

[5] Kubota J 1990 Mechanism of level fluctuation and mold powder casting in slab continuous casting at High Speed *Proc. of 6th Int. Iron and Steel Congress (Nagoya)* 356-362

[6] Yuan Q, Vanka S P and Thomoas B G 2004 Computational and experimental study of turbulent flow in a 0.4-scale water model of a continuous steel caster *Metallurgical and Materials Transaction* **35B**(10) 967-982

[7] Tianfei M, Guoqi L and Xinfu W 2008 Water model investigation of fluid field in mold of wide slab continuous caster *Refractories* **42**(5) 341-344

Development of Cu Reinforced SiC Particulate Composites

Harshpreet Singh*, Lailesh Kumar, Syed Nasimul Alam

Department of Metallurgical and Materials Engineering, National Institute of Technology Rourkela, Orissa, Pin-769008, India

*Email- harshpreet00@gmail.com

Abstract. This paper presents the results of Cu-SiCp composites developed by powder metallurgy route and an attempt has been made to make a comparison between the composites developed by using unmilled Cu powder and milled Cu powder. SiC particles as reinforcement was blended with unmilled and as-milled Cu powderwith reinforcement contents of 10, 20, 30, 40 vol. % by powder metallurgy route. The mechanical properties of pure Cu and the composites developed were studied after sintering at 900°C for 1 h. Density of the sintered composites were found out based on the Archimedes' principle. X-ray diffraction of all the composites was done in order to determine the various phases in the composites. Scanning electron microscopy (SEM) and EDS (electron diffraction x-ray spectroscopy) was carried out for the microstructural analysis of the composites. Vickers microhardness tester was used to find out the hardness of the samples. Wear properties of the developed composites were also studied.

1. Introduction

Composite materials offer several applications in the field of aerospace, automotive and ship building industry as it has different advantages over other conventional metals.Metal matrix composites (MMCs) combine both metallic properties such as toughness and ductility and ceramic properties such as modulus and high strength.They possesshigh shear strength and provide service at high temperatures. MMCs offer the possibility to tailor the properties of a metal by adding an appropriate reinforcement to develop materials having superior mechanical properties.Copper has high thermal conductivity and used as a structural material for cooling. In order to increase its high temperature properties different reinforcements are being used. The objective is to develop Cu based metal matrix composites by mechanical alloying (MA) route having high strength and microstructural stability at elevated temperatures [1-4].

Metal matrix composites are becoming an important structural material in engineering application. These composites find application in the area of aerospace, aircraft and automotive industries because of their excellent mechanical properties such as high specific strength, improved wear resistance and high specific modulus. Apart from this they also have very good thermophysical properties, such as low coefficient of thermal expansion and high thermal conductivity. Recently, metal-ceramic composites with high ceramic contents have become an area of focus for thermal management applications. Copper is widely used in thermal and electronic applications, due to the high electrical(5.96×10^7 S/m) and thermal conductivity (401 W/m K), good corrosion resistance and high melting point (1357.77 K). However, the low mechanical property of Cu at both room and high temperatures limits the extensive application of pure Cu. Discontinuously reinforced metal matrix composites are a class of materials that exhibit superior properties. In the present study, the focus was

Content from this work may be used under the terms of the Creative Commons Attribution 3.0 licence. Any further distribution of this work must maintain attribution to the author(s) and the title of the work, journal citation and DOI.
Published under licence by IOP Publishing Ltd

towards development of Cu based metal matrix composites with SiCp reinforcement by powder metallurgy. It is well known that the particle reinforced metal matrix composites have excellent mechanical properties and light weight due to the addition of the high strength and high modulus particles like TiC, Al_2O_3, SiC, TiB_2 etc. Silicon carbide (SiC) is widely studied for its high temperature stability, high thermal conductivity, high hardness and high heat capacity. Its melting point is above 2600°C, and the thermal conductivity is 16.7 W/m K. These properties present SiC the possibility as potential additive to copper based metal matrix composites [5-8].

2. Experimental

Pure Cu powder procured form Loba Chemie India and SiC powder from Search Chem having 99% purity was used for the current research work.Initially Cu was milled for 20 h in a Fritsch P-5 planetary ball mill using hardened chrome steel vials and balls in order to synthesize nanostructured elemental Cu. Green pellets of unmilled Cu and as-milled Cu with different volume fractions of SiCp (10, 20, 30 and 40%) are formed by blending the constituents followed by cold compaction of the samples in an uniaxial compaction machine under a load of 665MPa. Sintering of the samples were done at 900°C for 1 h in Ar atmosphere. Vickers microhardness test was carried out over all the samples to determine the hardness. Densities of the prepared composites were also determined by Archimedes principle. Optical micrographs of the samples were taken using Carl Zeiss microscopes.A JEOL JSM-6480LV scanning electron microscope (SEM) equipped with an INCAPentaFET-x3 x-ray microanalysis system with a high-angle ultra-thin window detector and a 30 mm^2 Si (Li) crystal was used for microstructural characterization of the samples. A JEM - 2100 JEOL high resolution transmission electron microscope (HRTEM) with point to point resolution 0.194 nm was used for analysing the milled powder. A Phillips Pananalytical PW 3040/00 x-ray diffraction (XRD) was also used to characterize the milled Cu-powder and the various phases in the Cu-SiCp composites. Cu Kα radiation having λ =1.5409 Å was used for all x-ray diffraction analysis.A DUCOM ball on plate tribometer was used for study the wear characteristics of all the composites. Wear test were performed on 15 mm diameter samples of the various Cu-SiCp composites in a ball-on-plate wear tribometer having a diamond indenter under ambient environment. Dry sliding wear tests were performed on all the samples. A normal load of 20 N was applied for a period of 10 minutes.

3. Result and Discussion

The results for HRTEM analysis of 20 h milled powder is shown in figure 1(a), which indicates that the powder has been reduced to nanometric dimension after high energy milling for 20 h. Figure1(a) is the bright field TEM image of the 20 h milled Cu sample. The SAD pattern in figure 1(b) shows complete ring pattern which suggest that the 20 h milled Cu powder has nanometric dimension. The ring patterns could be indexed to the various crystallographic planes of Cu.

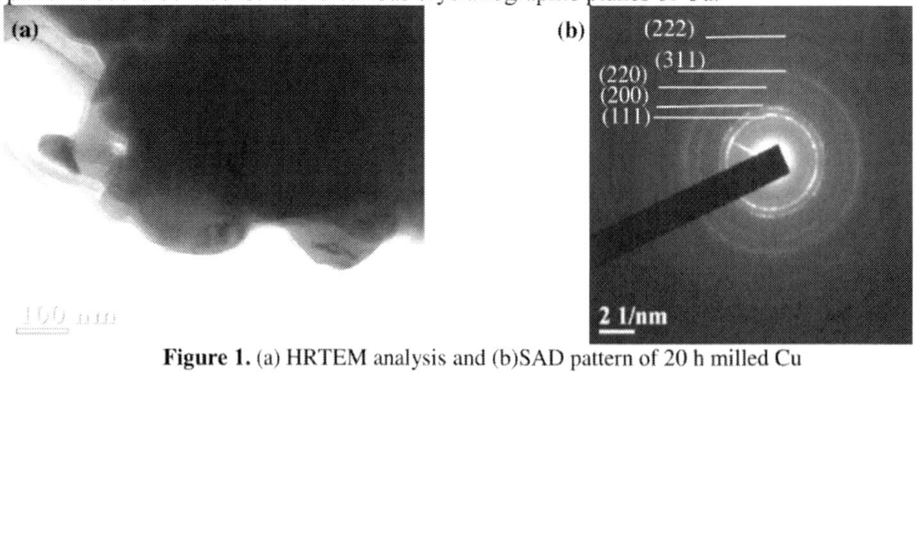

Figure 1. (a) HRTEM analysis and (b)SAD pattern of 20 h milled Cu

The x-ray diffraction of Cu powder milled for various durations of time are shown in figure 2(a). The various peaks in the x-ray diffraction pattern could be indexed to the different crystallographic planes of Cu. No other peaks could be detected in the x-ray pattern suggesting that there is no contamination from the milling media. Voigt's method was used for calculating the crystallite size of the milled Cu powder. After 20 h of milling the crystallite size of Cu was found to be 18 nm (figure 2(c)). There is a gradual drop in the crystallite size with milling time. From figure 2(c) it is evident that Cu could be reduced to nanometric dimension within 5 h of milling. Cold welding was also evident between 5 to 10 h of milling which led to the increase in crystallite size. There is a gradual increase in the lattice strain due to severe deformation of the milled powder and strain reaches a maximum value after 20 h of milling when the crystallite size is smallest (figure 2(d)). Lattice parameter of milled Cu was also determined using the Nelson-Riley function. The lattice parameter of Cu shows a slight increase during milling for 20 h [9-13].

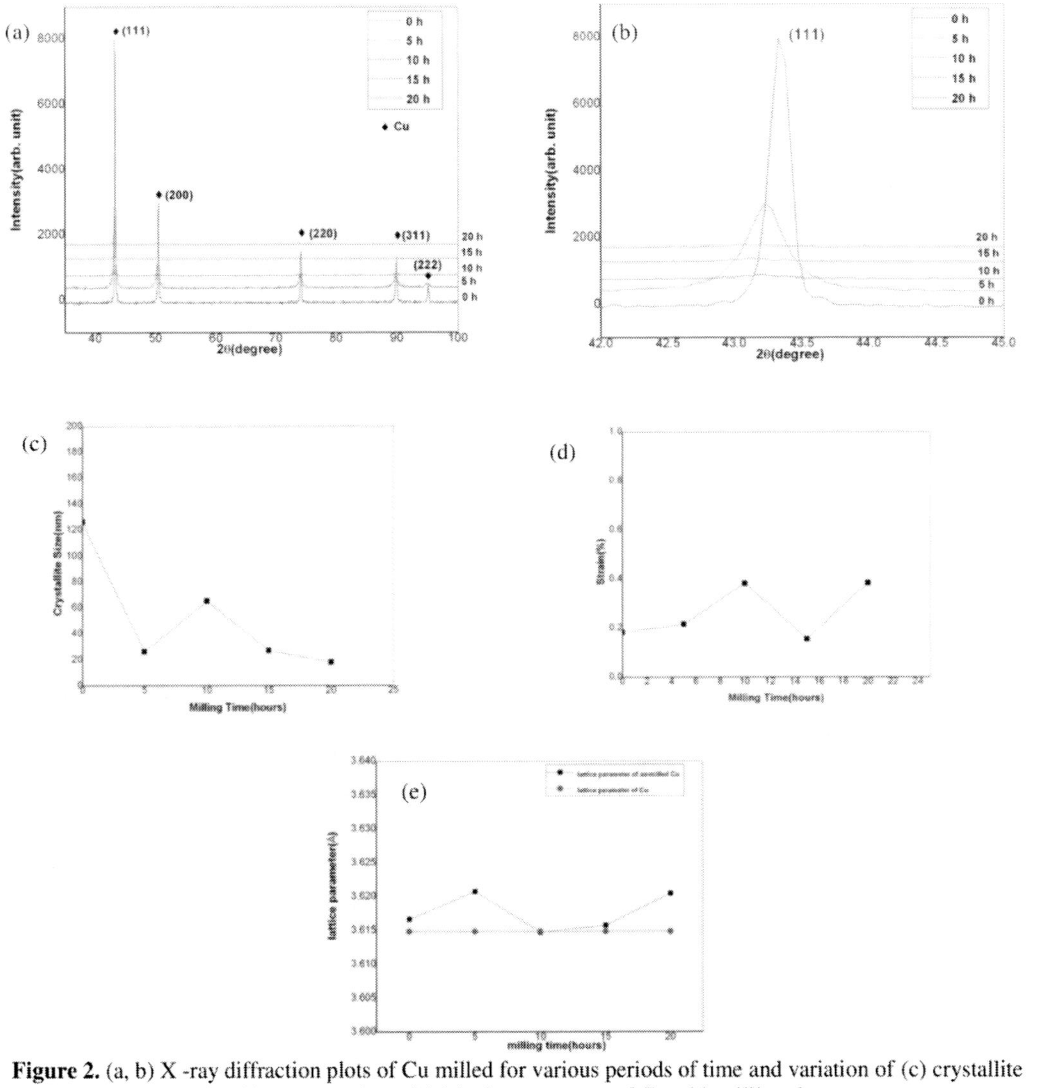

Figure 2. (a, b) X -ray diffraction plots of Cu milled for various periods of time and variation of (c) crystallite size (d) r.m.s. strain and (e) lattice parameter of Cu with milling time.

IOP Conf. Series: Materials Science and Engineering **75** (2015) 012007 doi:10.1088/1757-899X/75/1/012007

The optical micrographs of various Cu-SiCp composites in figure 3 (a-d) shows that SiCp is uniformly distributed in the Cu matrix. The SiCp are found to be less than 100 μm in size. The particles are discontinuously dispersed in the Cu matrix. Figure 4 is the SEM image of the various Cu-SiC composites developed using for different vol. % of SiC particle reinforcement. From the microstructure it is clear that the SiC particulates are homogenously distributed over the Cu matrix.

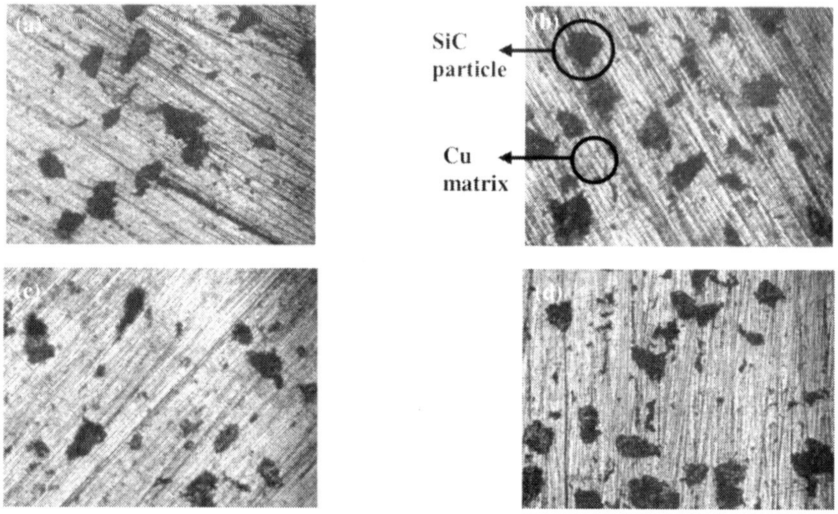

Figure 3. (a-d) Optical micrographs of unmilled Cu- SiC composites (10, 20, 30 and 40 vol. % of SiC).

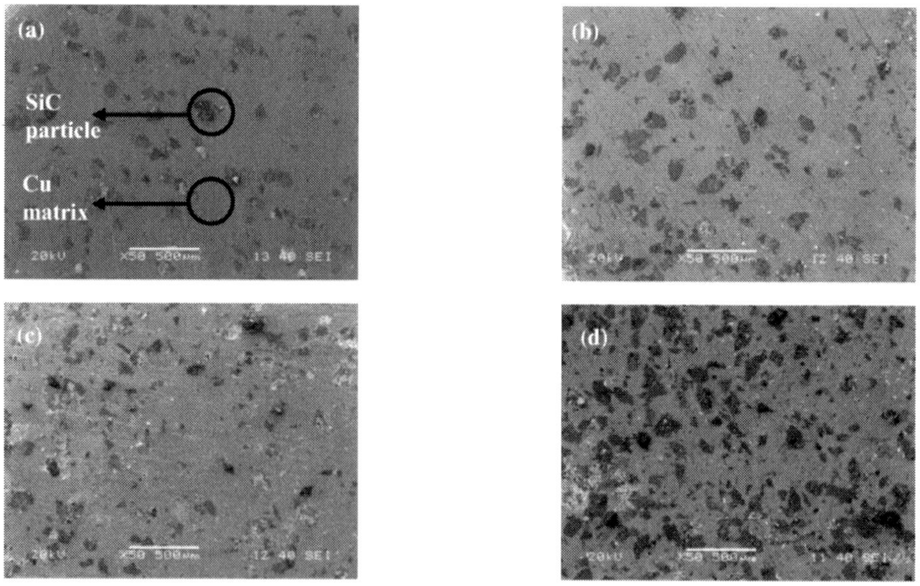

Figure 4. SEM images of unmilled Cu-SiC composites (a) 10 vol. %SiCp (b)20 vol. %SiCp(c) 30 vol. % SiCp and (d) 40 vol. %SiCp.

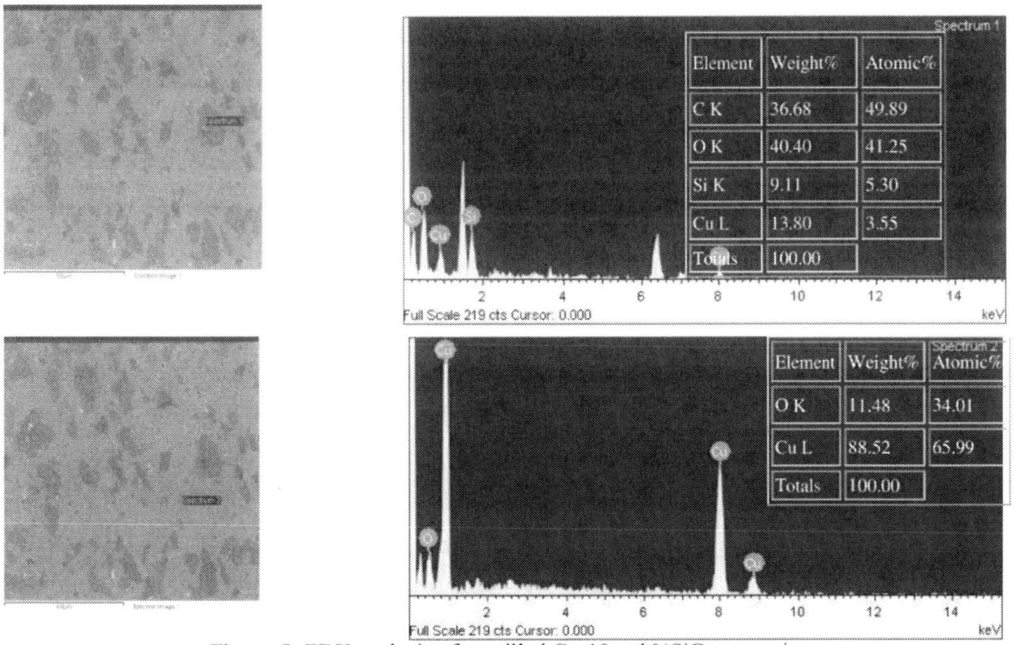

Figure 5. EDX analysis of unmilled Cu-10 vol.%SiC composite

The EDS analysis in figure 5 suggests that a dark region in the SEM image corresponds to the SiC particles and light coloured region corresponds to the Cu matrix. The dark coloured SiCp are surrounded by the light coloured Cu rich region. EDX analysis shows the presence of oxygen in the sintered sample. This is due to the undesirable oxygen present during sintering in the argon atmosphere. The x-ray diffraction analysis shown in figure 7 confirmed the formation of Cu_2O phase.

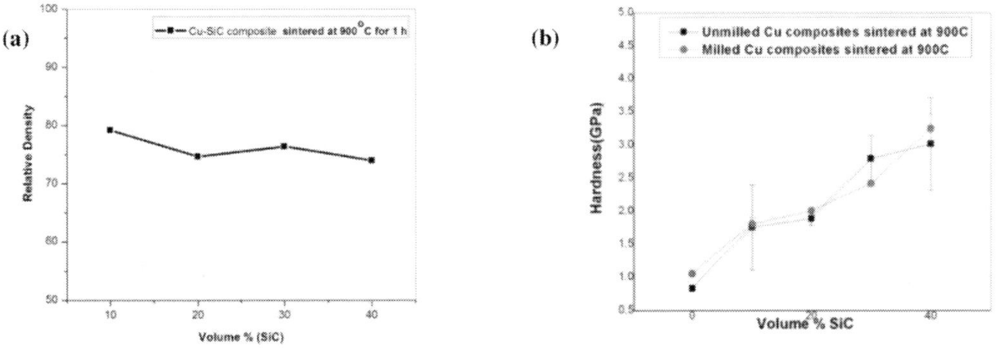

Figure 6. (a) Variation of density of unmilled Cu-SiCp composites and (b) variation of micro hardnessof as-milled and unmilled Cu-SiCp composites for different volume percent of SiCp

The figure6(a) shows the density plot for unmilled Cu SiC composites for different vol. % of reinforcement. The density plots show the gradual decrease in relative density with increase in vol. % of SiCp reinforcement in Cu matrix. The hardness plot shows the comparison between the unmilled Cu-SiCp composites and as milled Cu-SiCp composites for different vol. % of reinforcement. The hardness increases with the increase in vol. % of reinforcement. It can be seen from the figure 6(b) that as milled Cu-SiCp composites shows better hardness than unmilled Cu-SiCp composites. Milled Cu due to its smaller particle size leads to better sinterability and as a result gives higher hardness of the composites [14-16].

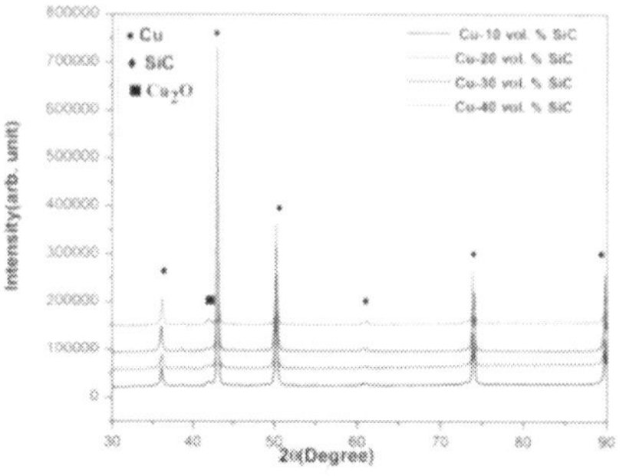

Figure 7. XRD plots for unmilled Cu-SiCp composites.

The figure 7 shows the x-ray diffraction and for unmilled Cu SiC composites for different vol. % of reinforcement. The higher content of SiCp in the composites leads to more intense x-ray diffraction peak corresponding to SiC. The x-ray diffraction shows peaks for Cu and SiC confirming the matrix and reinforcement material. The peaks confirm some formation of copper oxide (Cu_2O) layer in the composites.

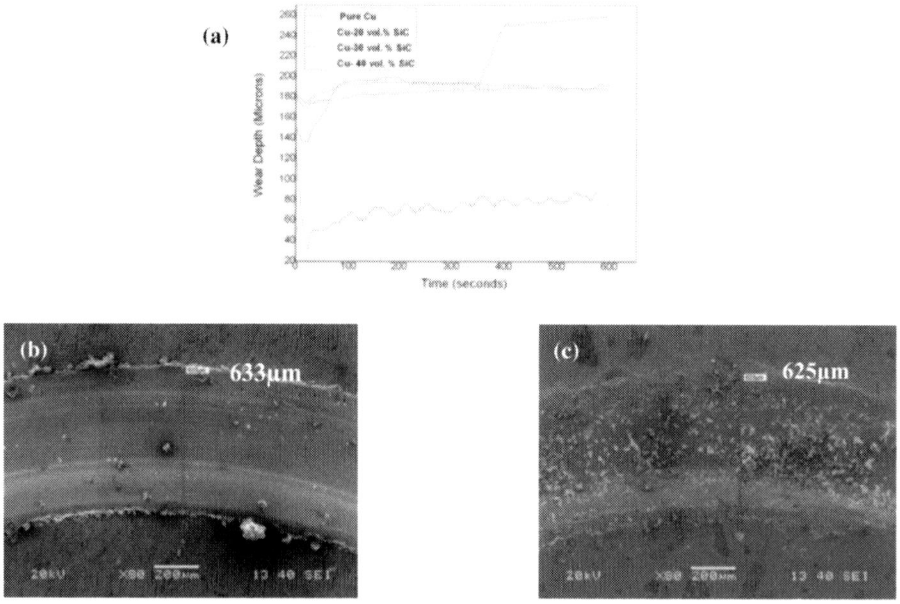

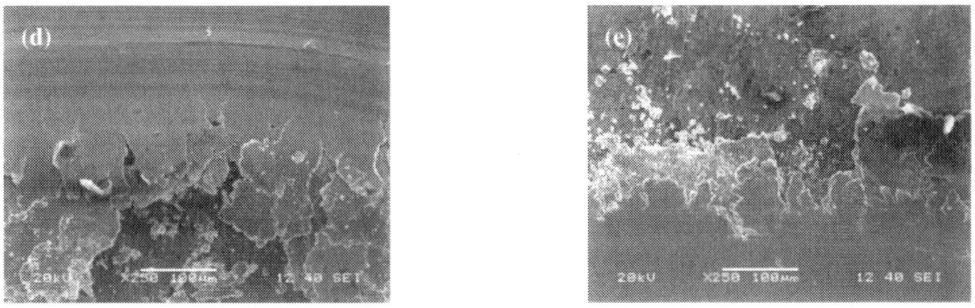

Figure 8. (a)Wear Characteristic of unmilled Cu- SiC composites. SEM images of the wear track of (b) unmilled Cu (c) unmilledCu-40 vol. % SiCp composite. (d-e) High magnification SEM images of the wear track of unmilled Cu-40 vol. % SiCp composite.

Figure 8 (a) the shows wear characteristic for unmilled Cu-SiCp composites. The plot shows that the wear resistance of the Cu-SiCp composites are higher than native Cu. It can also be observed that as the vol. % of the SiC particles is increased the wear resistance of the Cu-SiCp is increased. Cu-40 vol. % SiCp shows the least wear depth (figure 8(a)). The increase in hardness due to the addition of higher vol. % of hard SiCp phase leads to the decrease in wear depth. The drastic reduction in wear rate may be attributed to the enhancement in hardness of the composite reinforced by SiC particles and greater reduction of direct load contact between the Cu-SiCp composite surface and ball in comparison with the pure Cu due to the load bearing component action of hard SiC particles. Figure 8(b) shows the SEM images for wear track of the pure Cu and unmilled Cu-40 vol. % SiCp composite. From the images it is evident that Cu-40 vol. % SiC composites show less width of wear track as compared to pure Cu [17-20].

Figure 8(a) shows the wear behaviour of pure Cu and Cu-SiCp composites. It was found that the hardness of Cu-SiCp composites increased with the addition of SiCp due to the presence of the harder ceramic particles of SiC. The hard SiC particles support the stresses between the contact surfaces preventing large plastic deformations and abrasions. This reduces the amount of worn material. The Cu-SiCp composites containing the hard SiC particles show better wear resistance due to the formation of a tribolayer at the interface. During wear rapid removal and regeneration of the tribolayer at the interface takes place.The interface temperature also increases during the wear test and this causes oxidation of the sample. The increase of interface temperature during the wear test significantly enhances the loss of mass by oxidation, matrix softening, cracking and delamination wear. In the initial stages of wear, fine particles are generated from the two surfaces in contact by the micro-cutting and rubbing effects. Sharp SiC particle edges protrude out of the Cu-SiCp composite surface. The wear particles entrapped by the contacting surfaces undergo a mechanical mixing process, which is very similar to the mechanical alloying (MA) process. During milling some original particles fracture further and expose atomically clean surfaces that come in contact with each other. With further mixing, cold welding and particle fracturing takes place which leads to a steady state particle size distribution in the mixture. Few fine particles are dislodged from the interface whereas some agglomerate and pile up. Due to the pressing and flattening effect of the normal load and frictional forces a compact mechanically mixed layer on the stationary and relatively soft composite surface is formed. This layer acts as a protective layer for the composite. It should be noted that due to the fine size of the debris particles and high temperature at the contact the metal components like Cu can be oxidized very rapidly [21-24].

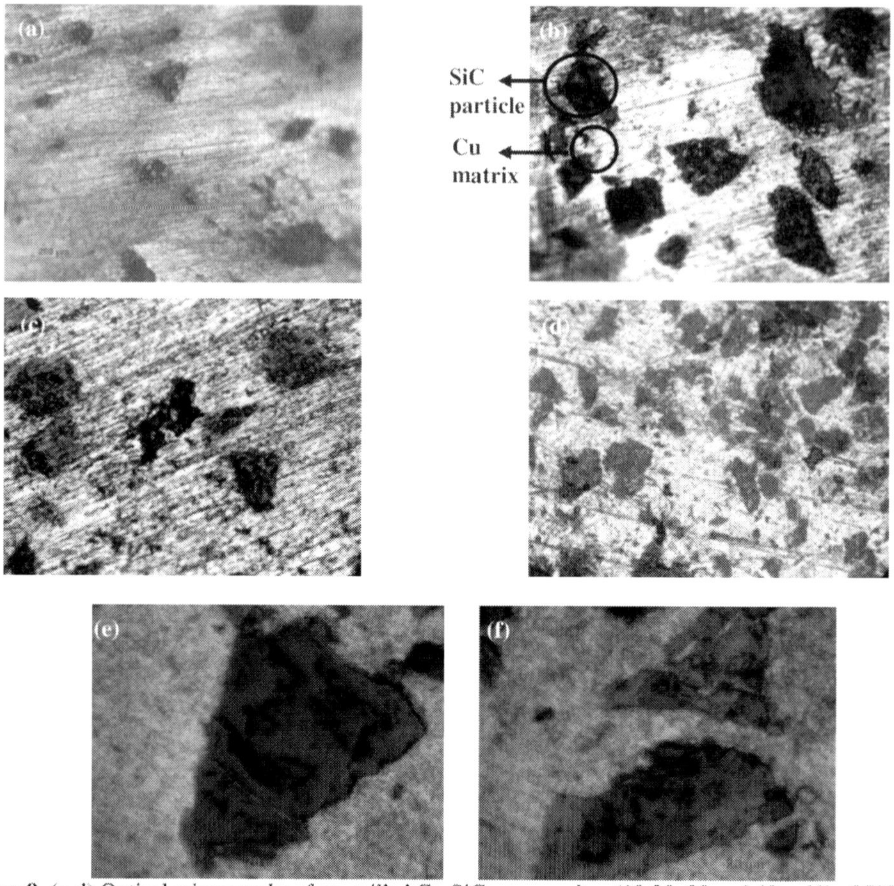

Figure 9. (a-d) Optical micrographs of as- milled Cu-SiCp composites (10,20, 30 and 40 vol% of SiCp) (e, f) Higher magnification optical micrographs of as-milled Cu-40 vol. % SiCp composites.

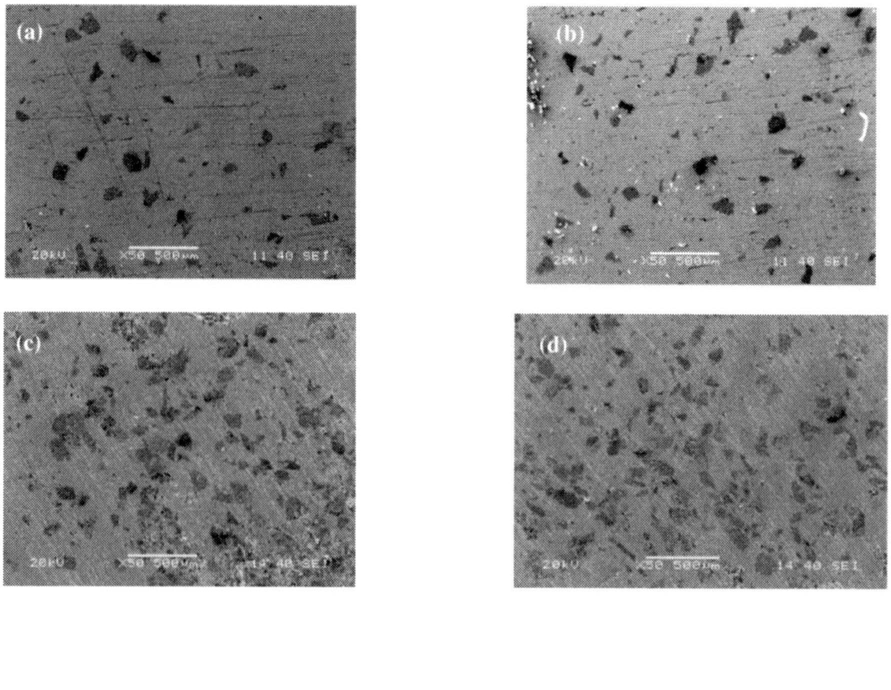

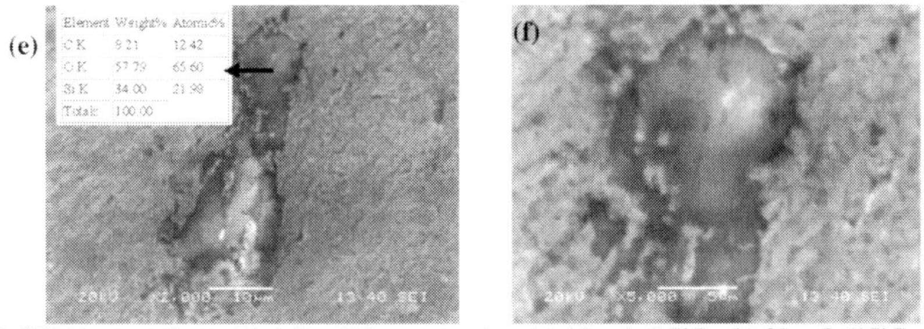

Figure 10. SEM image of as-milled Cu- SiCp composites (a) 10 vol. %SiCp (b) 20 vol. %SiCp (c) 30 vol. % SiCp and (d) 40 vol. %SiCp (e, f) Higher magnification SEM images of as-milled Cu-40 vol.% SiCp composites. The inset image in (e) shows the EDX analysis of SiC particle.

Figure 9 shows optical micrographs for as-milled Cu-SiCp composites for different vol. % of SiCp. From the micrographs it is evident that SiC particles are homogenously distributed all over the matrix. The high magnification optical micrographs of as-milled Cu-40 vol. % SiCp composites in figure 9 (e, f) shows the interface between the SiC particle and the Cu matrix. Figure 10 shows SEM for as milled Cu-SiCp composites for different vol. % of reinforcement respectively. The SEM images of the composites also show homogeneous dispersion of reinforcement over the Cu matrix.The scanning electron micrographs given in Figure 10 shows the typical microstructural features of a composite. In these micrographs of the various Cu- SiCp composites a range of SiC particulate sizes could be seen. Good interfacial integrity between the SiCp and the Cu matrix could also be observed. High magnification SEM images of as-milled Cu-40 vol. % SiCp composites in figure 10 (e, f) shows the interface between the SiC particle and the Cu matrix. EDX analysis inset in figure 10 (e) confirms that the dark particles are SiC. The SiC particles are embedded in the Cu matrix. The interface between the reinforcement and the matrix plays a crucial role in determining the mechanical properties of the composites.The SEM images the Cu-SiC interface suggests that the bonding between the Cu matrix and the SiC particles is very strong.

Figure 11 shows the density plots for as milled Cu-SiCp composites for different vol. % of reinforcement. The relative density gradually decreases with the increase in the content of SiC as the incorporation of hard and brittle material in soft Cu matrix which provides void in the matrix. From the plot it is evident that as-milled Cu- SiCp composites shows better densifications values as compared to unmilled Cu-SiCp composites. Figure 12 shows the XRD plots for as-milled Cu-SiCp composites for different vol. % of SiC particles. Peak corresponding to Cu_2O could be detected in the x-ray diffraction plot.

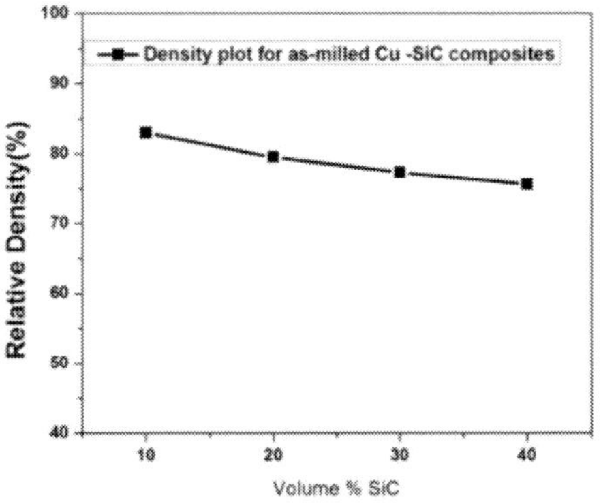

Figure 11. Density plots for as-milled Cu- SiCp composites.

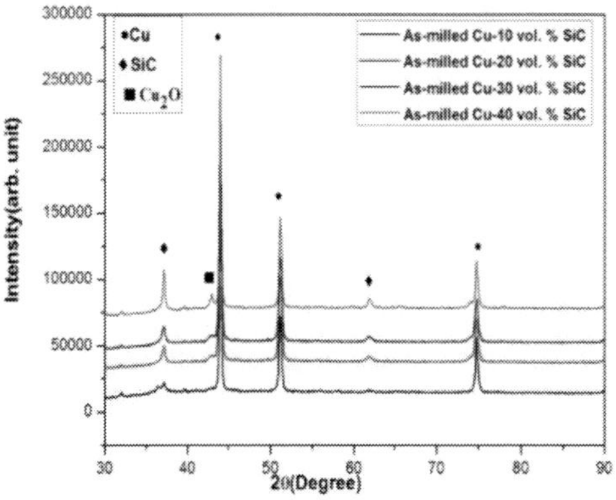

Figure 12. XRD plots for as-milled Cu-SiCp composites.

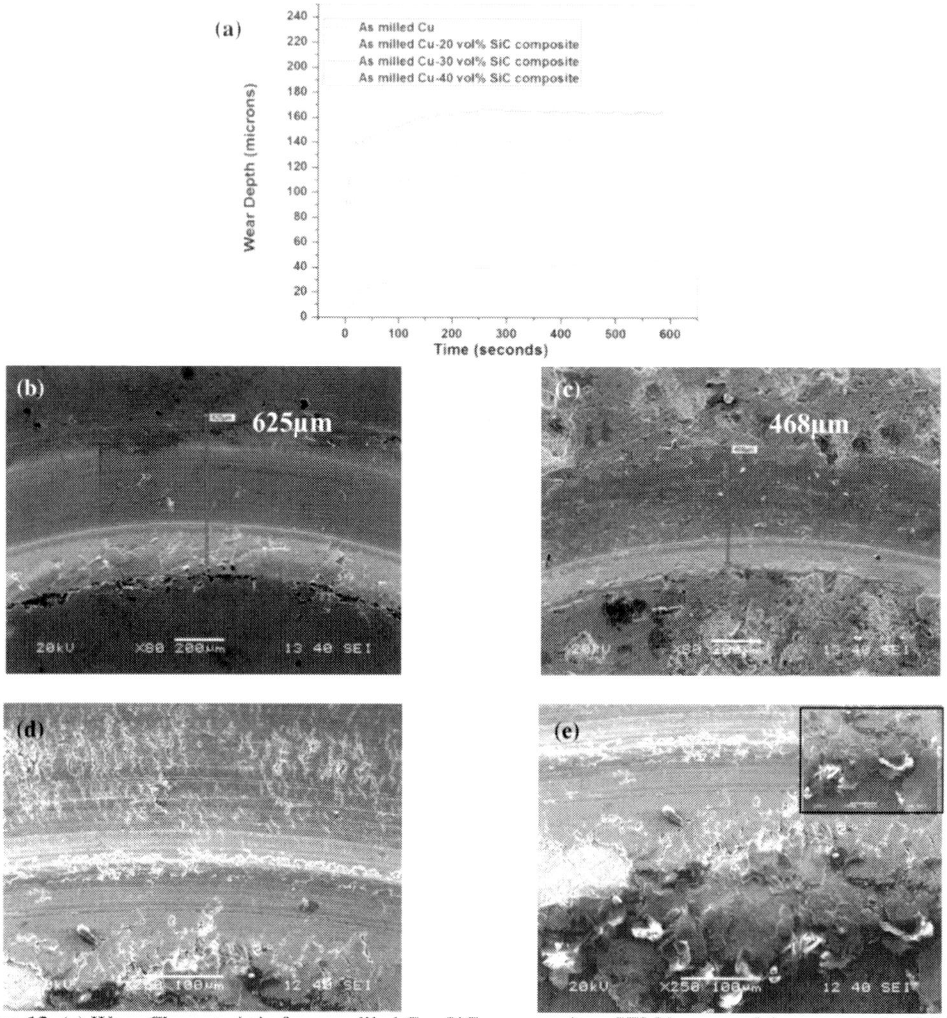

Figure 13. (a) Wear Characteristic for as-milled Cu- SiCp composites. SEM images of the wear track of (b) as-milled Cu sample (c) as-milled Cu-40 vol. %SiCp sample (d, e) High magnification SEM images of the wear track of as-milled Cu-40 vol. %SiCp sample. Inset image in (e) shows the edge of the wear track of as-milled Cu-40 vol. %SiCp sample

It is evident from figure 13(a) that there is a drastic reduction in wear rate with the addition of SiC particles to the Cu matrix. The dispersion of the SiC particles, as a hard ceramic phase in the Cu matrix, improves the wear resistance significantly. This can be attributed to the enhanced hardness of the composite reinforced by the SiC particles. It is known that improvement in hardness enhances wear resistance. The enhancement in the wear resistance of the composites can also be attributed to the good bonding between the Cu and the SiC particles. Bonding between the matrix and the reinforcement is known to play an important role in the wear resistance. Reduction of direct load at the contact between the Cu-SiCp composite surface and the ball due to load bearing component action of the hard SiC particles as compared to that of pure Cu also enhances the wear resistance of the Cu-SiCp composites.

To evaluate the wear resistance the as-milled Cu-SiCp composites were compared with pure Cu. It has been found that as-milled Cu-SiCp composites exhibits better wear behaviour in comparison to pure Cu. The enhancement of wear resistance is possibly due to the Orowan strengthening mechanism because of the fine dispersion of the SiC particles in the Cu matrix.The wear was found to be due to the combination of abrasion and delamination. From figure 6(b) it is evident that there is an enhancement of micro hardness due to the dispersion of SiC particles in the Cu matrix and this consequently improves the wear resistance of the composite. Figure 13 shows wear chractersties for as-milled Cu-SiCp composites.The plot between wear depth and time in figure 13(a) shows that as-milled Cu- 40 vol. % SiCp composites shows the lowest wear depth leading to higher wear resistance.Figure 13(b) shows the SEM images of wear tracks for as- milled Cu pellet and as- milled Cu- 40 vol.% SiCp reinforced composites [21-24].

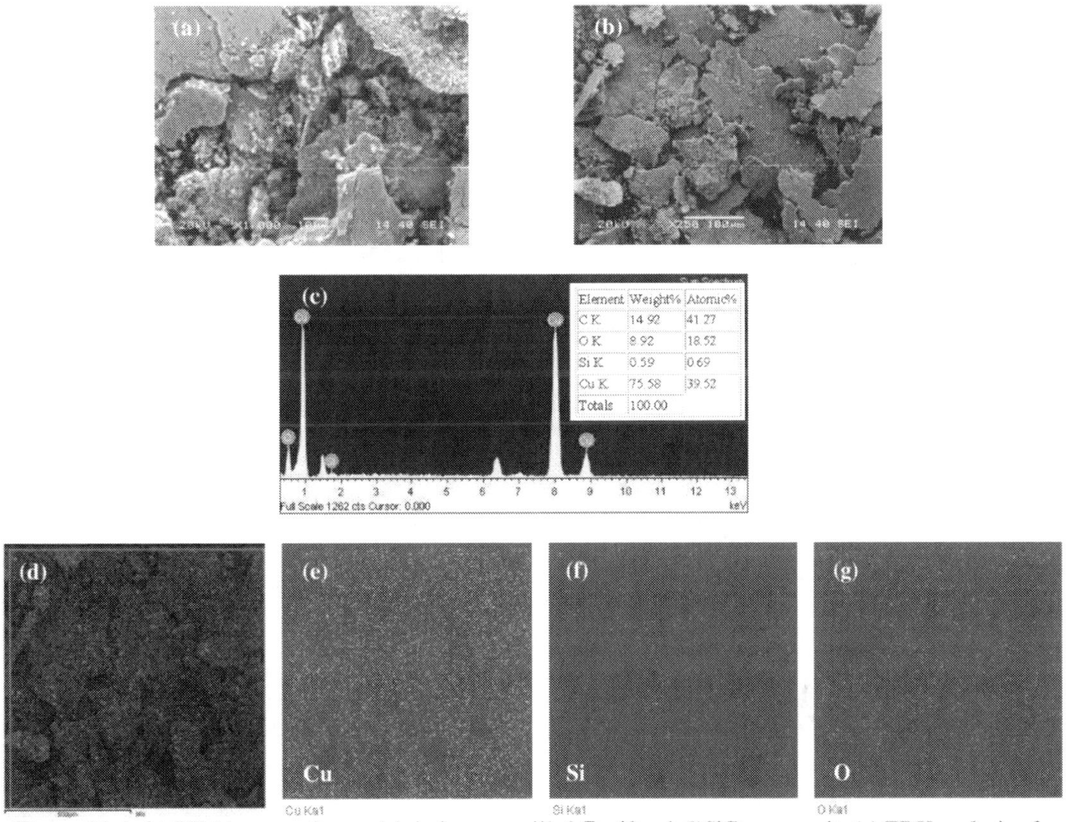

Figure 14. (a-b) SEM images of wear debris from as-milled Cu-40 vol. %SiCp composite (c) EDX analysis of the wear debris. (d) SEM image of the wear debris selected for elemental mapping. Elemental map of (e) Cu (f) Si (g) O.

Figure 14 (a, b) are the SEM images of the wear debris from as-milled Cu-40 vol. %SiCp composite. The EDX analysis of the wear debris in figure 14(c) shows mainly the presence of Cu and very small amount of Si. This suggests that the wear debris consists mainly of Cu and has very small amount of SiC in it. The elemental map of Cu and Si in the selected area shown in the SEM image in figure 14(d) also suggest that the wear debris mainly consists of Cu and there is only a very small amount of SiC in it. The oxygen in the wear debris is possibly due to the oxidation of the sample by the unavoidable oxygen during sintering. Oxidation of the wear debris could also take place during the wear test [25-28].

4. Conclusion

1. Milling of elemental Cu has led to the formation of nanostructured Cu. There has been no trace of contamination from milling media. Oxidation of Cu was also not found during milling.

2. In all the composites developed a homogenous distribution of the SiCp reinforcement in the Cu matrix was observed.

3. As-milled Cu-SiCp composites show better densification than unmilled Cu-SiCp composites.

4. The hardness of the Cu-SiCp composite increases with the increase in the content of SiC particles. As-milled Cu-SiCp composites showed higher hardness as compared to unmilled Cu-SiCp composites.

5. Cu-SiCp composites containing higher vol. % of the SiCp shows better wear resistance. All the Cu-SiCp composites developed show better wear resistance than pure Cu.

Acknowledgement

We would like to thank the staff in the SEM and XRD laboratories at the National Institute of Technology-Rourkela, India. We would also like to thank everyone involved at the National Institute of Technology Rourkela for their useful feedback and help. The authors would also like to thank the Metallurgical and Materials Engineering Department of Indian Institute of Technology Kharagpur, for carrying out HRTEM.

References

[1] Suryanarayana C 2001 Mechanical alloying and milling *Progress in MaterialsScience*461-184.
[2] Bloor D, Brook R J, Flemings M C, Mahajan S editors. 1994 *The encyclopaedia of advanced materials*(Oxford: Pergamon Press).
[3] Clyne T W and Withers P J 1993 *An Introduction to Metal Matrix Composites*University of Cambridge (CambridgeUnited Kingdom).
[4] Alam S N, Singh Harshpreet2014 Development of copper-based metal matrix composites: An analysis by SEM, EDS and XRD *Microscopy and Analysis***28(4)** 8-13.
[5] Tjong S C and Lau K C 2000 Tribological behaviour of SiC particle-reinforced copper matrix composites *Materials Letters***43** 274-280.
[6] Shu K M, Tu G C 2001 Fabrication and characterization of Cu-SiCp composites for electrical discharge machining applications Material *Manufacturing Process***16** 483-502.
[7] GwidonWS2009*Wear-Materials, Mechanisms and Practice*, Wiley Publication **ISBN: 978-0-470-01628-2**.
[8] Schubert T, Brendel A, Schmid K, Koeck T, Ciupinski L ,Zielinski W, Weißgärber T, Kieback 2007 Interfacial design of Cu/SiC composites prepared by powder metallurgy for heat sinks applications *Composites Part A* **38** 2398-2403.
[9] Raghu T 2001 Synthesis of nanocrystalline copper–tungsten alloys by mechanical alloying *Materials Science and Engineering A***304** 438-441.
[10] Alam Syed Nasimul 2006 Synthesis and characterization of W–Cu nanocomposites developed by mechanical alloying *Materials Science and Engineering A* **433**161-168.

[11] Ahamed H , Kumar V S 2011 A Comparative Study on the Milling Speed for the Synthesis of Nano-Structured Al 6063 Alloy Powder by Mechanical Alloying*Journal of Minerals & Materials Characterization & Engineering* **10** 507-515.

[12] Yusoff M, Othman R, Hussain Z 2011 Mechanical alloying and sintering of nanostructured tungsten carbide-reinforced copper composite and its characterization *Materials & Design* **32(6)** 3293-3298.

[13] Gilman P S, and Benjamin J S 1983 Mechanical Alloying *Annual Review of Materials Science***13** 279-300.

[14] Singh Harshpreet, Alam S N 2013 Development of Cu based metal matrix composites by mechanical alloying route *International Union of Materials Research Societies – International Conference in Asia.*

[15] Sundberg G 2006 Fabrication of Cu-SiC metal matrix composite *Journal of Materials Science* **41** 485-504.

[16] G Celebiefe, T Yener, I Altinsoy, M Ipek, S Zeytin, C Bindal 2010 The fabrication and properties of SiC particulate reinforced copper matrix composites *13th International Materials Symposium.*

[17] Zhang Y S, Han Z, Wang K, Lu K 2006 Friction and wear behaviors of nanocrystalline surface layer of pure copper *Wear* **260**942-948.

[18] Kennedy F E, Balbahadur A C, Lashmore D S 1997 The friction and wear of Cu based silicon carbide particulate metal matrix composites for brake applications *Wear* **203/204** 715-721.

[19] Pradhan A and Das S 2014 Dry Sliding Wear and Friction Behaviour of Cu-SiC Nanocomposites coating prepared by Pulse Reverse Electrodeposition *Tribology Transactions* **57**46-56.

[20] Dasgupta Rand, Meenai H 2003 Sliding wear properties of Al-Cu based alloys with SiC particle reinforced composites under varying experimental conditions *Journal of Materials Science Letters***22**1573-1576.

[21] Barmouz M, Asadi P, Basharati Givi M K, Taherishargh M 2011 Investigation of mechanical properties of Cu/SiC composite fabricated by FSP *Material Science and Engineering A* **528**1740-1749.

[22] Barmouz M, Basharati Givi M K, Seyfi J 2011On the role of processing parameters inproducing Cu/SiC metal matrix composites via friction stir processing: Investigating microstructure, microhardness, wear and tensile behavior *MaterialCharacterization***62**108-117.

[23] Barmouz M, Besharati Givi M K 2011 Fabrication of in situ Cu/SiC composites usingmulti-pass friction stir processing: Evaluation of microstructural, porosity, mechanical and electrical behavior*Composite Part AApplied Science Manufacturing* **42** 1445-53.

[24] Ramesh CS, Noor Ahmed R, Mujeebu MA, Abdullah MZ 2009 Development andperformance analysis of novel cast copper–SiC–Gr hybrid composites *Materials Design***30**1957–1965.

[25] Dhokey N B, Paretkar R K 2008 Study of wear mechanisms in copper-based SiC_p (20% by volume) reinforced composite *Wear* **265** 117-133.

[26] Zhan Y, Zhang G 2003 The effect of interfacial modifying on the mechanical and wear properties of SiC_p/Cu composites *Materials Letters* **57** 4583-4591.

[27] Prabhu T R, Varma V K, Vedantam S 2014 Tribological and mechanical behavior of multilayer Cu/SiC + Gr hybrid composites for brake friction material applications *Wear* **317** 201-212.

[28] Zhan Y, Zhang G and Zhuang Y2004 Wear Transitions in Particulate Reinforced Copper Matrix Composites *Materials Transactions***45** 2332-2338.

IOP Conf. Series: Materials Science and Engineering **75** (2015) 012008 doi:10.1088/1757-899X/75/1/012008

Development of High Capacity $Na_{0.7}(Ni_{0.4}Mn_{0.4}Co_{0.1}Fe_{0.1})O_2$ Cathode Material for Sodium Ion Batteries

Chandra Kant Kaithwas[1], **T. K. Kundu**
Department of Metallurgical and Materials Engineering,
Indian Institute of Technology Kharagpur, Kharagpur, WB, India -721302

[1]Email: ckaithwas14@gmail.com

Abstract. Sodium ion battery (SIB) has great potential as sustainable large scale energy storage application compared to lithium-ion battery due to abundance and cost effectiveness of sodium. $Na_{0.7}(Ni_{0.4}Mn_{0.4}Co_{0.1}Fe_{0.1})O_2$ as new cathode material for SIB is prepared by solid state reaction synthesis method. The structure of the new cathode material was characterized by X-ray diffraction using Co-Kα radiation. Morphologies and particle size range (0.37-1.9 microns) of the $Na_{0.7}(Ni_{0.4}Mn_{0.4}Co_{0.1}Fe_{0.1})O_2$ cathode material have been identified by scanning electron microscope. Electrochemical performance of the cathode material for coin cell battery using sodium as anode and $NaClO_4$ as electrolyte was examined in constant current mode. The material cycling performance showed very good reversibility between 2.0 - 4.3 V with reversible capacity of 202 mAh g^{-1} at 0.11 mA current. At C/10 reversible capacity of 191 mAh g^{-1} have been found. The prepared material shows considerable (40%) retention capacity after 45 cycle of charging and discharging with retention capacity of 79 mAh g^{-1}. Electrochemical impedance spectroscopy analysis has been performed between 100 kHz to 10 mHz frequency range and after 10 cycles the resistance for grain and grain boundaries are found to be 26.20 Ω and 354.7 Ω respectively. $Na_{0.7}(Ni_{0.4}Mn_{0.4}Co_{0.1}Fe_{0.1})O_2$ can be a promising cathode material for SIB as it shows very good charging and discharging characteristics with high reversible capacity.

1. Introduction

World-wide sodium ion battery is considered a better promising alternative to present Li-ion battery as it has great potential due to high abundance and cheap availability of sodium compound [1]. Efforts are being made to develop sodium batteries for electric vehicles and grid energy storage applications. Last few years most interest in Na-ion batteries have been in development of high capacity reversible cathode materials. Layered transition-metal oxides ($NaMO_2$, M = Co, Cr etc.) were proposed to be the cathode materials for Na-ion batteries. Cobalt and chromium are costly and toxic component. Moreover these materials usually show poor life cycle and low thermal stability [2]. Many alternatives cathode material for sodium ion batteries replacing (partly/completely) Co and Cr by other cheap and benign transition metals (Fe, Mn, Ni etc) have been searched like $Na(Ni_{1/3}Co_{1/3}Fe_{1/3})O_2$ [3], $Na_{0.45}Ni_{0.22}Co_{0.11}Mn_{0.11}O_2$ [4], $Na_{2/3}Ni_{1/3}Mn_{2/3-x}Ti_xO_2$ [5], $Na_3V_2(PO_4)_3$ [6], $Na_4Co_{2.4}Mn_{0.3}Ni_{0.3}(PO_4)_2P_2O_7$ [7], $Na[Ni_{1/3}Fe_{1/3}Mn_{1/3}]O_2$ [8], $Na[Li_xNi_{0.34-x}Mn_{0.47}Co_{0.19}]O_2$ where (x = 0.10,0.20.0.30) [9]. Sodium ion battery cathode materials are generally of P2 and O3 types. In P2 type structure Na as prismatic coordination with oxygen with 2 transition metal layer in repeat unit perpendicular to the layering. In O3, Na is as octahedral coordination with oxygen with 3 transition metal layer in repeat unit perpendicular to the layering. The P2 crystal structure of $Na_{2/3}[Ni_{1/3}Mn_{2/3}]O_2$ [10] could be well maintained for long term cycling in 2.0–4.0 V. When the $Na_{2/3}[Ni_{1/3}Mn_{2/3}]O_2$ cell was cycled between 1.6 V and 3.8 V, the first discharge capacity increased to the 135 mA g^{-1} at C/10. To make the material cheaper and environmentally, manganese has been found as an alternative source. Mn-based materials, such as P2-$Na_{0.67}[Fe_{0.5}Mn_{0.5}]O_2$ [12], $NaNi_{1/3}Mn_{1/3}Co_{1/3}O_2$[13] etc. possess some advantages. Addition of transition metals such as Ni and Co shows that the synthesized material possess better reversible capacity of 165 mA g^{-1} at C/20. In this paper we show that, addition of another transition metal Fe in the base material (Mn, Ni) improves the reversible capacity of the cathode material to 191 mA g^{-1} at C/10.

Content from this work may be used under the terms of the Creative Commons Attribution 3.0 licence. Any further distribution of this work must maintain attribution to the author(s) and the title of the work, journal citation and DOI.
Published under licence by IOP Publishing Ltd

2. Experimental method

2.1 Synthesis method

$Na_{0.7}(Ni_{0.4}Mn_{0.4}Co_{0.1}Fe_{0.1})O_2$ was prepared by a solid-state synthesis method from Na_2CO_3 (99.5% pure form Merck), $Ni(OH)_2$ (98% pure form Aldrich), Mn_2O_3 (99% pure form Aldrich), $CoCO_3$ (95% pure form Aldrich), Fe_2O_3 (90% pure form Merck) mixed in mass ratio of 0.30:0.27:0.25:0.09:0.06. The mixture were thoroughly ball milled using Retsch 67 (PM100) for 30 hours at 415 rpm using acetone as the liquid medium. The slurry sample obtained was dried in the open air muffle oven for evaporation of acetone. The dried ground material then pressed into pellets using 5 ton hydraulic press. The pellets were heated at 860°C for 24 hours for calcination in air. Then the calcine was cooled to room temperature and stored in an argon-filled glove box until use.

2.2 Characterization (XRD)

The structure of the prepared materials has been evaluated with X-ray diffractometer (XRD BRUKER, D8 Advance) using Co-Kα radiation. The XRD spectra were collected with diffraction angle (2θ) value ranging from 10° to 120° with 0.02° step size. The morphologies of the cathode material were studied by a scanning electron microscope (ZEISS, EVO 60).

2.3. Methods of Electrochemical Measurements

Sodium chlorate and glass fibre have been used as electrolyte and separator respectively to prepare the Na-ion battery. The synthesised cathode material is mixed with polyvinylidene fluoride (PVDF) binding agent and Acetylene Black (to increase the conductivity) in the ratio of 8:1:1. This mixture is then coated on an aluminium foil. The coated aluminium foil is dried in vacuum oven at 80°C for 24 hours. The foil is punched into circular cross sections of diameter 15 mm to act as cathode. This cathode is then assembled inside a glove box with spacer, separator, spring, Na anode of same circular cross section, spring, top case and bottom case to form the CR2032 coin cell. The weight of the active material in the prepared battery is found to be 0.0032 grams.

A potentiostat-galvanostat system (Gamery instrument, G-750) using (PITE 200) software and electrochemical analysis using (EIS 300) software for cyclic voltammetry and impedance spectroscopy measurements respectively. Electrochemical impedance spectroscopy was performed range 100 KHz to 10 MHz with 5mV. Electrochemical cyclic voltammetry was carried out one cycle charging and discharging at a scan rate of 0.1 mV s^{-1} for 2 to 4.3 volt potential range. Charging and discharging performance of the cell was evaluated at different current rates between 2 to 4.3 V.

3. Results and discussion

The powder X-ray pattern is shown in Fig 1. From X pert High Score analysis it has been revealed that the main phase of the material can be indexed as R-3m(166) space-group with the rhombohedral structure. Lattice parameters are found to be a = 2.95510 Å, c = 16.56300 Å. Some minute quantities of NiO are found in the synthesized cathode material. The scanning electron microscopy (SEM) image of the synthesized $Na_{0.7}(Ni_{0.4}Mn_{0.4}Co_{0.1}Fe_{0.1})O_2$ are shown in Fig (2-5). Irregular agglomerated particles are seen and particle size ranges between 0.37 - 1.9 μm are observed of the cathode material. As the particles are agglomerated to each other, they form several micro-sized secondary particles. According to the X-ray energy dispersive spectra (EDS), as shown in Fig (6-7), the estimated Na : Ni :

Mn : Co : Fe : O atomic ratios are found to be 18.9 : 10.4 : 10.3 : 2.7 : 2.8 : 54.9 respectively in the cathode material as given in Table 1.

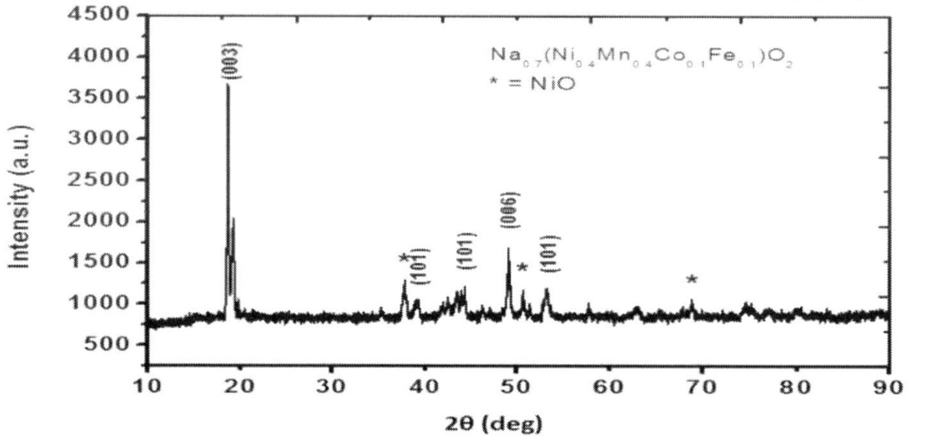

Figure 1. XRD pattern of $Na_{0.7}(Ni_{0.4}Mn_{0.4}Co_{0.1}Fe_{0.1})O_2$ material

Figure 2. SEM image at 15.00 KX magnification.

Figure 3. SEM image at 15.00 KX magnification.

Figure 4. SEM image at 10.00KX magnification.

Figure 5. SEM image at 10.00KX magnification.

IOP Conf. Series: Materials Science and Engineering **75** (2015) 012008 doi:10.1088/1757-899X/75/1/012008

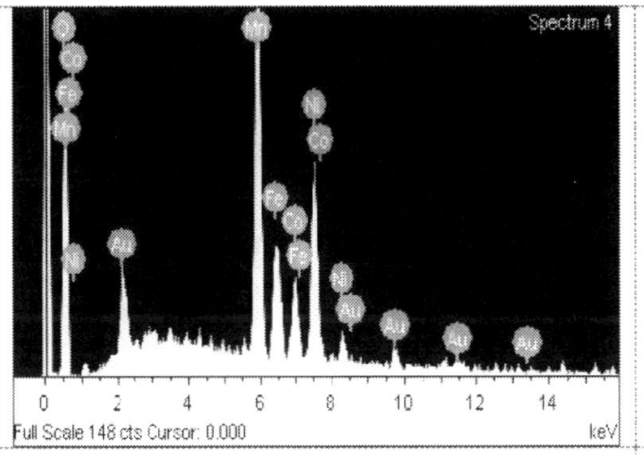

Figure 6. X-ray energy dispersive spectra of $Na_{0.7}(Ni_{0.4}Mn_{0.4}Co_{0.1}Fe_{0.1})O_2$.

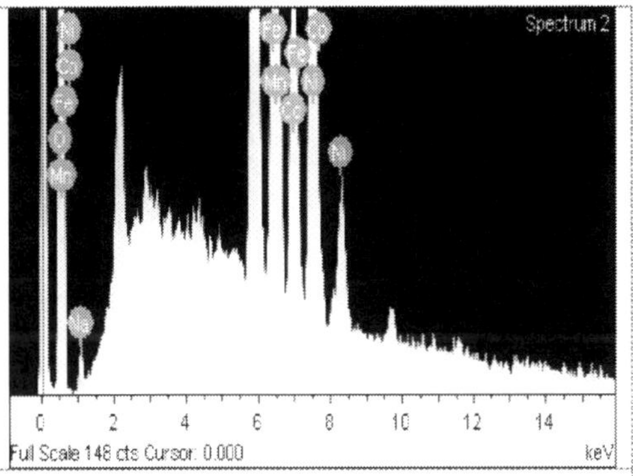

Figure 7.X-ray energy dispersive spectra of $Na_{0.7}(Ni_{0.4}Mn_{0.4}Co_{0.1}Fe_{0.1})O_2$.

Table 1. EDS analysis of the synthesized material

Element	Weight %	At %
Na	15.82	18.9
Ni	21.43	10.4
Mn	20.45	10.3
Co	5.80	2.7
Fe	5.50	2.8
O	31	54.9

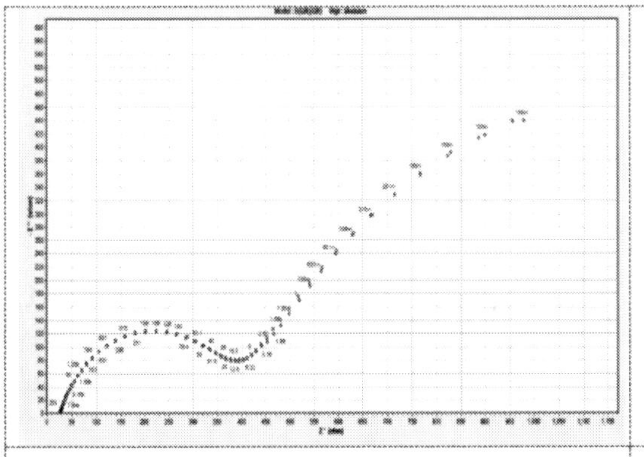

Figure 8. Nquist plot before 10 cycles of charging and discharging.

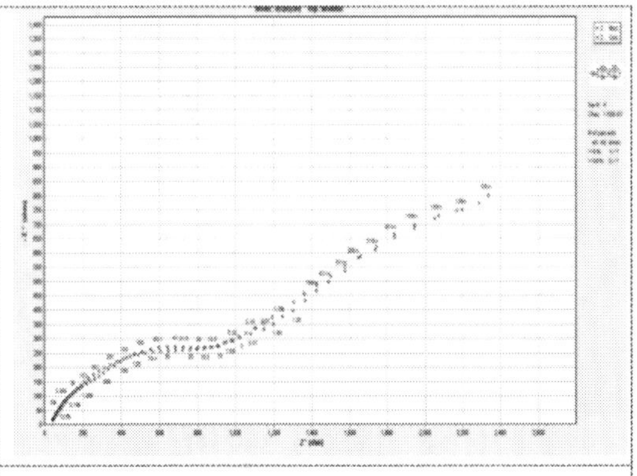

Figure 9. Nquist plot after the 40 cycles of charging and discharging

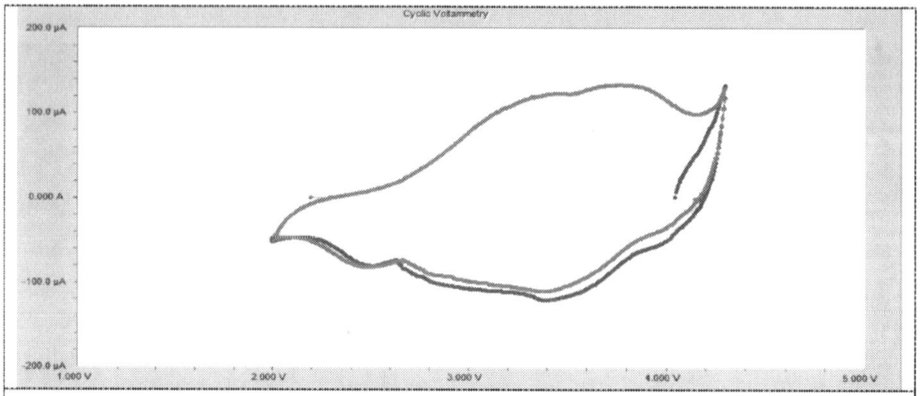

Figure 10. Cyclic voltammogram profiles of the $Na_{0.7}(Ni_{0.4}Mn_{0.4}Co_{0.1}Fe_{0.1})O_2$ cathode material at scan rate of 0.1 mV/s.

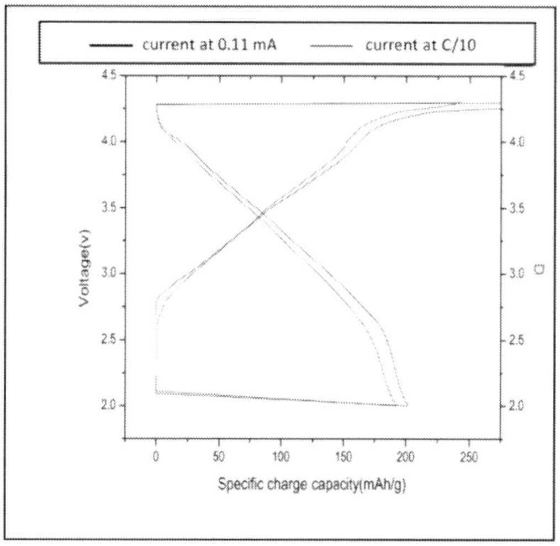

Figure 11. Full specific discharge capacity

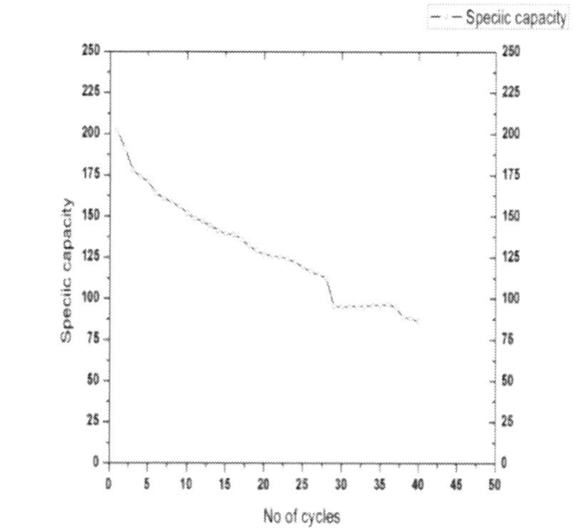

Figure 12. Specific charge capacity vs. Number of cycles

Electrochemical performance of the synthesized $Na_{0.7}(Ni_{0.4}Mn_{0.4}Co_{0.1}Fe_{0.1})O_2$ cathode material was evaluated by making a CR2032 type coin cell and then galvanostatic cyclic discharge-charge characteristics was performed in the 2.0 to 4.3 V voltage range at 0.11 mA current. Electrochemical impedance spectroscopy analysis has been performed and the corresponding Nquist plot has been shown in Fig 8. It is found that after 10 cycles the resistance for grain and grain boundaries were 26.20Ω and 354.7Ω respectively. After continuous charging and discharging for 40 cycles it was noticed that the resistance values for grains and grain boundaries were 26.18Ω and 836.7Ω respectively as shown in Fig 9. This drastic increase in grain boundary resistance at higher discharge cycle can be attributed to the degradation of the cathode material. Electrochemical cyclic voltammetry were performed for one cycle charging and discharging at a scan rate of 0.1 mV s^{-1} between 2 to 4.3 Volt potential range is shown in Fig 10. CV was performed to determine the extent of oxidation and reduction of the transition metals in the cell, and after CV test we observed that broad peaks, due to presence of a number of transition metals. Similar peaks position both during charging and discharging and complete reversibility indicate good stability and minimal polarization of the cathode material during charging and discharging. The material cycling performance was carried out between cyclic 2.0 - 4.3 V and reversible capacity of 202 mAh g^{-1} at 0.11 mA current was observed as shown in Fig. 11. At C/10, reversible capacity of 191 mAh g^{-1} have been found and shown in Fig 11. The prepared material shows considerable retention capacity (40%) after 45 cycle of charging and discharging with retention capacity of 79 mAh g^{-1}.

4. Conclusion

$Na_{0.7}(Ni_{0.4}Mn_{0.4}Co_{0.1}Fe_{0.1})O_2$ cathode material for sodium ion battery has been prepared by solid state synthesis with uniform size range with desired structure. Electrochemical charging-discharging characteristic using the synthesized material as cathode in a coin cell battery indicate it to be a promising material for sodium ion battery. The material cycling performance showed very good reversibility between cyclic ranges 2.0 - 4.3 V with reversible capacity of 202 mAh g^{-1} and 191 mAh g^{-1} at 0.11 mA and C/10 current respectively. The prepared material shows considering retention capacity (40%) after 45 cycle of charging and discharging with retention capacity of 79 mAh g^{-1}. $Na_{0.7}(Ni_{0.4}Mn_{0.4}Co_{0.1}Fe_{0.1})O_2$ can be a promising cathode material for SIB as it shows very good charging and discharging characteristics with high reversible capacity.

Acknowledgement

The authors are sincerely thankful to Prof. S. B. Majumdar, Materials Science Centre and Prof. S. Ghosh, Dept. of Metallurgical and Materials Engineering, Indian Institute of Technology Kharagpur, for allowing their laboratory to carry out this work.

References

[1] Dingding Yuana, Xiaohong Hub, Jiangfeng Qianb, Feng Peic, Fayuan Wuc, Rongjun Maoc 2014 P2-type $Na_{0.67}Mn_{0.65}Fe_{0.2}Ni_{0.15}O_2$ Cathode Material with High-capacity for Sodium-ion Battery, Electrochimica Acta **116** 300– 305

[2] Yujie Zhu, Yunhua Xu, Yihang Liu, Chao Luo and Chunsheng Wang 2013 Comparison of electrochemical performances of olivine $NaFePO_4$ in sodium-ion batteries and olivine $LiFePO_4$ in lithium-ion batteries, RSC publishing **5** 780

[3] Plousia Vassilaras, Alexandra J. Toumar, Gerbrand Ceder 2014 Electrochemical properties of $Na(Ni_{1/3}Co_{1/3}Fe_{1/3})O_2$ as a Cathode material for Na-ion batteries, Electrochemistry Communications **38** 79–81

[4] Daniel Buchholz, Luciana Gomes Chagas, Martin Winter, Stefano Passerini 2013 P2-type layered $Na0_{.45}Ni_{0.22}Co_{0.11}Mn_{0.66}O_2$ as intercalation host material for lithium and sodium batteries, Electrochimica Acta **110** 208– 213

[5] Hiroaki Yoshida, Naoaki Yabuuchi, Kei Kubota, Issei Ikeuchi, Arnd Garsuch, Martin Schulz-Dobrick and Shinichi Komaba 2014 P2-type $Na_{2/3}Ni_{1/3}Mn_{2/3-x}Ti_xO_2$ as a new positive electrode for higher energy Na-ion batteries, Royal Society of Chemistry **244** DOI: 10.1039/c3cc49856e

[6] Guangqiang Li, Danlu Jiang, Hui Wang, Xinzheng Lan, Honghai Zhong 2014 glucose-assisted synthesis of $Na_3V_2(PO_4)_3$ composite as an electrode material for high-performance sodium-ion batteries, Journal of Power Sources **26** 325-334

IOP Conf. Series: Materials Science and Engineering **75** (2015) 012008 doi:10.1088/1757-899X/75/1/012008

[7] Masafumi Nose, Shinya Shiotani, Hideki Nakayama, Kunihiro Nobuhara, Shinji Nakanishi, $Na_4Co_{2.4}Mn_{0.3}Ni_{0.3}(PO_4)_2P_2O_7$ 2013 High potential and high capacity electrode material for sodium-ion batteries, Electrochemistry Communications **34** 66–269

[8] Donghan Kim, Eungje Lee, Michael Slater, Wenquan Lu, Shawn Rood, Christopher S. Johnson 2012 Layered $Na[Ni_{1/3}Fe_{1/3}Mn_{1/3}]O_2$ cathodes for Na-ion battery application, Electrochemistry Communications **18** 66–69

[9] XinhuiZhanga, Chuang Yub 2012 Novel composites $Li[LixNi0_{.34-x}Mn_{0.47}Co_{0.19}]O_2$ ($0.18\leq x\leq 0.21$): Synthesis and application as high-voltage cathode with improved electrochemical performance for lithium ion batteries, Electrochimica Acta **8** 233– 238

[10] Hong Wang, Bingjian Yang, Xiao Zhen Liao, Jing Xu, Dezhi Yang, Yu-Shi He, Zi-Feng Ma 2013 Electrochemical properties of P2-$Na_{2/3}[Ni_{1/3}Mn_{2/3}]O_2$ cathode material for sodium ion batteries when cycled in different voltage ranges, Electrochimica Acta **13** 200– 204

[11] N. Yabuuchi, M. Kajiyama, J. Iwatate, H. Nishikawa, S. Hitomi, R. Okuyama,R. Usui, Y. Yamada, S. Komaba 2012 P2-type $Na_{.67}[Fe_{1/2}Mn_{1/2}]O_2$made from earth-abundant elements for rechargeable Na batteries, Nature Materials **11** 512

[12] Ying Zhang, Ke Ye, Kui Cheng, Guiling Wang, Dianxue Cao 2014 Three-dimensional lamination-like P2-$Na_{2/3}Ni_{1/3}Mn_{2/3}O_2$ assembled with two-dimensional ultra thin nanosheets as the cathode material of an aqueous capacitor battery, Electrochimica Acta **148** 195–202

[13] M. Sathiya, K. Hemalatha, K. Ramesha, J. M. Tarascon, A. S . Prakash 2012 Synthesis, structure, and electrochemical properties of the layered sodium insertion cathode Material: $NaNi_{1/3}Mn_{1/3}Co_{1/3}O_2$, Chemistry of Materials **24** 1846

[14] D. Kim, S. H. Kang, M. Slater, S. Rood, J. T. Vaughey, N. Karan, M. Balasubrama-nian, C. S. Johnson 2011 Enabling sodium batteries using lithium-substituted sodium layered transition metal oxide cathodes, Advanced Energy Materials **1** 333

Dry sliding wear system response of ferritic and tempered martensitic ductile iron

V K JHA, Y H MOZUMDER, S SHAMA, R K BEHERA[1], A PATTANIAK, SINDHOORA L P, S C MISHRA and S SEN
Dept. of Metallurgical and Materials Engineering, NIT Rourkela, India-769008

[1]Corresponding author e-mail: ranjanbehera.2419@gmail.com

Abstract. Spheroidal graphite cast iron (SG iron) is the most preferable member of cast iron family due to its strength and toughness along with good tribological properties. SG iron specimens with annealed and martensitic matrix were subjected to dry sliding wear condition and the system response was correlated to matrix microstructure. Respective microstructure was obtained by annealing and quench and tempering heat treatment process for an austenitizing temperature of 1000°C. Specimens were subjected to Ball on plate wear tester under 40N, 50N, 60N load for a sliding distance of 7.54m. Except for quench and tempered specimen at 50N, weight loss was observed in every condition. The wear surface under optical microscope reveals adhesive mechanism for as-cast and annealed specimen whereas delaminated wear track feature was observed for quench and tempered specimen.

Keywords: SG cast iron, dry sliding wear, microstructure, wear morphology

I. Introduction

Spheroidal graphite cast iron (SG iron) or ductile iron (DI) unlike every other cast iron has graphite in the form of spheroids which act as crack arrester, due to which it possesses higher strength and toughness along with better wear resistance. Furthermore it has been proved that the properties of SG iron can be improved by application of suitable heat treatment process leading to transformation of as-cast ferritic or ferritic/pearlitic matrix into pearlitic, martensitic and bainitic. Austempered ductile iron is one of the most favorable materials among all other type because of its excellent strength, toughness, wear resistance, fatigue strength and fracture toughness [1-8]. Tribological investigation carried out on austempered SG iron with varying austempering time and temperature, reported increased wear resistance with increasing austempering temperature and time along with increased hardness during wear due to the bainitic ferrite which is less prone to thermal instability than martensite, might undergo strain hardening [9, 10]. Tempering treatment on ductile iron with boron increased the wear resistance but with increasing boriding time wear rate of boro-tempered ductile iron decreased [11]. Apart from alloying element and heat treatment graphite nodule size and distribution affects the wear resistance of SG iron. According to a study by Sugishita and Fujiyoshi [12] and Zimba et.al [13] presence of large size graphite nodules reduces the wear rate by acting as lubricating agent.

Content from this work may be used under the terms of the Creative Commons Attribution 3.0 licence. Any further distribution of this work must maintain attribution to the author(s) and the title of the work, journal citation and DOI.
Published under licence by IOP Publishing Ltd

The past results reported are mainly leaned towards austempered ductile iron and the studies were focused on the effect of austempering time and temperature on wear behavior. Hence the present investigation is concentrated to study the wear behavior of as-cast ferritic ductile iron compared with annealed ferritic matrix and quenched and tempered ductile iron subjected to dry sliding condition at higher loads.

II. Experimental details

2.1. Specimen Preparation

In order to investigate the structure-property relationship, ductile iron test blocks with different alloying elements were brought from L&T Kansbahal, India. The chemical composition test blocks by weight percentage is presented in Table 1. In order to carry out the experiment specimens of 25×10×5 mm^3 were machined from the test block. Specimens were then austenitized at 1000°C for 90 minutes and then it was cooled down to 700C and kept there for 5hr 30 minutes followed by furnace cooling to room temperature for annealing heat treatment. On the other hand specimens after austenitization were quenched in mineral oil maintained at 100°C and immediately after quenching, tempered at 500°C for 120 minutes followed by air cooling to room temperature for quenching and tempering heat treatment process. After heat treatment oxide layer from each of the specimen was removed by conventional filling and emery paper polishing method.

Table 1: Chemical composition of specimens in wt. %

Elements	Content (Wt. %)
C	3.61
Si	2.1
Mn	0.2
S	0.007
P	0.022
Cr	0.03
Ni	0.47
Mo	0.001
Cu	0.009
Mg	0.043
Ce	0.004
Fe	Balance

2.2. Vickers hardness and wear testing

Vickers hardness was measured by applying a load of 20Kg and dwell time being 10seconds on each heat treated and as-cast specimen. Ducom TR-208-M1 Ball on plate type wear monitor [14-17] with specimen as flat plate and a spherical tipped diamond cone of 120° angle 0.4mm tip diameter was used in order to investigate the wear system response of the as-cast heat treated specimens. The mechanism of Ball on plate wear test is very much similar to that of Pin on disc wear tester. However the minor difference is that in pin-on-disc specimen is in the form of cylindrical pin, held stationary in specimen holder and disc is the counter body which rotates against the pin, whereas in case of Ball on plate wear test instead of pin specimen is a flat one rotates against a counter ball which is fixed. Also, in pin on disc machine only the disc rotates whereas in Ball on plate mechanism both the specimen and indenter rotate at same relative speed. The Ducom TR-208-M1 Ball on plate type wear monitor along with schematic diagram of pin on disc wear monitor is presented in Fig. 1 (a) and Fig. 1 (b) respectively. Test was conducted at 40N, 50N, 60N loads for a sliding distance of 7.54m at linear velocity of 0.063m/s. The weight loss for corresponding specimens was measured with the help of electronic balance of 0.1mg accuracy, prior to the weight measurement specimens were cleaned ultrasonically with acetone.

Fig. 1: (a) Ducom TR-208-M1 Ball on plate type wear monitor

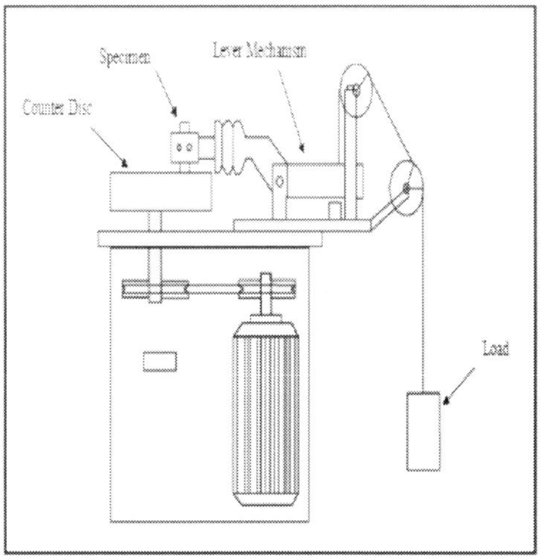

Fig. 1: (b) Schematic diagram of pin on disc wear test machine [18]

2.3. Microstructural and wear morphology investigation

In order to correlate the wear response with matrix structure standard metallographic technique was followed for microstructural investigation. Specimens were first polished with belt polisher followed by 1/0, 2/0, 3/0, 4/0 grades of emery paper and finally cloth polishing was done with alumina slurry followed by diamond polishing. Metallographic images were taken with the help of computer integrated optical microscope at 100X magnification. The worn surfaces of each specimen under various loading conditions were observed under optical microscope at 200X.

III. Results and Discussions

3.1. Microstructural investigation

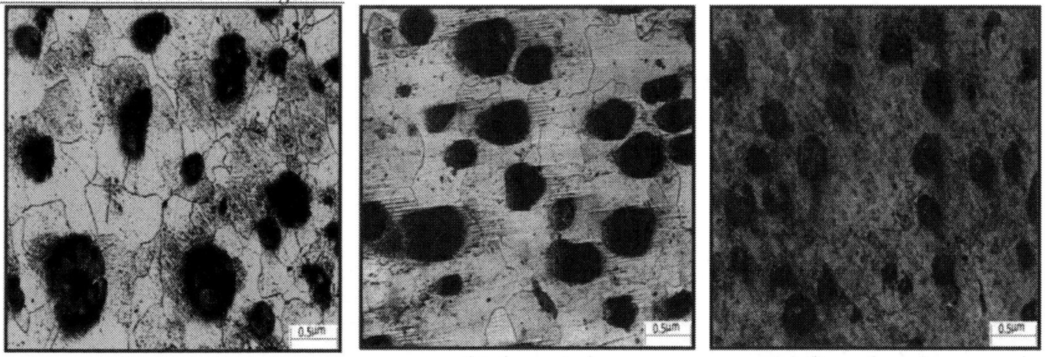

(a) As-cast (b) Annealed (c) Quench and tempered

Fig. 2: Microstructures of respective specimens

Fig. 2 illustrates the microstructures of respective as-cast and heat treated specimens. Both the as-cast (Fig. 2(a)) and annealed (Fig. 2(b)) specimens have graphite nodules embedded in ferritic matrix whereas quench and tempered (Fig. 2(c)) specimen had tempered martensitic matrix with graphite nodules lodged in it.

3.2. Hardness and Wear properties

The Vickers hardness for respective specimens was presented in Fig. 3(a). It can be noticed that although the as-cast and annealed specimen, both were having ferritic microstructure, the annealed specimen was softer than the as-cast one. This can be attributed to the fact that, the stresses developed in as-cast specimen during solidification and specimen machining had been removed during annealing treatment. The quench and tempered specimen had the highest hardness value due to the hard tempered martensite matrix. Fig. 3(b) depicts the wear system response of respective as-cast and heat treated specimen with respect to their hardness values. The as-cast ferritic matrix observed to lose weight marginally when load is increased to 50N than that at 40N. On further increase in load to 60N no weight loss was observed and can be due to the work hardening effect as the specimen was ductile one. On the other hand annealed specimen which is free from stresses developed during machining and solidification appeared to lose weight continuously at every load. The rate of wear in annealed specimen as compared to the as-cast specimen was less due to the higher hardness value of the former specimen and work hardening effect when operated under 50N load. At the same time the quench and tempered specimen gained some weight at 50N load and on further increasing the load to 60N again there was weight loss at the same rate. As compared to the ferritic matrix wear rate of quench and tempered specimen was very less due to the hard tempered martensitic matrix [19].

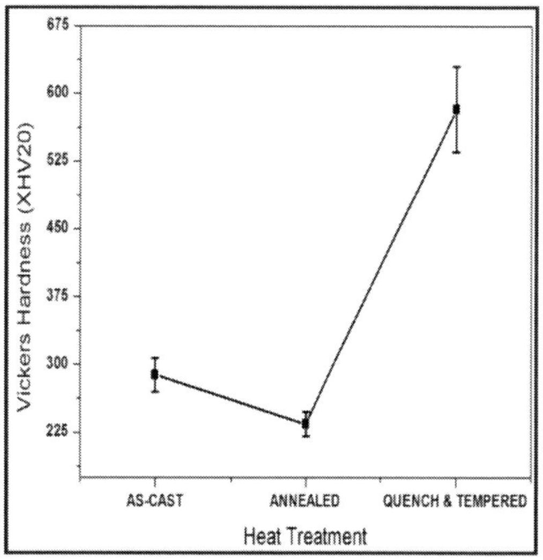

Fig. 3(a): Vickers hardness of respective as-cast and heat treated specimens

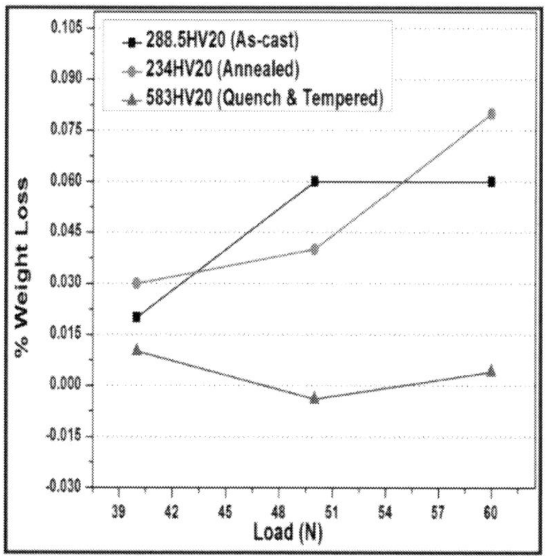

Fig. 3(b): % Weight Loss vs. Load for respective specimen's w.r.t. hardness

3.3. Wear morphology

In order to investigate the wear mechanism, micrographs of worn surfaces of as-cast and heat treated specimens were taken with the help of optical microscope at 200X, and presented in Fig. 3. Direction of wear track was shown by the black arrows in each case. Micro cracks along with broken wear platelets were observed in as-cast specimen at 40N (Fig. 4(a)), due to the surface work hardening. Similar kind of phenomenon was also observed for annealed specimen at 40N (Fig. 4(b)). The graphite nodules appeared to expand in the transverse direction as well as shear deformation in the direction of wear. The features noticed in case of as-cast and annealed specimen suggest the adhesive wear principle. On the other hand the quench and tempered specimen at 40N (Fig. 4(c)) adhesive mechanism of wear was observed indicated by the particle pull out phenomenon and continued to remove particles from the specimen and eventually deposited on the surface leading to broken delaminated wear track [20], however no micro cracks were observed in this case. Further increase in load i.e., to 50N no difference in wear mechanism was found for as-cast and annealed specimen, Fig. 4(d) and (e) respectively, except more micro cracks formation due to increase in surface hardness while operating under such high load. Similarly no change in wear phenomenon was observed for quench and tempered specimen, Fig. 4(f), except formation of micro cracks. Further increment of load to 60N did not affect the wear principle but only to surface hardness which can be clearly seen in Fig. 4(g) for as-cast, Fig. 4(h) for annealed and Fig. 4(i) for quench and tempered specimen. Fig. 5 showed EDAX analysis of worn surface of the quench & tempered specimen at 50N, taken under NOVA NANOSEM 450, field emission scanning electron microscope. It was observed that oxide layers were formed on the wear track, which led to the gain in weight as mentioned in the previous section.

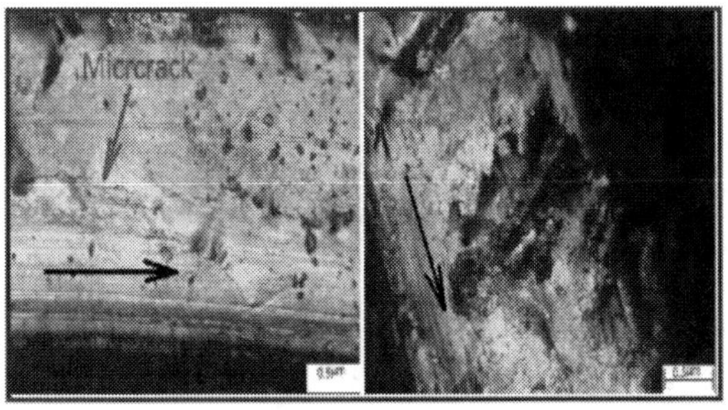

(a) As-cast 40N

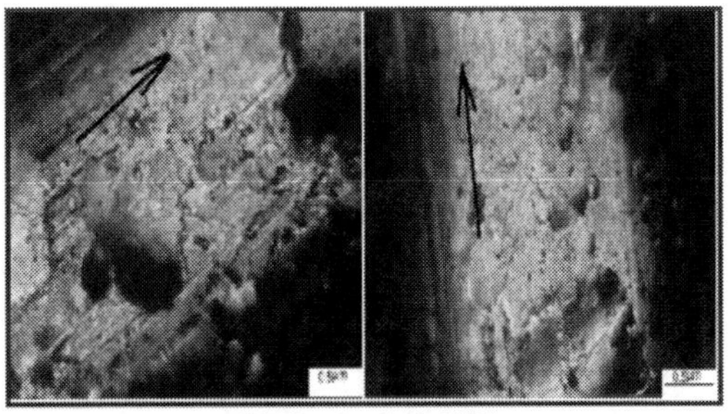
(b) Annealed 40N

(c) Quench and Tempered 40N

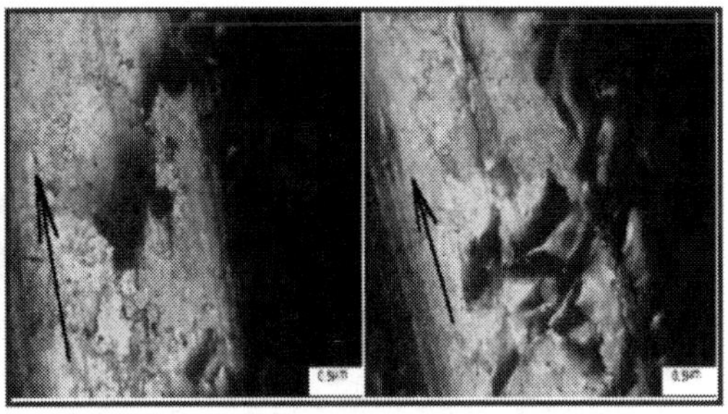
(d) As-cast 50N

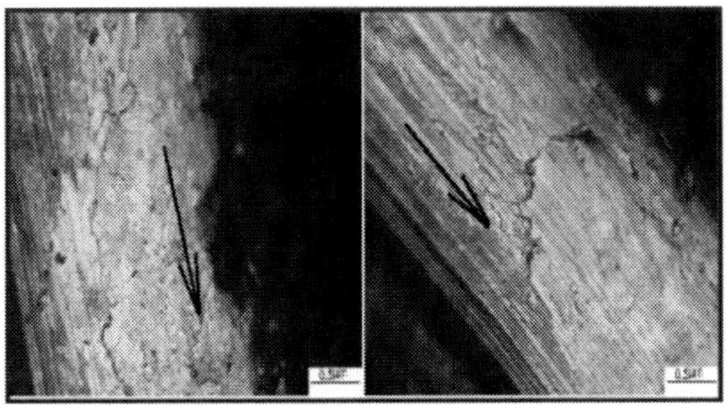
(e) Annealed 50N

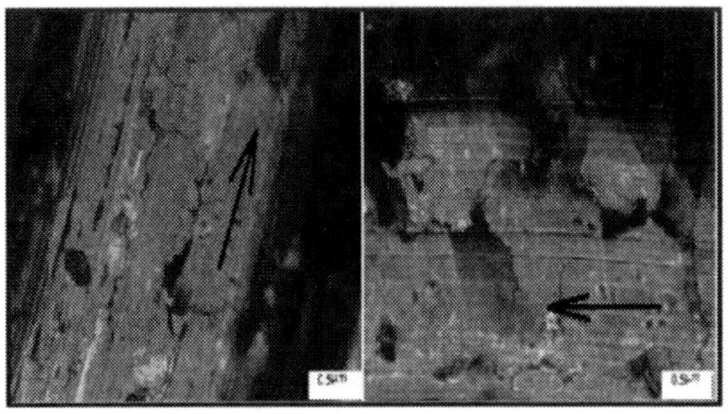
(f) Quench and Tempered 50N

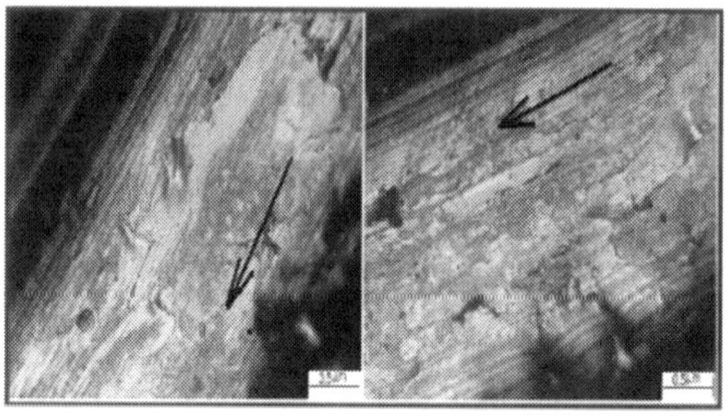

(g) As-cast 60N

(h) Annealed 60N

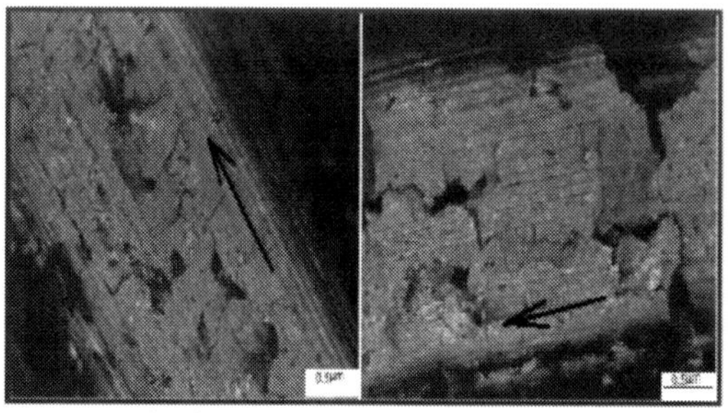

(i) Quench and Tempered 60N

Fig. 4: Worn surface micrographs of respective specimens at 200X

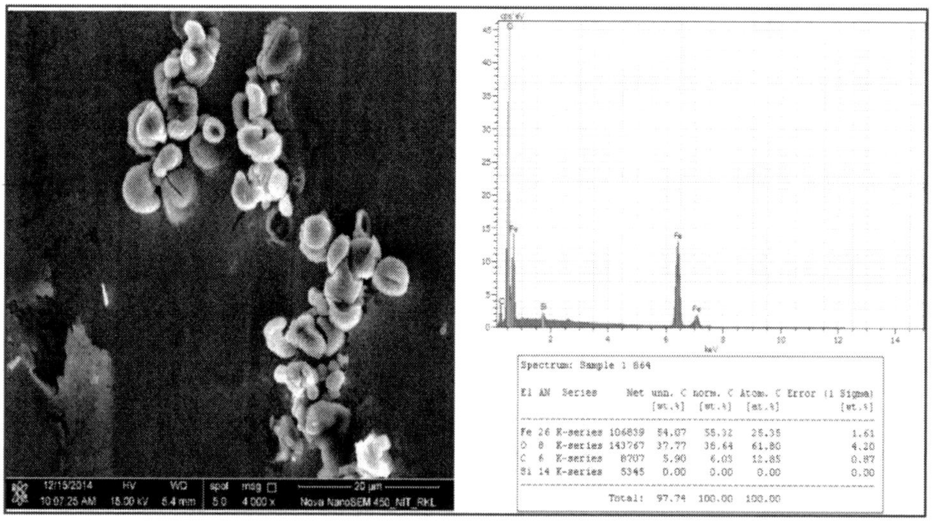

Fig. 5: EDAX of quench and tempered specimen showing oxides when operated under 50N load

IV. Conclusion

Ductile iron specimens with ferritic and tempered martensitic matrix were undergone Ball on plate wear system. The results of the study can be concluded as follows.

1. Soft ferritic matrix was less resistant to wear due to lower level of hardness, whereas tempered martensitic matrix with higher hardness value was more resistant to wear.
2. Adhesive wear phenomenon was observed for specimens with soft ferritic matrix along with micro cracks due to the hardening of surface while operating under high load and kept on increasing with increase in load.

3. Hard quench and tempered specimen showed adhesive wear mechanism featured by particle pull out from the wear surface leading to formation of another layer over the specimen and finally breakage of the wear plate.

V. Acknowledgement

The authors also express their gratitude towards the help of L&T Kansbahal, India for providing test blocks for this investigation.

VI. References

[1]. Gudlanch RB, Janowak JF 1983 *AFS Trans* **94** 377–88.
[2]. Harding RA, Gilbert GNJ 1986 *Brit Foundryman* **79** 489–96.
[3]. Harding RA 1993 *The Foundryman* 97–208.
[4]. Yang J, Putatunda SK 2005 *Mater Sci Eng A* **406** 217–28.
[5]. Putatunda SK, Bartosiewicz L, Krause AR, Singh I, Alberts FA 1993 *Mater Charact* **30** 221–34.
[6]. Harding RA 1985 *Mater Des* 177–84.
[7]. Hughes ICH 1985 *Mater Des* **6** 124–6.
[8]. Putatunda SK 2001 *Mater Sci Eng A* **315** 70–80.
[9]. S. M. A. Boutorabi, J. M. Young and V. Kondic, M. Salehi 1993 *Wear* **165** 19-24.
[10]. A.S.M.A. Haseeb, Md. Aminul Islam, Md. Mohar Ali Bepari 2000 *Wear* **244** 15–19.
[11]. Yusuf Kayali, Yılmaz Yalcin, Sukru Taktak. *Mater. Des 2011* **32** 4295–4303.
[12]. Junji Sugishita and Shonoshin Fujiyoshi 1982 *Wear* **77** 181 – 193.
[13]. J. Zimba, M. Samandi, D. Yu, T. Chandra, E. Navara, D.J. Simbi. 2004 *Mater Des* **25** 431–438.
[14]. A. Roy Choudhury, Tamer Ezz, Satyajit Chatterjee, Lin Li 2008 *Surface & Coatings Technology* **202** 2817–2829.
[15]. L. Steiner, V.Bouvier, U.May, N.Huber 2001 *Tribology International* **43** 1410–1416.
[16]. A. Lyubicheva, I.Soldatenkov 2012 *Tribology International* **55** 77–80.
[17]. M. Alvarez-Vera, A.Juarez-Hernandez, C.E.Gonzalez-Rivera, R.D.Mercado-Solis, M.A.L. Hernandez-Rodriguez 2013 *Wear* **301** 243–249.
[18]. Sandeep Kumar, Sant Ram Chauhan 2012 *International Journal of Composite Materials*, **2(5)** 101-114.
[19]. M.A. Islam, A.S.M.A. Haseeb, A.S.W. Kurny 1992 *Wear* **188** 61-65.
[20]. H.R. Abedi, A. Fareghi, H. Saghafian, S.H. Kheirandish 2010 *Wear* **268** 622–628.

IOP Conf. Series: Materials Science and Engineering **75** (2015) 012010 doi:10.1088/1757-899X/75/1/012010

Effect of band-overload on fatigue crack growth rate of HSLA steel

S.V.Abhinay*, Om Prakash Tenduwe, Ajit Kumar, K. Dutta, B. B. Verma and P.K.Ray

Department of Metallurgical and Materials Engineering, National Institute of Technology, Rourkela –769008, India

E-mail: smrutin9999@gmail.com

Abstract. Fatigue crack growth behavior is important parameter of structural materials. This parameters can be used to predict their life, service reliability and operational safety in different conditions. The material used in this investigation is an HSLA steel. In this investigation effect of single overload and band-overload on fatigue crack growth of same steel are studied using compact tension (CT) specimens under mode-I condition and R=0.3. It is observed that overload and band-overload applications resulted retardation on the fatigue crack growth rate in most of the cases. It is also noticed that maximum retardation took place on application of seven successive overload cycles. Application of ten and more overload cycles caused no crack growth retardation.

1. Introduction

Fatigue and fracture are common cause of service failure of engineering components and structures. To study about fatigue and fracture related problem is very important of any kind of machine parts, components and engineering structure that is related to various type of loading condition during their operation, so realistic fatigue crack growth and fatigue life prediction is one of the most importance part in terms of economic and safety point of view.

The fluctuating load nucleates a crack, which then propagates slowly and finally the crack growth rate per cycle is accelerated and subsequently comes to the stage when the crack-length is long enough to be considered critical for a catastrophic fracture failure. The fatigue crack growth rate may be significantly affected by application of overload cycles [1].In fatigue crack growth, load applied in the form of a single or band overloads may follow either in mode I or mixed-mode (mode I and II) [2]. It has been evidenced that a pure mode-I overload and multiple overloads leads to maximum crack growth retardation, however in mode-II overload has least effect on fatigue crack growth retardation [2,3]. Most of engineering machine parts and structures are failed by fatigue and fracture causes problem [4]. We need to understand how materials fail, how crack start and propagatehow we control it and our ability to prevent such failures. Several investigators [5- 9] observed that changes in magnitude of cyclic load may result in retardation or acceleration in fatigue crack growth rate. The overload induced crack tip plastic zone and crack closure level have been found to control the delay behaviour significantly [6]. Extensive published data show that the rate of fatigue crack growth rate

Content from this work may be used under the terms of the Creative Commons Attribution 3.0 licence. Any further distribution of this work must maintain attribution to the author(s) and the title of the work, journal citation and DOI.
Published under licence by IOP Publishing Ltd

under constant amplitude cyclic load fluctuation can be retarded significantly as a result of application of single or multiple tensile overload cycle having peak load greater than that of the constant amplitude loading cycles. The ductile solid generally exhibits a temporary accelerated crack growthrate during and immediately after the application of overload. Due to the application of overload the sharp edge of crack tip is get curved. This reducessevere stress concentration that would present at sharp edge of crack tip [10].The assessment of life under those complex situations is certainly tedious because of the lack of proper understanding of micro-mechanisms of retardation. Based on various mechanisms, a number of retardation models have been proposed [11-13].However, each model has its own merits and demerits as a result; significant ambiguities and disagreements still exist in terms of the exact mechanism of retardation. Due to the number and complexity of the mechanisms involved in this problem, no universal model exists yet.

2. Materials and experimental details:

2.1.Specimen Preparation

Fatigue crack growth tests, were conducted on compact tension (CT) specimens with a narrow notch and reduced thickness, which is fabricated from 12 mm thick plate. The CT specimens were made in the L-T orientation, both sides of the specimen surfaces were given mirror-polish with the help of different grades of emery papers with the loading aligned in the longitudinal direction and notch given in the transverse direction; ASTM standard E647-13 [14] is followed for specimen design. The dimensional details of specimen are presented in Fig. 1.

2.2. Chemical composition

The material studied in current investigation is HSLA steel, collected from Rourkela steel plant, Rourkela. The chemical composition of material is provided in Table 1.

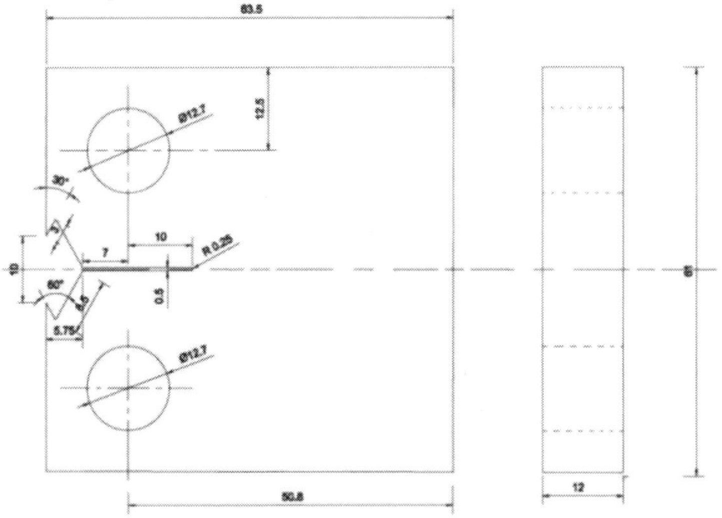

Fig.1. CT specimen with reduced thickness

Table 1. Chemical composition of HSLA steel

Elements	C	Mn	Si	P	S	Al	V	Nb	Mo	Fe
Wt. %	0.2	1.27	0.25	0.021	0.014	0.05	0.001	0.005	0.001	Balance

2.3.Microstructural investigation

Well-polished and etched metallographic specimens were studied using an optical microscope (Carl Zeiss). Typical optical micrograph of as-received material is shown in Fig. 2. The white portion of microstructure refers to ferrite and light black portion refers to pearlite. The dark black portion appears as martensitic along with carbide precipitate throughout structure in this steel. The ferrite matrix gives ductility and toughness to the investigated steel.This optical microstructure illustrates the alignment and grain structures of the rolled plate in three mutually orthogonal directions. The microstructures of all three directions were superimposed to obtain the 3-D view and shown in Fig. 2.

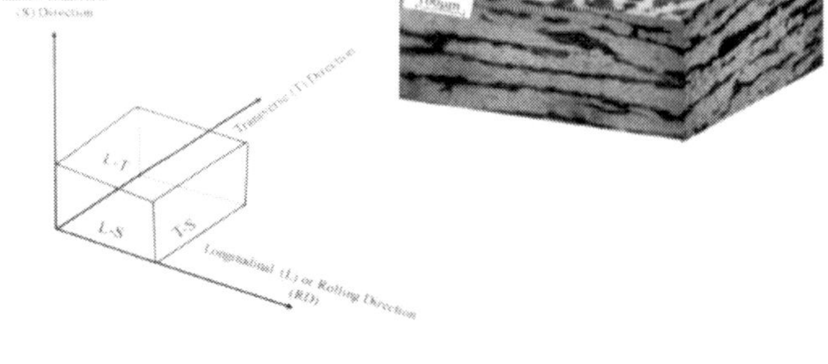

Fig. 2. 3-D view of the microstructure of the investigated steel.

2.4.Fatigue crack growth rate (FCGR)test

The FCGR tests were done in a 100kN universal test machine (BiSS)using variable amplitude crack propagation (*VAFCP*) software. The software permitted on–line monitoring of the crack length (a), compliance, ΔK, load range and the crack growth rate per cycle (*da/dN*). All tests were conducted at constant load mode at stress ratio of (R) 0.3 and using 10Hz frequency at room temperature. A crack opening displacement (COD) gauge was used to measure the displacement. The gauge was mounted on knife edges on the specimen. The specimen surfaces were stickered by graph paper for manual examination of the crack extension during the test as well.Fatigue pre-cracking was done under mode-I loading (crack opening mode) at constant amplitude loading mode to an *a/W*ratio of 0.24. Following three different case of crack growth tests were performed in this investigation:

(i) Constant amplitude loading with constant stress ratio (R).
(ii) Constant amplitude loading with single overload in mode-I.
(iii) Constant amplitude loading with band (multiple) overload in mode-I.

The stress intensity factor range (ΔK)[22] for CT specimen were calculated by following equation.

$$\Delta K = \frac{\Delta P}{B\sqrt{W}} \frac{(2+\alpha)}{(1-\alpha)^{1.5}} \left(0.886 + 4.6\alpha - 13.32\alpha^2 - 14.72\alpha^3 - 5.6\alpha^4\right) \quad (1)$$

Where, $\alpha = a/W$;expression valid for $a/W>0.2$

In case (I) CT specimens were tested under constant amplitude load mode maintaining a fixed load ratio, $R = 0.3$. In case (II)the specimens were tested under same loading conditions with single tensile overload are applied in mode I at $a/W = 0.28$ with overload ratio (R_{ol}) were applied 1.25, in 1 Hz frequency;the overload ratio is $R_{ol} = K_{ol} / K^B_{\max}$, where K_{ol} is overload stress intensity factor, and $K^B_{\max}$ is the maximum stress intensity factor for base line test. The specimens were subsequently subjected to mode-I constant amplitude load cycles after overload.

In case (III), constant amplitude loading with band (multiple) tensile overload were applied in mode-I. After band overload on the subsequent constant amplitude fatigue crack growth test were allowed for continue the test. The crack was allowed to grow up toa/W = 0.67, band-overload tests ware followed by multiple tensile overload at a/W = 0.28, and overload ratio (R_{ol}) were applied 1.25, in 1 Hz frequency, The number of band overload were applied during test are 3,5,7,10,100, in the same crack opening mode.

3. Results and discussion

3.1. Constant amplitude loading interposed with mode-I overload and band overload

Crack retardation behaviour of specimens under investigation are illustrated in the form of crack length (a)vs. number of stress cycles (N) in Fig.3. The figure clearly shows that the application of overload cycle retards a growing crack. It is also observed that the magnitude of retardation increases with increasing number of overload cycles.

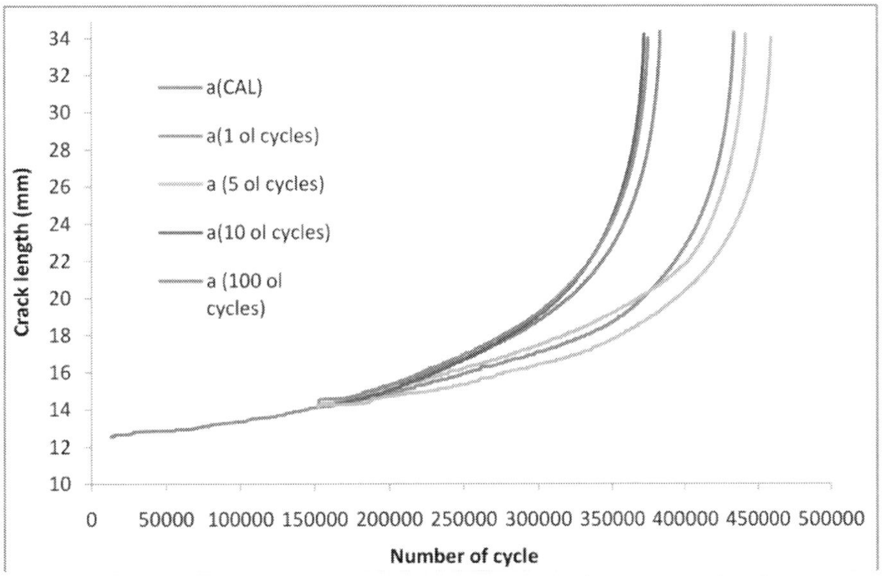

Fig. 3. Superimposed crack length vs. numberintensity factors rangeof cycle curve

This increase in magnitude of retardation is noticed up to 7 number of overload cycles. Application of overload cycles 10 and 100 has nullified the retardation effect and resulted little crack growth crack growth acceleration. To visualise the effect of overload cycle on the magnitude of retardation, number of stress cyclesrequired to attend a fixed crack length (a=15mm) following overload cycles from 1 to 100 can be viewed in Fig.4. Similar observations are also reported in literature[15].

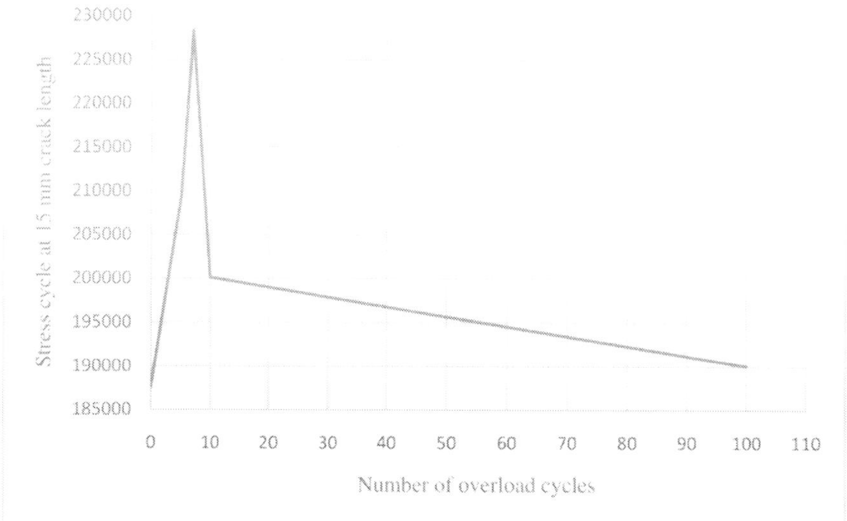

Fig.4. Number of stress cycles vs. number of overload cycles

The overload application is known to retard a growing crack. The retardation observed following overload cycles are explained on the basis of development of plastic zone in elastic enclave resulting compressive residual stresses, crack closure crack tip blunting etc. [10, 16] Compressive residual stresses and crack tip blunting[11, 12] play significant role in retarding the propagating crack. The application of an overload or a few overloads appears to introduce a plastic region embedded in elastically deformed enclosure. Repetition of the overload cycles beyond a certain number (7 in the present case) may have developed plastic regions at the crack tip larger than the critical to produce further retardation. It may be noted that the size of elastic enclave reduces with increasing plastic zone resulting reduced constrain and reduced retardation effect.

Fatigue crack growth rate as a function of applied stress intensity factor is presented on log-log scale in Fig.5. The magnified view of plots for a few overloading conditions are presented in Fig.6.This conforms that drop in crack growth rate occurs on application of 7 numbers of overload cycles. The repetition of same overload to 100 cycles caused acceleration in crack growth rate.

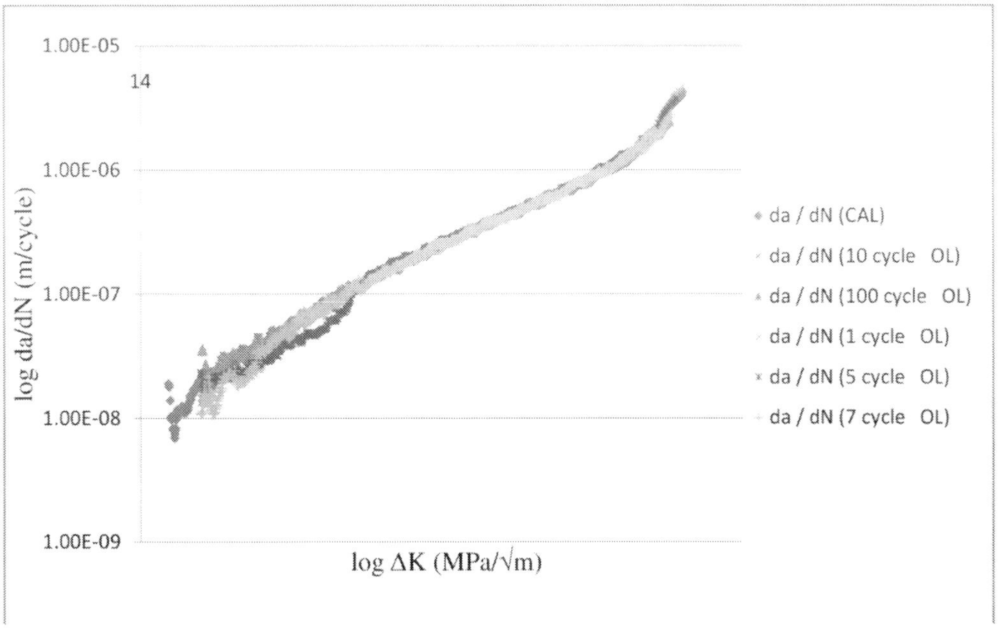

Fig. 5.Log-log plot of crack growth rates vs. stress of cycle curve.

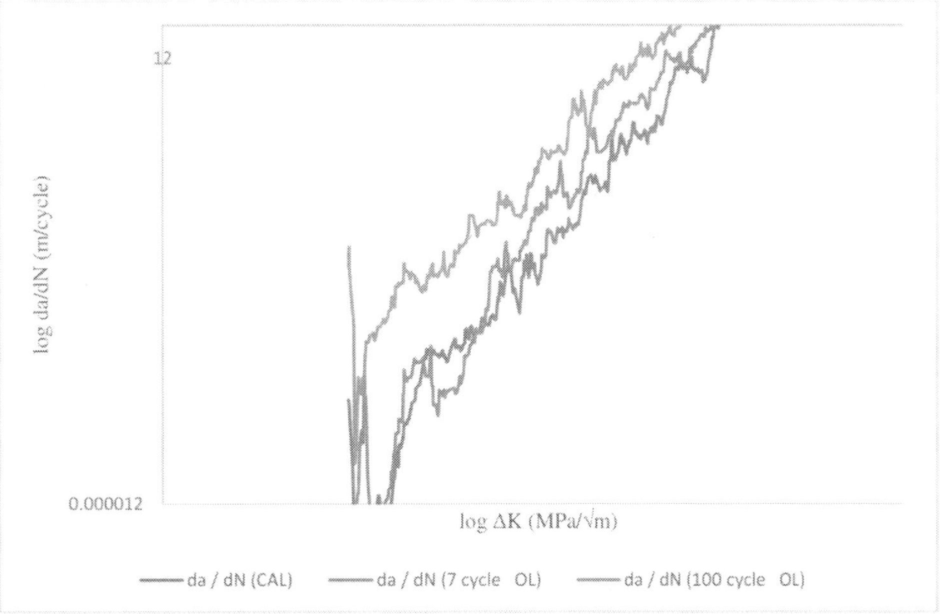

Fig. 6.Log-log plot of crack growth rates vs. stress of cycle curve

The reduction in the magnitude of retardation on application of 10 and 100 overload cycle may be due to development of plastic zone large enough to upset the constrain by the reduced elastically deformed region. It may also be possible that repeated overload application had sharpened the crack resulting suitable condition for rapid crack growth [7]. However the proposed explanations need further investigation.

3.2 Photograph of fatigue fracture surface

Various regions of a fracturedspecimen showing fatiguecrack growth, imposed to 7 cyclesof overload of an HSLA steel tested at R_{ol}= 1.25 are presented in Fig. 7.

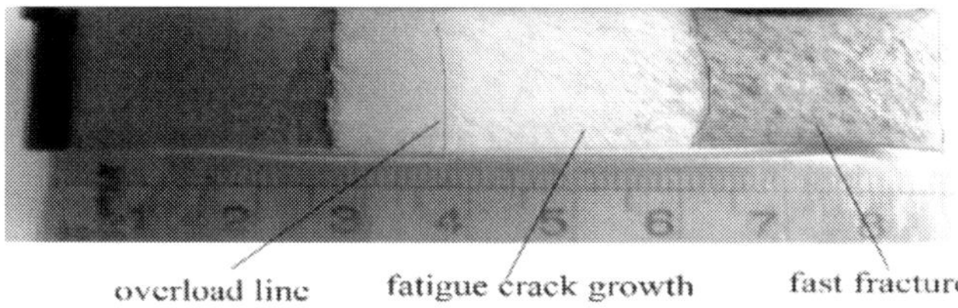

Fig. 7. Various regions of a fracturedspecimen showing fatiguecrack growth, imposed to 7 cyclesof overload.

4. Conclusions

In fatigue crack growth study three different loading conditions were applied: constant amplitude loading with fixed stress ratios, constantloading interspersed with single spike overload, and constantamplitude loading interspersed with multiple (band) spike overload. Effect of overload and band overload on fatigue growth life is determined.

The conclusions drawn from the present work are summarized as follows:

- The application of overload and band overload reduces the crack growth rate. However, the extent of retardation is little on application of single overload cycle.
- Maximum retardation was observed on application of 7 numbers of consecutive overload cycles. This may be due to further enlargement of plastic zone and other retardation inducing factors.
- Reduced retardation effect on application of overload cycles 10 and 100 is possibly due to development of plastic zone large enough to upset the constrain due to elastic enclave.

References

1 Brog T, Jones J, Was GS 1984 *Eng Fract Mech* **20(2)** 313.
2 Sander M, Richard H 2005 *Int J Fatigue* **27(8)** 905.
3 Sander M, Richard H 2006 *Int J Fatigue* **28**(5) 583.
4 Anderson TL 2005: *Fracture Mechanics: Fundamentals and Applications*. CRC press.
5 Sun Y 2006 *The Study of Overload Effects on Fatigue-Crack-Growth Behavior by Neutron Diffraction*, MS thesis, University of Tennessee – Knoxville, USA.
6 Verma BB, Kumar A, Ray PK 2000, Trans Indian Inst Metals **53(6)** 591.
7 **Schijve J, 1972, DTIC** Document.
8 Dubensky R 1974 *Fatigue Crack Growth Delays*. SAE Technical Paper. **doi:10.4271/740223**.
9 Wei RP, Shih TT 1974 Int J of Fract, **10(1)** 77.
10 Shin C, Fleck N 1987 Fatigue Fract Eng Mater Struct, **9(5)** 379.
11 Willenborg, J., R. Engle, and H. Wood, 1971, DTIC Document.
12 Wheeler OE. June 30, 1970. General Dynamics Report FZM-5602, Fort Worth.
13 Elber W 1971 *ASTM STP 486, American Society for Testing and Materials*, Philadelphia, 230.
14 E 647-13, *Standard Test Method for Measurement of Fatigue Crack Growth Rates, ASTM Annual Book of Standards*, **03.01**, West Conshohocken, PA, 2013.
15 Corbly D, Packman P 1973 *Eng Fract Mech*, **5(2)** 479.
16 Ramos MS, Pereira MV, Darwish FA, Motta SH, Carneiro MA 2003 Fatigue Fract Eng Mater Struct **26(2)** 115.

IOP Conf. Series: Materials Science and Engineering **75** (2015) 012011 doi:10.1088/1757-899X/75/1/012011

Effect of mechanical mixing on the microstructure and mechanical properties of Al-based metal matrix composites reinforced with nanocrystalline Al-Ca intermetallics

A. K. Chaubey, Rohit Kumar, Tapas sahoo

CSIR- Institute of Minerals and Materials Technology (IMMT), Bhubaneswar-751013

Email: anil.immt@gmail.com

Abstract. Al-based metal matrix composites reinforced with 40 vol.% of nanocrystalline Al-Ca intermetallic particles were synthesized by hot pressing followed by hot extrusion process and the effect of manual blending as well as mechnical mixing on microstructure and mechanical properties was studied. Microstuture reveals that mannual blending leads to the agglomeration of Al-Ca intermetallic particles and on the other hand, the composites prepared by milling display a more homogeneous distribution of the reinforcing particles. Mechanical mixing has a strong impact on the mechanical properties. The strength increases from 112MPa for pure Al to 250 and 415MPa for the composites produced by blending and milling respectively. This behavior is linked to the reduced matrix ligament size characterizing the milled composites.

1. Introduction

In the category of light weight advanced engineering materials, Al-based metal matrix composites (MMCs) have been shown its importance due to their remarkable mechanical properties, including low density, high strength and good fatigue and wear resistance [1,2]. In addition, MMCs offer the possibility to tailor their properties to meet specific requirements, which renders this type of materials quite unique in comparison to conventional unreinforced materials [1-4]. Among the reinforcements intermetallics in MMCs have emerged as possible substitutes for ceramic reinforcements [3-6]. This appears to be a highly promising option in view of high hardness of intermetallics and their compatibility with the metal matrix. In addition, thermal expansion coefficients of intermetallics are much closer to the metals than those of ceramic reinforcements.

Among all the intermetallics in the Al-Ca system Al_3Ca_8 is one of the important intermetallic phases that is associated with important mechanical properties. It is known to be hard, brittle and stable up to 850 K temperature [7].The mechanical properties of intermetallics are closely related to their structure and stability [8]. Hence, the complete knowledge of these aspects is a necessary condition for their application. The Al-Ca system has been investigated mostly around the Al-rich corner. However, three new Ca-rich intermetallic compounds (AlCa, Al_3Ca_8 and $Al_{14}Ca_{13}$) have been recently reported in addition to the already established Al_4Ca and Al_2Ca compounds [9, 10].

For most engineering applications, the composite powders have to be compacted to produce highly dense samples [11]. Among the different consolidation techniques, hot extrusion, which involves the simultaneous application of temperature and pressure, shows a large potential for achieving fast and full densification of powders avoiding possible detrimental phase transformations and grain growth) during compaction. Hot extrusion of powders allows a high shear strain rate, which promotes high

Content from this work may be used under the terms of the Creative Commons Attribution 3.0 licence. Any further distribution of this work must maintain attribution to the author(s) and the title of the work, journal citation and DOI.
Published under licence by IOP Publishing Ltd

strength bonding between the particles. In the case of Al- and Mg-based alloys and composites, hot extrusion breaks the typical oxide layers that coat the powder particles, and hence it gives improved bonding between the particles [12]. Hence, in order to improve the density and interfacial characteristics of the consolidated bulk samples, the hot pressed samples were extruded under argon atmosphere at 673 K and 500 MPa and extrusion ratio was 6:1. Effect of dispersion in extruded samples with respect to microstructure and mechanical property of the composite reinforced with 40 vol.% of Al-Ca intermetallics was studied.

2. Experimental

Al metal matrix was prepared by mannual blending as well as mechnical mixing of Al-Ca intermetallic reinforcement using a Retsch PM400 planetary ball mill with hardened steel balls and vials. The powders were milled for 3 h with a ball-to-powder mass ratio (BPR) of 10:1 and at rotational velocity of 100 rpm. The Al-Ca intermetallic reinforcement was prepared by crushing of pre-alloyed $Ca_{65}Al_{35}$ ingots (prepared by arc melting techniques) followed by 100h milling and details are reported elsewehre [13]. Handling of samples was carried out in a glove box under purified argon atmosphere (less than 0.1 ppm O_2 and H_2O). In order to produce Al based metal matrix composites (MMCs) reinforced with Al-Ca intermetallic particles, the Al alloy powder was manuual blended as well as mechanically mixed in the planetary ball mill with 40 vol. % of Al-Ca intermetallic powder. The composite powders were consolidated by hot pressing at 673 K and 400 MPa. Structural characterization was performed by X-ray diffraction (XRD) using a Philips PW 1050 diffractometer (Co Kα radiation; $\lambda = 0.17889$ nm) and by scanning electron microscopy (SEM) using a Gemini 1530 microscope. The Rietveld method was applied for the profile-fitting structure refinement using the WinPlotR software package [14]. The density of the hot pressed samples was measured using Archimedes principle. Cylinders having a length/diameter ratio of 2.0 (6 mm length and 3 mm diameter) were prepared from the hot consolidated samples for compression testing in accordance with ASTM standard [15]. The specimens were tested at room temperature using universal testing machine (Instron model 8562) under quasistatic loading (strain rate ~ 1×10^{-4} s^{-1}).

3. Result and discussion

Fig. 1 shows the XRD patterns of the Al–Ca reinforcement and extruded composites. The initial powder (Fig. 1(b)) shows sharp Bragg peaks of the Al_3Ca_8 intermetallic phase. After 200 h of milling the Al_3Ca_8 intermetallic transforms into the $Al_{14}Ca_{13}$ phase. In addition, significant peak broadening can be observed. This can be attributed to the reduced crystallite size to the nanometer regime. Indeed, Rietveld structure refinement reveals a crystallite size of about 10 nm. The pattern also shows the presence of a small amount of CaO, which may be ascribed to oxidation of the highly reactive Ca during handling and sample preparation. The XRD patterns of the composites mixed with 40 vol.% of the Al-Ca reinforcement by blending (B20) and my milling (M40) reinforcement (Fig. 1(a)), show sharp crystalline peaks due to pure Al along with broad peaks of the intermetallic reinforcement.

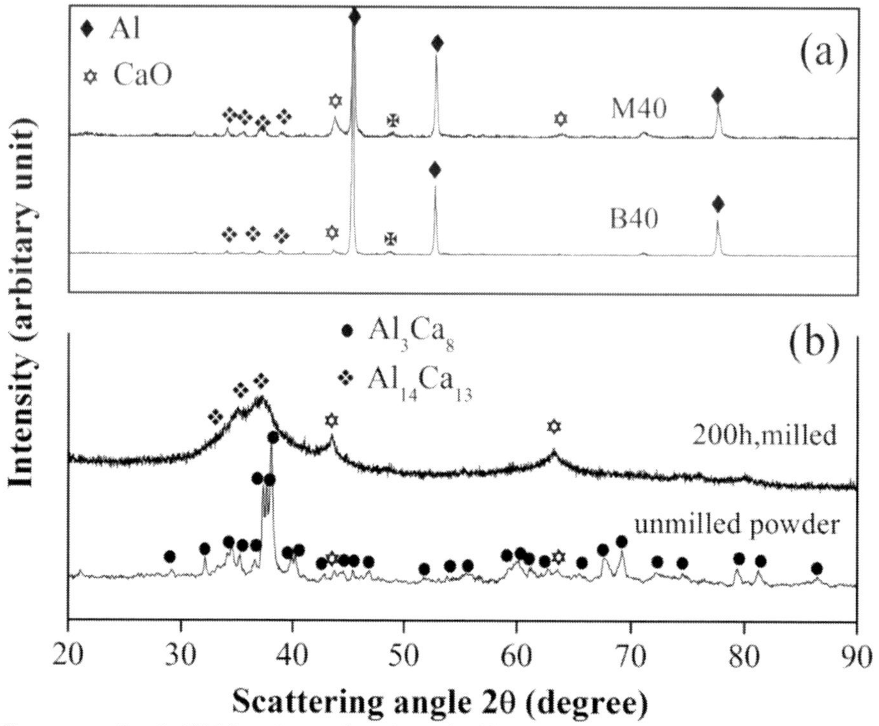

Fig. 1 XRD patterns (λ= 0.17889 nm) as a function of milling time for the mechanically milled Al-Ca intermetallic particles (a) composites with 40 vol.% of intermetallic reinforcement by blending (B20) and by milling (M40) (b) the initial pre-alloy and Al-Ca powder milled for 200 h.

The microstructure of the composites along with base material was investigated by SEM and the corresponding micrographs taken from the cross-section of the extruded specimens are shown in Fig. 2. The base material of the composite (pure Al sample) (Fig. 2(a) shows homogeneous microstructure without any defect in the microstructure. The composite sample prepared by manual blending of the reinforcement (Fig. 2(b)) shows agglomeration of the particles and these agglomerated particles cracked during extrusion(as marked by arrow, crack length= 40-120 μm). But the samples prepared by milling of the reinforcement (Fig. 2(c)) shows small size intermetallics particles (dark area) homogenously distributed within the Al matrix without any measure carks in the particle or at interface.

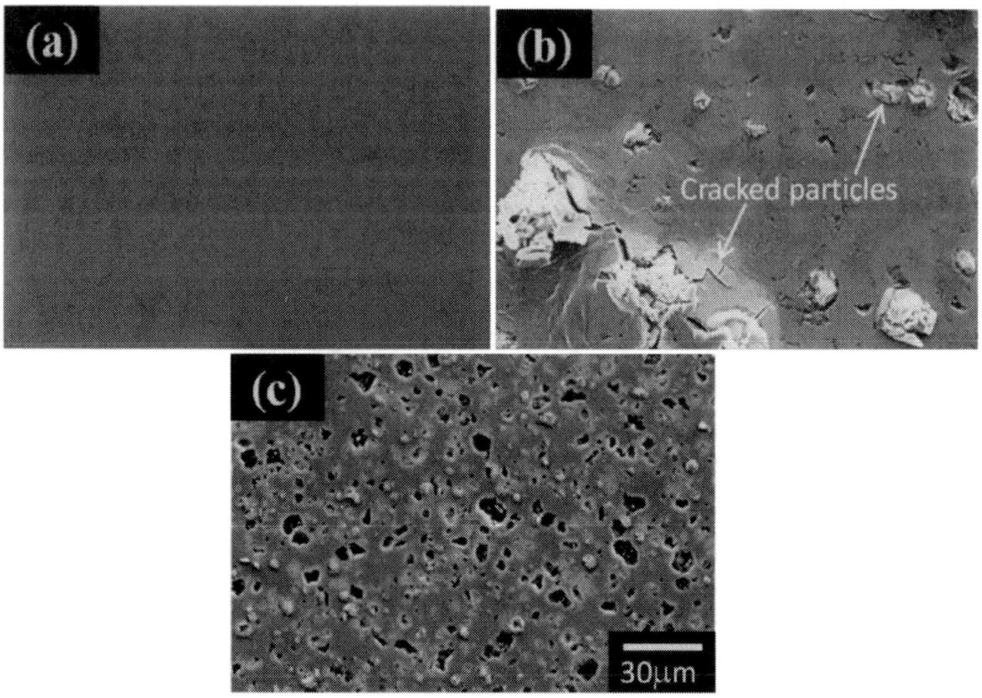

Fig. 2 SEM micrographs for the consolidated samples (a) pure Al (b) composites reinforced with 40 vol.% of Al-Ca intermetallic particle (a) blended (b) milled.

The cracking of the particles in the blended composite during extrusion is due to brittle nature of the intermetallics particles and after agglomeration it fails to transfer load to the matrix. Fig. 3 displays the SEM micrograph and the corresponding elemental distribution maps, Al, Ca & O combine (Fig. 3 (b)), Al (Fig. 3(c)) and Ca (Fig. 3(d)) obtained by SEM-EDX for the extruded composites reinforced with 40 vol.% Al-Ca intermetallic reinforcement. It reveals oxidation of the small intermetallics particles and small cracks in the particles, this is because more brittle and fragile nature of CaO.

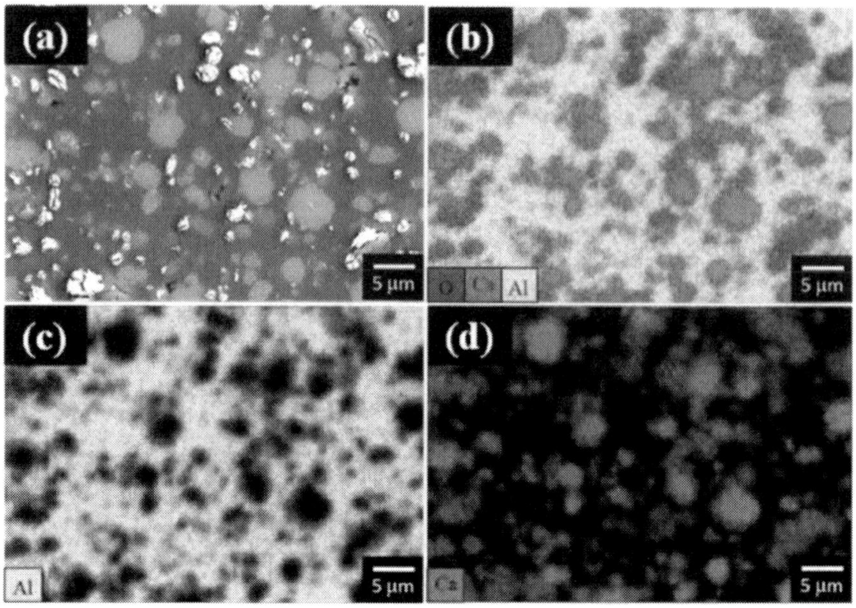

Fig. 3 EDX-SEM elemental distribution maps of the consolidated composites with 40 vol.% of Al-Ca intermetallics.

The mechanical properties of the extruded samples under compressive loading of the composite are shown in Fig.4. Extruded samples shows increase in strength in both blended as well as milled condition as compared to pure Al. The compressive strength increases from 250 MPa for the blended composites (B40) to 415 MPa for milled composite (M40) but the ductility remains 10% in both cases. Further the milled sample shows 1.7 times strength as compared to the blended samples, exhibiting the effectiveness of the milling on particle dispersion. Milling leads to homogeneous particle dispersion in the matrix leads to decrease in inter particle spacing (λ) as it was previously also seen in case of hot pressed sample, which is directly related to the strength of the material. This increase in compressive strength could be attributed to well-known factors (pertaining to reinforcement) such as: (i) the dislocation strengthening in the matrix, which is related to the nucleation of additional dislocations in the matrix due to the introduction of the reinforcement [16,17], (ii) Orowan strengthening mechanism [18], and (iii) load transfer from the matrix to reinforcement [17].

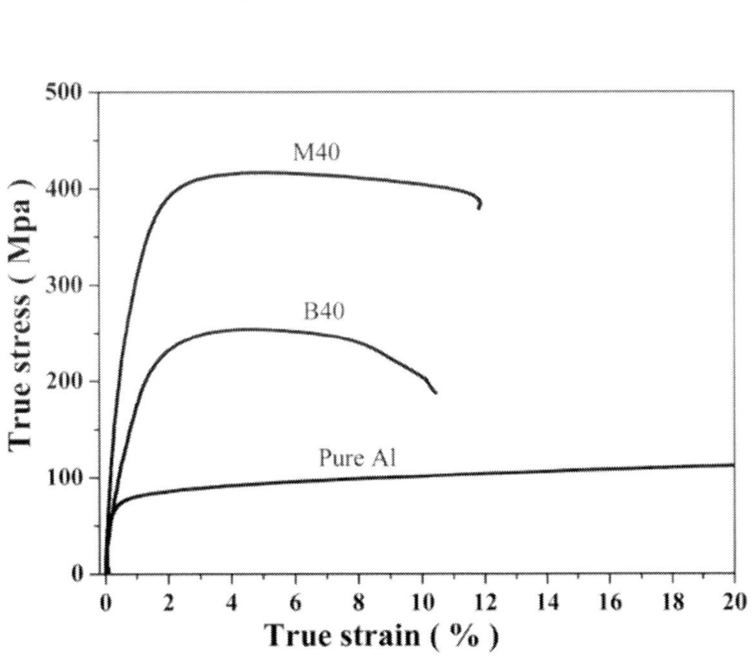

Fig. 4 Room temperature compressive stress-strain curves for the hot pressed and hot extruded pure Al, reinforced with 40 vol.% Al-Ca intermetallic particles prepared by blending (B40) and milling (M40) respectively.

4. Conclusions

Al-Al_3Ca_8 composite was prepared successfully by manual blending as well as milling of the reinforcement in Al matrix followed by hot consolidation process. Manual blending leads to the agglomeration of the Al_3Ca_8 particles to form a cell network throughout the sample. On the other hand, the composites prepared by milling display a more homogeneous distribution of the reinforcing particles. The distribution of reinforcing particles has a strong impact on the mechanical properties. The strength increases from 112 MPa for pure Al to 250 and 280 for the corresponding composites produced by blending as well as milling respectively. This behavior is due to the significant decrease of the matrix ligament size in the composites produced by milling similar to the strengthening by grain refinement. Thus milling leads to higher strength as compared to the composites produced by manual blending.

REFERENCES

[1] Clyne T W and Withers P J 1993 *An Introduction to Metal Matrix Composites* (Cambridge University Press Cambridge)

[2] Kainer K U 2006 *Metal Matrix Composites. Custom-made Materials for Automotive and Aerospace Engineering* (WILEY-VCH Weinheim)

[3] Lee J M, Kang S B, Sato T, Tezuka H and Kamio A 2003 *Mater. Sci. Eng. A* **343** 199

[4] Muñoz-Morris M A, Rexach J I and Lieblich M 2005 *Intermetallics* **13** 141

[5] Torralba J M, Velasco F, Costa C E, Vergara I and Cáceres D 2002 *Composites A* **33** 427

[6] Velasco F, Lima, W M, Antón N, Abenójar J and Torralba J M 2003 *Tribol. Int.* **36** 547

[7] Matsuyama K 1928Sci. Reports *Tohoku University* **17** 783

[8] Ozturk K. Chen L-Q and Liu Z-K 2002 *J. Alloys Comp.* **340** 199

[9] Kevorkov D and Schmid-Fetzer R 2001 J. *Metallkde.* **92** 946

[10] Huang B Q and Corbett J D1998 *Inorganic Chem.* **27** 5827

[11] Underwood E E 1985 *Metals Handbook, 9th edition* (ASM International **34** 12)

[12] Wang S H 2011 *Nanocrystalline metals and alloys: Processing, microstructure, mechanical properties and application.* (Wood head Publishing Ltd. Cambridge, USA)

[13] Chaubey A K , Scudino S, Prashanth K G, Stoica M, Vaughan G, Mukhopadhyay N K, Mishra B K and Eckert J 2012 *Mater. Lett.* **79** 145

[14] Roisnel T and Rodríguez-Carvajal J 2001 *Mater. Sci. Forum* **118** 378

[15] ASTM E9-89a. 2000 *Standard test methods for compression testing of metallic materials at room temperature*, (West Conshohocken) pp. 1-9.

[16] Hassan S F and Gupta M J, 2006 *Mater. Sci.* **41** 2229

[17] Shi N, Wilner B and Arsenault R J 1992 *Acta. Metall. Mater.* **40** 2841

[18] Szaraz Z, Trojanova Z, Cabbibo M and Evangelist E 2007 *Mater. Sci. Eng. A* **462** 225

IOP Conf. Series: Materials Science and Engineering **75** (2015) 012012 doi:10.1088/1757-899X/75/1/012012

Effect of post-curing on thermal and mechanical behavior of GFRP composites

D S Kumar*, M J Shukla, K K Mahato, D K Rathore, R K Prusty and B C Ray

Composite Materials Group, Department of Metallurgical and Materials Engineering, National Institute of Technology, Rourkela, India-769008

*Email: dskumar343@gmail.com

Abstract. Curing cycle has a strong impact on the thermal and mechanical behavior of thermosetting polymers. The extent of cross-linking which is a strong function of curing temperature and time is directly linked to the glass transition temperature (T_g) of the thermosetting polymer. This transition temperature speaks about the transformation of the polymer from glassy state to rubbery state, hence decides the applicability of the material at certain temperature with certain degree of safety and reliability. Hence assessment of T_g and its possible improvement is quite essential from material point of view. The present study is emphasized on the impact of post curing parameters on thermal as well as mechanical behavior of glass fiber reinforced polymer (GFRP) composite. Post curing was carried out at 3 different temperatures (80°C, 110°C and 140°C) for different time periods (2h, 4h, 6h, 8h and 12h). Short beam Shear (SBS) test was performed on each of the post cured samples to determine the apparent Interlaminar Shear Strength (ILSS) and the corresponding T_g was also evaluated using differential scanning calorimetry (DSC) analysis. The results revealed that the ILSS and T_g are significantly affected with post curing parameters. No significant change in ILSS was obtained at 80°C over the entire curing time. In case of 110 °C a smooth increment in ILSS was observed with time (even till 12 hrs). For samples post cured at 140 °C a rapid improvement in ILSS takes place with time followed by saturation. With all the possible combinations of curing temperature and time, optimum values are noticed at 140°C for 6 hrs.

1. Introduction

Composite materials can be defined as a combination of two or more materials that results in better properties than those of the individual components used alone. These materials have high strength and stiffness combined with low density compared to the conventional metals. Polymers are structurally more complex than metals. Polymer – matrix composites are viscoelastic materials and their mechanical properties are significantly influenced by temperature. They are used mostly in areas from aerospace, automobiles and boats to cryogenic equipment such as cryogenic fuel tanks, cryogenic fuel delivery lines, cryogenic wind tunnels and parts of the cryogenic side of turbo-pumps because of their ease of handling, low fabrication cost and excellent mechanical properties [1].Curing is an irreversible reactionwhere chemical covalent cross-links are formed whichare thermally and mechanically stable. The curing process plays a major role in achieving the final mechanical properties and chemical resistance of the material. State of polymer resin is liquid (soft) before the fabrication of composite, which then changes to solid matrix (hard) after curing.For the fabrication of composite, two types of resins are used i) primary resin (matrix) ii) secondary resin (hardener). Most commonly used primary resins are epoxy, unsaturated polyester and polyurethane. Secondary resin i.e. hardener is added for curing purpose. Most commonly used secondary resin includesamines or peroxides. During cross-linking, the stateof matrix changes from liquid to gel and then transforms into solid. Curing can be done at room temperature as well as at elevated temperature [2].This depends on composition of resin and hardener. Optimum curing results in a perfectly cross-linked polymer network which leads to increased T_g and mechanical properties.The T_g depends on different factors, including composition of the resin molecule, curing agent, curing time, cross-linking density and temperature [3]. Generally, during post curingthe $T_g$increases with increasing post curing temperature but it will not exceed the cure temperature. There are several parameters that define the post-cure process. Two biggest variables are temperature and time, but also the time between initial curing and post curing and temperature profile gradient play a role. So, post curing process can play crucial role in obtaining optimum mechanical and thermal properties of polymer matrix composites [4]. Present investigation

Content from this work may be used under the terms of the Creative Commons Attribution 3.0 licence. Any further distribution of this work must maintain attribution to the author(s) and the title of the work, journal citation and DOI.
Published under licence by IOP Publishing Ltd

IOP Conf. Series: Materials Science and Engineering **75** (2015) 012012 doi:10.1088/1757-899X/75/1/012012
includes the effect of two most primary post curing parameters, post curing temperature and time on the ILSS and T_g of GFRP composite and an attempt has been made to mathematically describe the variation of ILSS and T_g with post curing time at different temperatures.

2. Materials and experimental methods

2.1 Materials
Diglycidyl ether of Bisphenol A (DGEBA) based epoxy resin was used as matrix and triethylene tetra amine (TETA) was used as hardener. Resin and hardener were provided by Atul Industries Ltd, Gujarat under the trade name of Lapox L-12 and K-6 respectively.E-Glass fibres used were manufactured by Saint-Gobain. The reinforcement was used in the form of glass fibre sheets (3k, plain weave).

2.2 Fabrication of GFRP Composite
Woven fabric E- glass fibre were cut into the size of 25 cm X 25 cm to form 14 layer sheets and weighed by using electronic weighing machine. Epoxy resin was weighed to be 40% of the total weight of the fibre. Then hardeneris added which is equal to 10% of the weight of epoxy resin. Glass fibre/epoxy laminate have been prepared by hand lay-up method and cured in a hydraulic press by at 60°C temperature and 1MPa pressure for 20 minutes. Table 1 represents the properties of epoxy resin and glass fibre.

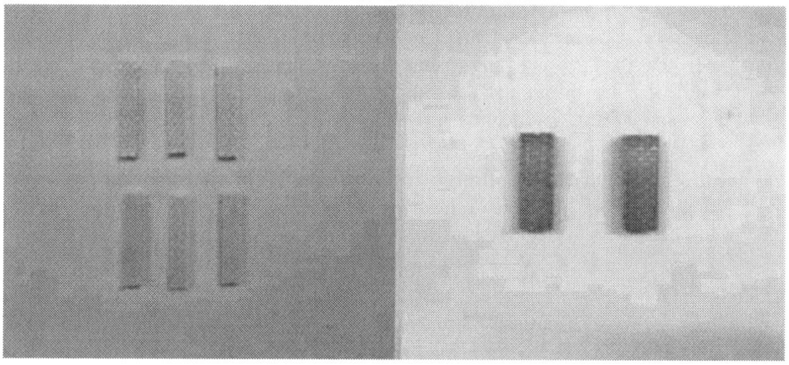

Figure 1. Glass/fibre composite before post curing and after post curing

Table 1. Properties of epoxy resin and glass fibre

Property	Epoxy	Glass Fibre
Tensile Strength (GPa)	0.11	3.4
Tensile Modulus (GPa)	4.1	72.3
Poisson's Ratio	0.3	0.2
Strain at Failure %	4.6	4.8
Density g/cm^3	1.162	2.58

2.3 Experimental methods

2.3.1.Post Curing schedule
The dried GFRP specimens were post cured in an oven at 80°C, 110°C and 140°C for 2h, 4h, 6h, 8h and 12h. The heating rate was 2°C/min. The specimens were cooled in oven atmosphere.

2.3.2. Short Beam Shear (SBS) test

The test was performed in 3 point bending fixture of Instron 5967 UTM using ASTM D2344-84 standard specimens to evaluate the apparent interlaminar shear strength.The tests were carried out at ambient temperature and at 1mm/min loading rate. The span to thickness ratio was kept 6.

2.3.3. Thermal analysis

DSC is a thermal analytical technique in which the difference in the amount of heat required to increase the temperature of a sample and reference is measured as a function of temperature or time. DSC wascarried out in a nitrogen atmosphere. The weight of the samples was taken around 15 mg. Heating was performed from 30°C to 150°C with heating rate of 10°C/min. The glass transition event is observed as an endothermic step wise increase in the heat flow. Glass transition temperature represents the region in which the resin transforms from a hard glass solid to a viscous liquid.

3. Results and discussion

3.1 Effect of temperature and time on ILSS

Post cured samples showed change in colour in contrast to the non-post cured samples as shown in figure 1. A substantial increment in maximum stress value is observed in case of post curing the GFRP sample at 140 °C for 2 hrs in comparison to the control sample, followed by a small increment with time which can be observed from figure 2 (a). After certain time period, the stress-strain curves overlap on each other indicating saturation in the strength value. Figure 2(b) represents stress-strain curves of samples cured at 80,110and 140°C for 6 hrs. No significant change in maximum stress values was observed at 80°C curing temperature. Gradual increment in maximum stress was further noticed on increasing curing temperature. Variation in ILSS forGFRP composite at different time periods is shown in figure 3. At 80°C temperature, no significant improvement in ILSS was noticed even after 12hrs of conditioning,because the ILSS mainly depends oncross-linking density at interface region.

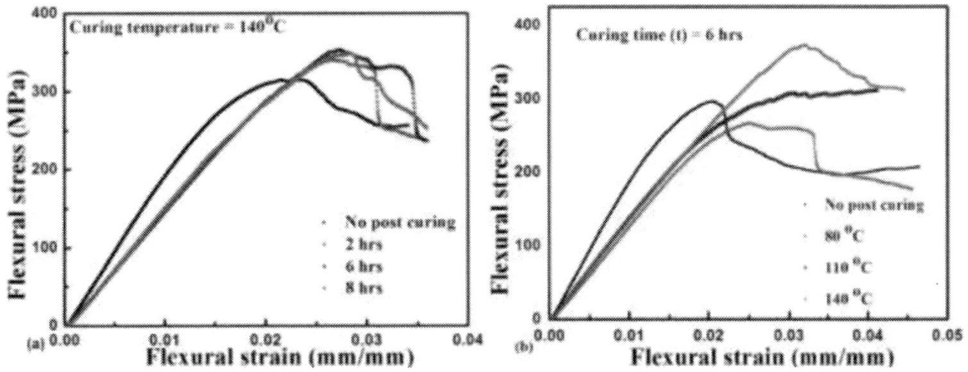

Figure 2. Effect of temperature and time on flexural stress.a) at 140°C for 2hrs, 6hrs and 8 hrs. b) at 80°C, 110°C and 140°C for 6hrs.

The energy available at these post curing parameters may not be sufficient to further enhance the wettability at the fibre/matrix interface, which in turn reflects no improvement in ILSS.At 110°C the ILSS increases with time steadily at a very slower rate. Here the variation in ILSS with time may be represented with a linear equation (i) as

$$ILSS\,(MPa) = 23.96 + 1.24t\,(hrs) \text{-------- (i)}$$

At 110°C, the heat energy may be sufficient for the movement of polymeric chains at the interface for formation of certain cross-links and the density of the crosslinks at the interface gradually increases with time. Hence steady increment in ILSS is noticed with time.At 140°C, a very drastic increment was noticed inILSS at an earlier stage.

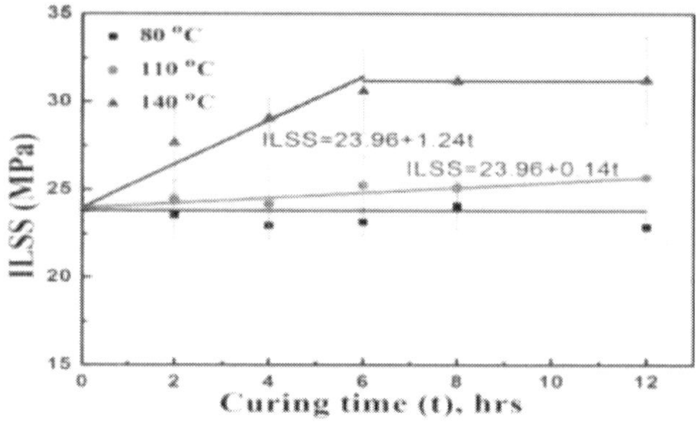

Figure 3. Effect of temperature and time on ILSS

A linear trend is noticed as shown in equation (ii) as

$$ILSS\,(MPa) = 23.96 + 1.24t\,(hrs) \text{-------- (ii)}$$

The crosslink density at the interface increases rapidly with time and gets saturated after certain time period. Hence a drastic improvement in ILSS at the rate of 1.24 MPa/hr is noticed till 6 hrs of curing and then onwards saturation follows with an ILSS of approximately 31.45 MPa, which accounts for 25%higher than that of without post-cured GFRP.

3.2 Effect of temperature and time on T_g

The DSC curves for various curing temperature and time is plotted in figure 4. T_g is obtained from the onset of change in slope of the heat flow vs. temperature. Hence the variation in T_g with different curing time at different curing temperatures is reported in figure 4(a) and 4(b). Primarily the T_g of the composite is governed by the bulk polymer matrix. The variation of T_g when plotted against post curing time for various temperatures, it followed sigmoid curve as per the equation (iii).

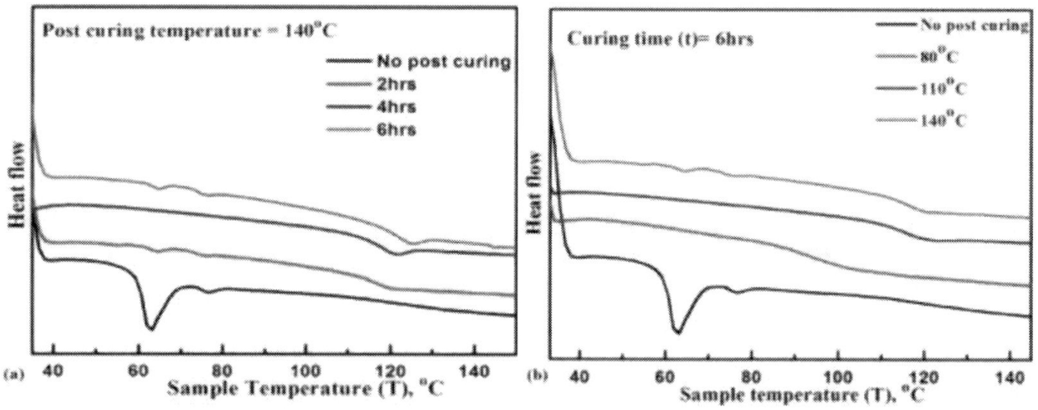

Figure 4. DSC curves for various post cured GFRP samples a) at 140°C for 2hrs, 4hrs and 6 hrs. b) at 80°C, 110°C and 140°C for 6hrs.

$$T_g = T_g^0(°C) + A\frac{t^n}{B+t^n} \qquad \text{-------- (iii)}$$

Where, $T_o = T_g$ for no post curing
t = Post curing time (hrs)
A, B and n are constants

Table 2. Values of constants for 80, 110 and 140°C for equation (iii)

Temperature (°C)	A	B	n
80	29.45	14.9	2.67
110	36.51	7.29	0.15
140	186.03	2.57	0.13

From table 2 it can be seen that the time coefficient (A) drastically increases with increase in curing temperature. 'A' is the dominating factor at lower time and hence at the earlier stage of curing time. The T_g is strongly dependent on 'A.' As at 140°C, the A value is quite high an instantaneous increase in T_g is noticed at small't' values.The exponent (n) represents the dependency of T_g on time.The value of 'n' is quite high at 80°C, hence the T_g is strongly dependent on time. With increase in curing temperature, the 'n' value significantly reduces and at 140°C it achieves a value of 0.13. Hence at 140°C, T_g has a little dependency on time.

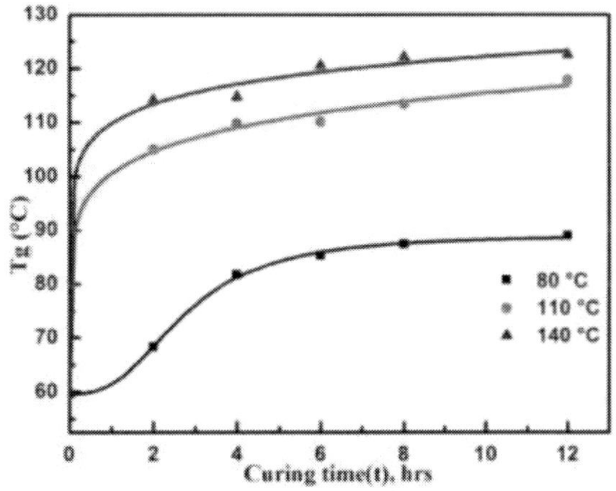

Figure 5. Variation in Tg with curing time at different curing temperatures

The figure 5 represents variation in T_g with curing time at different curing temperatures. At 80°C, initially the rate of increase in T_g for 2 hrs is higher as compared to other post curing time.Then the rate of increasing in T_g gradually slows down and after 6hrs the T_g get saturated. At high post curing temperatures i.e. 110°C and 140°C the rapid increase in T_g is observed for 2 hrs. After 2 hrs further exposing the samples for longer duration did not significantly altered the T_g and the T_g is almost saturated after 4hrs time for both 110°C and 140°C.

Figure 6(a) shows the SEM micrograph of fractured surface of a non-post cured GFRP specimen which was tested in 3-pont bend test. Polymer matrix without post curing is not sufficiently strong and under loading condition failure of composite primarily occurs due to the crushing of polymer matrix.Matrix crushing resulted in formation of small clusters of matrix which are adhered on the fibre surfaces as shown in figure 6(a). Whereas post curing of composite makes the polymer matrix significantly strong through the formation of well cross-linked polymer networks.The primary mode of failure in post cured GFRP is the failure at interface which leads to fibre/matrix debonding and further upon loading results in complete detachment of fibres from the matrix as shown in figure 6(b).

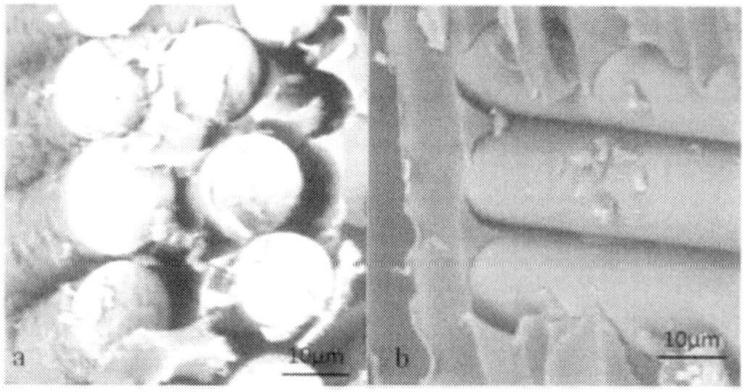

Figure 6. Scanning electron microscopy (SEM) images of fractured surface of glass fibre/epoxy composite a) without post curing. b) with post curing at 140°C for 6 hrs.

4. Conclusions

The following conclusions can be drawn from the present investigation

1. At 80°C temperature, no significant improvement in ILSS was noticed even after 12hrs of conditioning. The energy available at these post curing parameters may not be sufficient to further enhance the wettability at the fibre/matrix interface, which in turn reflects no improvement in ILSS.
2. At 110°C the ILSS value increases with time steadily at a very slower rate like mentioned as above equation (i).
3. At 140°C temperature a very drastic increment in ILSS value was observed at an earlier state. A linear trend is noticed as shown in equation (ii). This trend of increment is observed to be valid till a certain period of time i.e. 6hs after which the ILSS gets saturated.
4. High temperatures like 110°C and 140°C the energy received by the system is fairly high for activating the monomers to further cross-link.
5. From the present investigation it was observed that post curing at 140°C for 6 hrs gave better thermal and mechanical properties as compared to post curing at different temperatures and time periods.
6. The variation of T_g with time for different temperature is plotted and the governing mathematical expression is provided. The variation in T_g followed a trend as fitted with sigmoidal curve.

References

[1] Ray B C and Rathore D 2014 Durability and integrity studies of environmentally conditioned interfaces in fibrous polymeric composites: Critical concepts and comments *Adv. Colloid Interface Sci.* **209** 68–83
[2] Aruniit A, Kers J, Krumme A, Poltimäe T and Tall K 2012 Preliminary Study of the Influence of Post Curing Parameters to the Particle Reinforced Composite's Mechanical and Physical Properties *Mater. Sci.* **18** 256–61
[3] Furtos G, Silaghi-Dumitrescu L, Moldovan M, Baldea B, Trusca R and Prejmerean C 2012 Influence of filler/reinforcing agent and post-curing on the flexural properties of woven and unidirectional glass fiber-reinforced composites *J. Mater. Sci.* **47** 3305–14
[4] Stark W 2013 Investigation of the curing behaviour of carbon fibre epoxy prepreg by Dynamic Mechanical Analysis DMA *Polym. Test.* **32** 231–9

Effect of process parameters on solidification of Al-33Cu strip in high speed twin roll strip casting- A numerical study

Seshadev Sahoo
Department of Mechanical Engineering, Institute of Technical Education and Research, Siksha 'O' Anusandhan University, Khandagiri square, Bhubaneswar,OR, India

Email: seshadevsahoo@soauniversity.ac.in

Abstract. In twin roll strip casting process, the solidification behavior in the molten pool affects the process efficiency and the product quality which is influenced by the process parameters. Numerical simulations have been carried out to analyze the fluid flow, heat transfer and solidification in high speed twin roll strip caster at different pouring temperatures and roll gaps and the knowledge of heat flux distribution at the liquid metal-roll interface. In the present study a numerical simulation has been carried out for the solidification of Al-33wt% Cu in the twin-roll strip casting system at different pouring temperatures and roll gaps. FLUENT 6.3.16 was used to model fluid flow, heat transfer and solidification behavior during the casting of Al-33wt. % Cu alloy. From the simulation results, the initial pouring temperature of the liquid metal and the roll gap has been optimized for producing quality strips.

1. Introduction

The twin roll strip casting process has gained more attention due to its higher productivity and good surface quality of the as cast strips. This process is applied to produce thin strips whose thickness varies between 0.1-6 mm directly from the liquid metal by combining casting and rolling in a single step. During strip casting operation, the molten metal is introduced into the casting machine via the nozzle, which solidifies on the surface of the casting rolls by the removal of heat and the thin strip is obtained continuously [1-3]. A schematic diagram of a vertical twin roll strip caster is shown in Figure 1. The control of microstructure of the strip is of primary importance in twin-roll strip casting process because further significant modification of the microstructure, on which the properties of the strip depends, may not be possible. The microstructure of the strip depends on cooling rate and solidification front speed at various locations in the strand, which in turn depends upon the casting parameters like casting speed, roll gap, superheat of liquid metal, roll force, roll materials, etc. These casting parameters affect the fluid flow, heat transfer and solidification behavior in twin roll strip casting process. As the process looks very simple but in actual practice it is very difficult to control the process parameters experimentally during the casting operation because the solidification of liquid metal occurs within few seconds. For arriving at the right combination of design and process parameters to achieve high speed/productivity as well as right microstructure, mathematical modeling and optimization based on it are indispensible.

Corresponding author: S. Sahoo, Email: seshadevsahoo@soauniversity.ac.in

Content from this work may be used under the terms of the Creative Commons Attribution 3.0 licence. Any further distribution of this work must maintain attribution to the author(s) and the title of the work, journal citation and DOI.
Published under licence by IOP Publishing Ltd

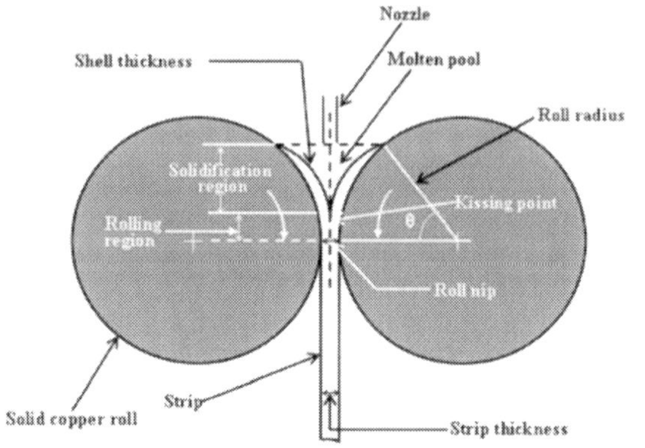

Figure 1. Schematic diagram of vertical twin roll caster.

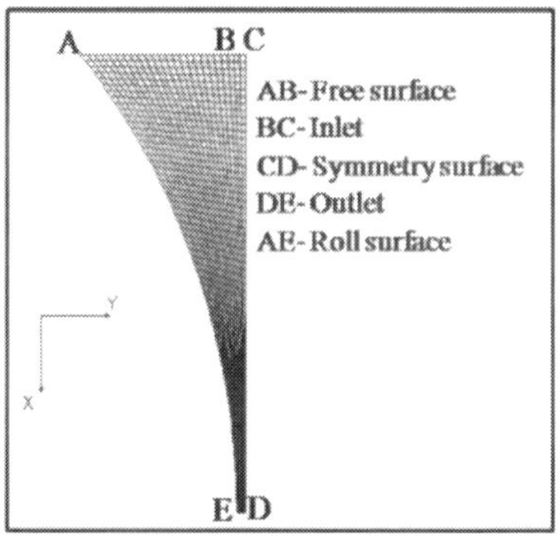

Figure 2. Computational domain

Santos et al. [4] presents a numerical model for the two-dimensional solidification process in twin roll continuous casting system by using finite difference technique and gives valuable information in to the thermal characteristics of solidification and processing for the strip casting. Kim et al. [5] numerically studied the fluid flow and heat transfer in wedge shaped pool during strip casting process and also investigated the effect of roll gap thickness and superheat of the melt on liquid metal velocity, variation in temperature distribution and solidification in the wedge shaped pool. But in their model the author considered temperature independent thermo-physical properties. The turbulent fluid flow, heat transfer, and solidification of a twin roll melt drag thin strip casting of steel simulated by Gupta and Sahai [6] and studied the effect of roll speed and melt superheat on thickness of the strip and found that by increasing the roll speed and melt superheat, thickness of the strip decreases. Zhao et al [7] developed a twin roll caster to cast aluminum alloys and investigated the process parameters including casting temperature, casting speed and thickness variation of the strip on microstructure and mechanical properties of the cast strip and found that twin roll caster has the ability to produce thin strips directly from the liquid metal with improved mechanical properties. The numerical investigation of turbulent fluid flow and solidification in twin roll caster was studied by Kim et al. [8] The authors studied the effect of roll gap thickness on the melt flow and temperature distribution in the melt pool and based on the results they estimated the patterns of the superheat removal. Zhang et al. [9] developed a FEM model to simulate twin roll strip casting process at 0.52 m/s casting speed and studied the influence of the process parameters, i.e., pouring temperature and height of liquid level to control the twin roll strip casting process and improve the quality of the strips. They also compared the variation of theoretically calculated and experimentally measured temperature of strip surface with the pouring temperature. A CFD model was developed by Zheng et al. [10] which provides better understanding of the melts flow characteristics and thermal exchange during rapid solidification of the Mg in twin roll casting process under different casting speed and gauge opening (roll gap) condition. They used constant thermo-physical properties like density, specific heat, latent heat, thermal conductivity and viscosity and measured the temperature of the casting strip both from the model and experiment and found that the calculated results for varying casting speeds are in good match with experimental determination. Fang et al. [11] simulated the temperature field of the strip in twin-roll casting method and studied the variation of temperature with different roll radii and roll gaps. In their model they simulated the model having casting speed of 0.5 m/s to 1.33 m/s and they did not consider the temperature dependency of thermo-physical properties. Miao et al. [12] developed a 3D model which describes the flow and temperature in molten pool and study the effect of process parameters on the flow and temperature field having casting speed 0.3-0.6 m/s. In their study they found the simulated results are in good agreement with the experimental results. Li et al. [13] developed a steady state finite element model to study the effect of processing parameters like casting speed and pouring temperature of liquid metal on the freezing point position of 304 stainless steel during twin-roll strip casting process. With the use of the model, the authors optimized the casting speed and pouring temperature for producing quality strips. But, in their model they also have considered temperature independent thermo-physical properties and casting speed of 0.5 m/s. Sahoo et al. [14, 15] developed a model for high speed twin roll caster having casting speed varies from 0.79 m/s to 3.98 m/s to produce layered structure material and compared with the experimental results.

Recently, numerical simulation has been developed rapidly, which can replace some experiments for the research of the twin roll strip casting process. However most of the above mentioned models have not taken into account the variation of thermo-physical properties with temperature and having low casting speed. The present research work is a numerical investigation of the characteristics of fluid flow, heat transfer and solidification in a wedge shaped pool during the high speed twin roll strip casting process and to study the effect of initial pouring temperature of liquid metal and roll gaps on solidification of Al-33wt. % Cu alloy in the molten pool.

2. Mathematical model

Liquid metal which is poured between the two counter rotating rolls is solidified as the rolls rotate and is machined to strips at the exit. Considering the practical process of casting operation the following assumptions are made in the development of the mathematical model: (1) The process is steady state, (2) Liquid metal is incompressible and Newtonian fluid, (3) There is no slip condition between the roll and the solidified strip, (4) Process is symmetric around the center line of roll gap, (5) The heat transfer is dominated by convection and conduction modes, (6) The value of heat transfer coefficient is constant along the strip/roll interface.

2.1 Governing equations

Numerical simulations of fluid flow and heat transfer in twin roll strip casting processes has been conducted by solving the two dimensional turbulent form of the Navier-Stokes equation which is fully coupled with a differential energy balance equation, which takes solidification into account.

The continuity equation is given by

$$\frac{\partial u}{\partial x}+\frac{\partial v}{\partial y}=0 \qquad (1)$$

The momentum equation in x-direction is given by

$$\frac{\partial p}{\partial x}+\frac{\partial(\rho uv)}{\partial x}=\frac{\partial}{\partial x}\left(\mu\frac{\partial u}{\partial x}\right)+S_x \qquad (2)$$

and that in y-direction is given by

$$\frac{\partial p}{\partial y}+\frac{\partial(\rho uv)}{\partial y}=\frac{\partial}{\partial y}\left(\mu\frac{\partial v}{\partial y}\right)+S_y \qquad (3)$$

The two-equation k-ε model of turbulence was used for the turbulence considerations.

Turbulent kinetic energy equation:

$$\left(u\frac{\partial k}{\partial x}+v\frac{\partial k}{\partial y}\right)\rho=\frac{\partial}{\partial x}\left(\frac{\mu_t}{\sigma_k}\frac{\partial k}{\partial x}\right)+\frac{\partial}{\partial y}\left(\frac{\mu_t}{\sigma_k}\frac{\partial k}{\partial y}\right)+\mu_t\phi-\rho\varepsilon \qquad (4)$$

Dissipation rate of turbulence equation:

$$\left(u\frac{\partial \varepsilon}{\partial x}+v\frac{\partial \varepsilon}{\partial y}\right)\rho=\frac{\partial}{\partial x}\left(\frac{\mu_t}{\sigma_\varepsilon}\frac{\partial \varepsilon}{\partial x}\right)+\frac{\partial}{\partial y}\left(\frac{\mu_t}{\sigma_\varepsilon}\frac{\partial \varepsilon}{\partial y}\right)+C_1\mu_t\phi\frac{\varepsilon}{k}-C_2\rho\frac{\varepsilon^2}{k} \qquad (5)$$

where ,

$$\mu_{eff}=\mu_0+\mu_t \qquad (6)$$

$$\mu_t = C_\mu \rho k^2 / \varepsilon \quad (7)$$

$$K_{eff} = K_0 + K_t \quad (8)$$

$$K_t = c_p \mu_t / \Pr \quad (9)$$

where , Pr is the turbulent Prandtl number.

According to Laundar and Spalding [16], the constants in the standard k-ε equation takes the following values: C_1=1.44, C_2=1.92, C_μ=0.09, σ_k=1.0, σ_ε=1.3.

The energy equation is given by

$$\left(\frac{\partial(\rho H)}{\partial t}\right) + \nabla.(\rho u H) = \nabla.(k \nabla T) + S_h \quad (10)$$

where ρ is the density, H is the enthalpy and S_h is the source term.

2.2 Boundary conditions

Because of the symmetry, the computational model adopted in this work is half of the real domain as shown in Fig. 2. The boundary conditions are as follows (refer to Figure 2):

(1) At the entrance of the liquid metal on the surface of the melting pool (BC), V_x=V_{in}, V_y=0, k=0.05(V_x^2+V_y^2), ε=$C_\mu k^{1.5}$/0.03D_{in}, T_{in}=851 K,
where V_x is the speed in x-direction, Vy the speed in y direction, V_{in} is the speed of the entrance, K the turbulence kinetic energy, ε the dissipation rate of the turbulent kinetic energy, and D_{in} the size of the inlet.

(2) On the free surface(AB), $\frac{\partial V_y}{\partial X} = \frac{\partial k}{\partial X} = \frac{\partial \varepsilon}{\partial X} = 0$, V_x=0

(3) At the symmetry interface (CD), $\frac{\partial V_x}{\partial Y} = \frac{\partial k}{\partial Y} = \frac{\partial \varepsilon}{\partial Y} = 0$, V_y=0

(4) On the outer surface of the roll, the tangential velocity of the roll is equal to the casting speed. So velocity components V_x and V_y on the roll boundary have the following values, V_x= $V_{roll}\cos\theta$, V_y=$V_{roll}\sin\theta$
where V_{roll} is the casting speed and θ is the angular position of the node on the roll surface. At the contact interface of liquid Al-33 wt.% Cu and the roll, a constant value of heat transfer coefficient was utilized in the model. The value of heat transfer coefficient was calculated based on the correlation given by Wang and Matthys [17]. The correlation is given by the following equation:

$$\overline{h_s} = 17.3 V_{roll}^{0.65} \quad (11)$$

where, $\overline{h_s}$ is the average heat transfer coefficient and V_{roll} is the casting speed. This correlation was validated by Guthrie et al. [18] experimentally.

(5) At the exit or nip of the roll (ED), fully developed boundary condition was used. The outflow conditions at the exit or nip of the roll are V_x= -V_{roll}, V_y=0.

3. Results and discussion

In the present work simulations were carried out for casting of Al-33 wt.% Cu strips having different initial pouring temperatures and roll gaps. The level of liquid metal pool was kept fixed at 0.0489 m. The liquid metal was fed through a nozzle of diameter 4 mm and the temperature drop during feeding was assumed to be negligible. Thus the temperature at the inlet was taken as the feeding temperature of the liquid metal. To know the effect of initial pouring temperature and roll gap on the solidification phenomena of Al-33wt. % Cu alloy during high speed twin roll strip casting, the thermal as well as the solidification behavior during twin roll strip casting process at high casting speed has been simulated on

FLUENT 6.3.16 platform. The parameters of the casting used for simulation are given in Table 1 and the detail procedure of the simulation are described elsewhere [14].

Table 1: Process parameters used in the simulation

Parameters	*Value*
Roll diameter (m)	0.1524
Roll width (m)	0.0254
Casting speed (m/s)	0.79
Roll gap (mm)	0.8 - 3
Pouring temperature of liquid metal, T_{in} (K)	831 - 941
Inlet diameter, D_{in} (m)	0.004

3.1 Initial pouring temperature of the liquid metal

Numerical simulations are carried out to study the effect of initial pouring temperature on solidification of Al-33wt. % Cu alloy during high speed twin roll casting. The simulations are carried out for casting of Al-33wt. % Cu strips having thickness of 2 mm in a twin roll caster at casting speed 0.7979 m/s (100 rpm). The initial pouring temperature of liquid metal varied from 831 K - 941 K and all other parameters is kept constant.

When other processing parameters are constant, the pouring temperature of liquid metal has a direct effect on the heat transfer between the casting roll and molten pool, which affects the flow and heat transfer in the molten pool. Figure 3 shows the solidification profile of liquid Al-33wt. % Cu in the molten pool for different pouring temperatures. From the solidification profile it is observed that, when the pouring temperature of the liquid metal is 831 K, the thickness of the solidified shell is 0.29 mm and the thickness of the solidified shell is 0.08 mm at pouring temperature of 941 K. Beyond 941 K the presence of solidified shell is not observed.

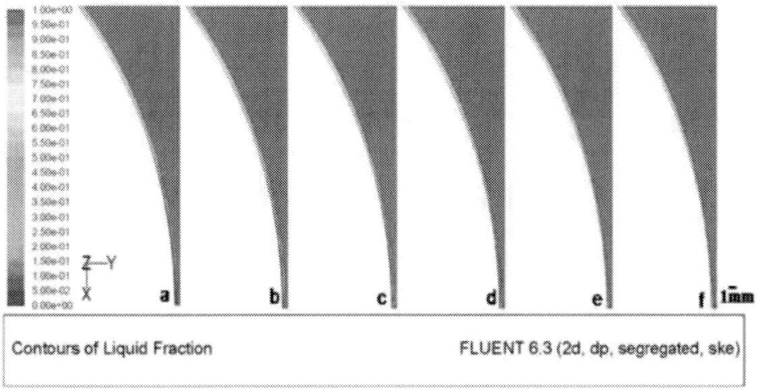

Figure 3: Solidification profile of Al-33Cu at different pouring temperatures (a) 831 K (b) 851 K (c) 871 K (d) 891K (e) 921 K (f) 941 K (casting speed 0.79 m/s).

The temperature of the strip at the outlet increases with increase in pouring temperature. Figure 4 shows that the variation of surface temperature of the cast strips at the exit section. It can be seen that the temperature of the strip at the exit section increases with increase in pouring temperature. However, at the surface of the strip the increase in the temperature as a result of 110 K increase in pouring temperature is not significant, i.e., ~5 K. At the plane of symmetry the increase is ~15 K. The higher pouring temperature results in higher temperature at the exit of the casting strip, which leads not only severe oxidation but also cracks of the casting strip. So considering the factors affecting the cast strip quality, the superheat should be as low as possible.

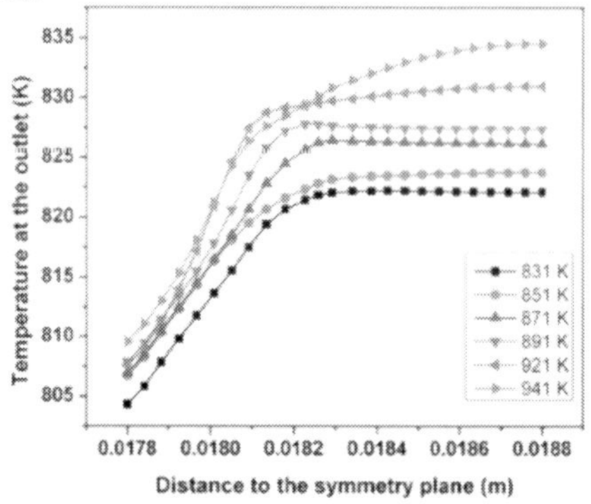

Figure 4: Temperature of the Al-33Cu strip at the nip position at different pouring temperatures

Figure 5 shows the effect of pouring temperature of melt on the solidification front speed. The solidification front speed decreases significantly with increase in pouring temperature of the melt. Similar trends can be observed for the variation of shell thickness at nip with the pouring temperature as shown in Figure 6. With increase in the pouring temperature, the superheat increases resulting in delay in the initiation of solidification and thus the shell thickness and front speed decreases.

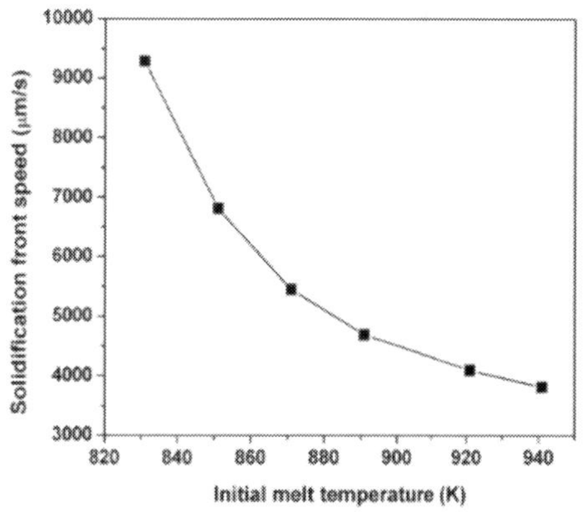

Figure 5: Effect of initial pouring temperature on solidification front speed

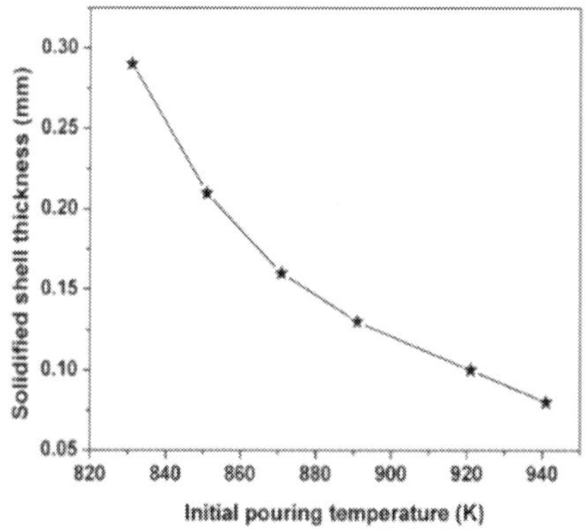

Figure 6: Effect of initial pouring temperature of the melt on solidified shell thickness

3.2 Effect of roll gap

The distance between the two rolls (roll gap) at the roll nip limits the maximum strip thickness and may affect transport phenomena in the molten pool. The simulations are carried out for casting of Al-33wt. % Cu strips at different roll gap, i.e., 0.8 mm, 1 mm, 2mm and 3 mm. The simulation results show the solidification pattern, temperature and velocity vector profile of Al-33wt. % Cu in the pool region respectively for the case when the roll speed is 0.79 m/s, pouring temperature of liquid metal is 851 K.

Figure 7 shows the solidification profile of the Al-33wt. % Cu alloy strip in the strand at different roll gaps. At roll gap thickness of 0.8 mm. the end point of solidification is above the roll nip position. When the roll gap thickness is 1 mm, the solidification end point is close to the exit. With increases in thickness of the roll gap from 2 mm to 3 mm, the solidification end point is below the roll nip position. As the roll gap thickness varies, the rate of heat transfer from the liquid metal to the roll surface remains constant because all other parameter like casting speed remains constant. However the mass of the liquid Al-33Cu enclosed between the rolls and thus the total heat to be removed increases with the roll gap. Therefore the solidification front speed and the shell thickness at nip decreases with increase in the roll gap as shown in Figure 8 and Figure 9. With decrease in roll gap, the solidification end point is close to the exit. This is because the quantity of liquid metal passing in and out of the molten pool is less than that with larger thickness and the heat brought into the melting pool is small. If the roll gap is too large and the end point of the solidification is lower than the exit, the strip may break off because of the incomplete solidification at the roll nip. If the roll gap is too small, then the roll pressure is too high which may break the strip.

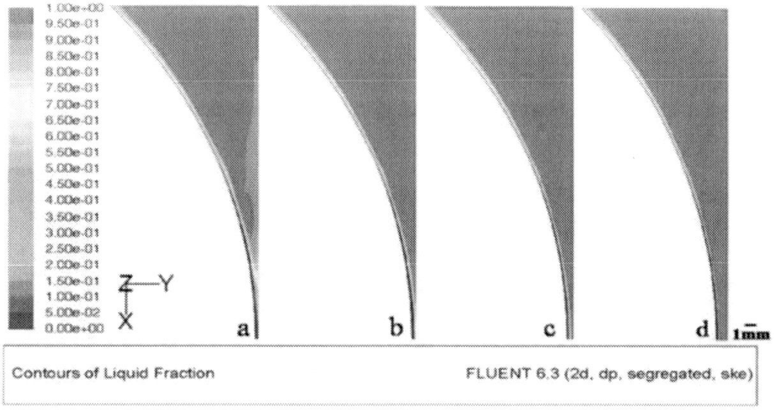

Figure 7: Solidification profile of liquid Al-33Cu at different roll gaps (a) 0.8 mm (b) 1 mm (c) 2 mm and (d) 3 mm

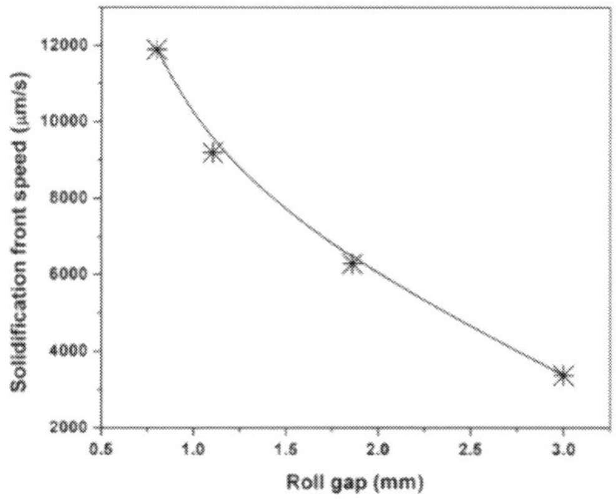

Figure 8: Effect of roll gap on solidification front speed at roll nip

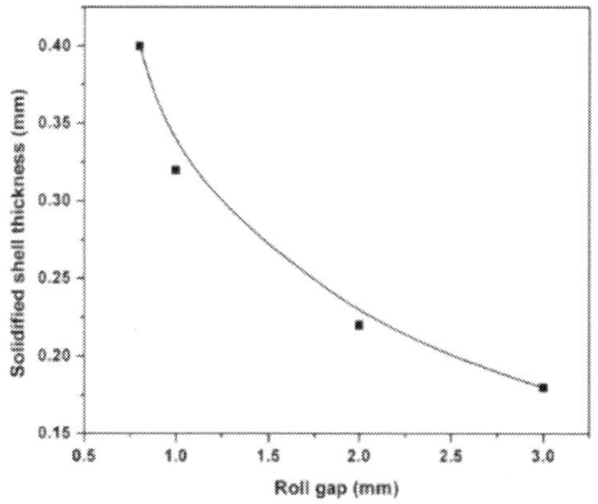

Figure 9: Effect of roll gap on solidified shell thickness at roll nip

Figure 10 shows the variation of the strip temperature from strip center to surface at the exit section with change of the roll gap. By varying the roll gap from 0.8 mm to 3 mm the surface temperature at the exit increases by ~9 K where as the temperature at the centre increases by ~15 K. The sensitivity of the temperature at surface and plane of symmetry (Figure 10) to roll gap is much higher when the roll gap is below 1 mm. It can be observed that the rise in temperature at surface when the gap is increased from 0.8 mm to 1.0 mm is approximately equal to that when the gap is increased from 1 mm to 2 mm.

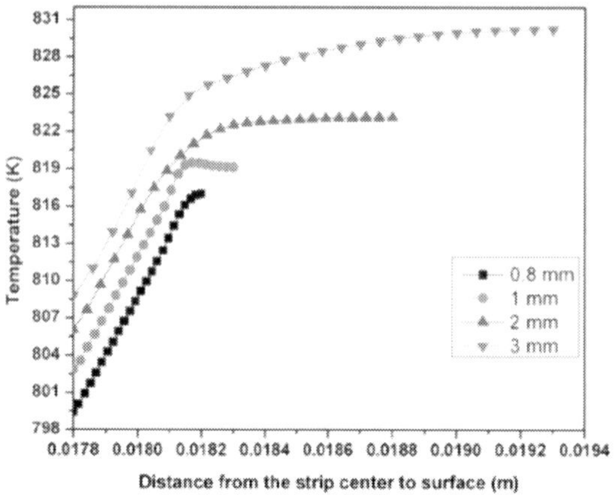

Figure 10: Temperature of the strip at the nip position from strip center to roll surface at different roll gaps.

Figure 11 shows the flow field in the molten pool at different roll gaps. This figure makes it possible to gain an insight into the effects of different roll gap on the flow pattern in the pool. As the roll gap increases, there is an increase in the intensity of the recirculation flow in the pool region. This is due to the fact that for a fixed inlet, with an increase in roll gap the vortex area in the pool becomes larger, which leads to an increase in mass flow within the pool. With increase in roll gap, the recirculation flow in the pool region becomes weaker due to the decrease in resistance at the nip. For roll gap of 0.8 mm recirculation is observed adjacent to the nozzle. With the increase in gap to 1 mm, significant shift in flow pattern can be observed. The recirculation in the top has shifted towards middle region and increased in size. The lower recirculation zone has also shifted to a lower position. As the gap increases to 2 mm the lower recirculation zone disappears and the size of the upper recirculation zone increases. With further increase in the gap no significant change is observed. These changes are observed because the recirculation zones are formed due to resistance to flow through the nip, which increases with decrease in the gap.

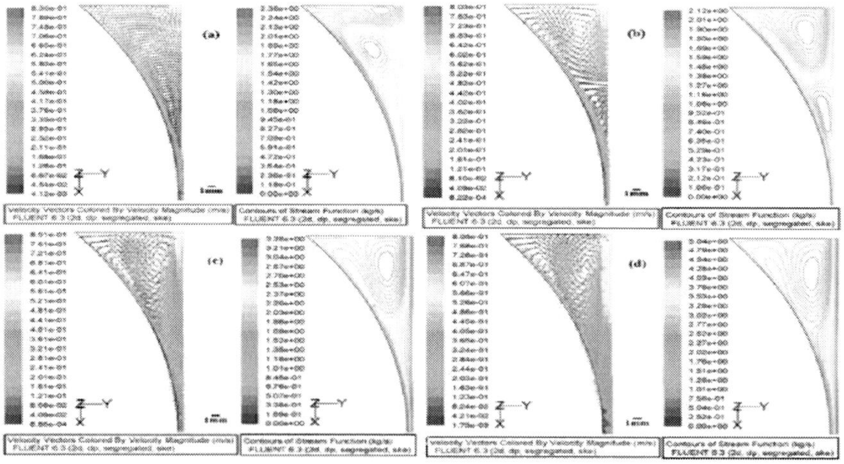

Figure 11: Contours of velocity vector and corresponding stream function of Al-33Cu in the molten pool at different roll gaps (a) 0.8 mm (b) 1 mm (c) 2 mm (d) 3 mm

4. Conclusions

When other processing parameters are constant, the pouring temperature and roll gap have a direct effect on the heat transfer between the casting roll and molten pool, which in turn influences the solidification of liquid metal in the molten pool. Simulations are carried out by varying the initial melt temperature from 831 K to 941 K (superheat 10 K to 120 K) and roll gap from 0.8 mm to 3 mm by keeping all other parameters constant. From the analysis of results it is found that

(i) The solidification front speed decreases from 9277.90 μm/s to 3815.42 μm/s and thickness of the solidified shell decreases from 0.29 mm to 0.08 mm with increase of initial pouring temperature of the melt from 831 K to 941 K.

(ii) The temperature of the strip at the exit section increases with increase in pouring temperature. However, at the surface of the strip the increase in the temperature as a result of 110 K increase in pouring temperature is not significant, i.e., ~5 K. At the plane of symmetry the increase is ~15 K.

(iii) With increase in roll gap, the recirculation flow in the pool region becomes weaker. The recirculation pattern significantly changes as the roll gap increases from 0.8 mm to 2 mm. Beyond a roll gap of 2 mm the change is not significant.

(iv) By varying the roll gap from 0.8 mm to 3 mm the surface temperature at the exit increases by ~9 K whereas the temperature at the centre increases by ~15 K. The sensitivity of temperature to roll gap is much higher for low values of roll gap.

(v) The solidification front speed and the shell thickness at nip decreases with increase in the roll gap.

Acknowledgement:

The authors would like to acknowledge the financial support provided by Council of Scientific and Industrial Research (CSIR) New Delhi, for generously supporting the research programme.

References

[1] Fiedler H, Jurisch M, Preiss P, Gobel R, Sickert G, Zimmermann H, Neumann W, Seilger R., 1991 J. Mater. Sci. Eng. A **133** 671
[2] Cook R, Grocock PG, Thomas PM, Edmonds DV, Hunt JD. 1995 J. of Mater. Process. Tech. **55** 5576
[3] Wechsler R. 2003 Scand. J. metall. **32** 58
[4] Santos CA, Spim Jr JA, Garcia A. 2000, J. of Mater. Process. Tech. **102** 33
[5] Kim DS, Kim WS, Kuznetsov AV. 2002, Numerical heat transfer Part A **41** 1
[6] Gupta M, Sahai Y. 2000, ISIJ International **40(2)** 144
[7] Zhao HY, Ju DY, Asakami T, Hu L. 2001, Materials and Manufacturing Processes **16(5)** 643
[8] Kim WS, Kim DS, Kuznetsov AK. 2000, International journal of heat and mass transfer **43** 3811
[9] Zhang XM, Jiang ZY, Yang LM, Liu XH, Wang GD, Tieu AK. 2007, J. of Mater. Process. Tech. **187-188** 339.
[10] Zeng J, Koitzsch R, Pfeifer H, Friedrich B. 2009, J. of Mater. Process. Tech. **209** 2321
[11] Fang Y, Wang Z, Yang Q, Zhang Y, Liu L, Hu H, Zhang Y. 2009, Int. J. of Minerals, Metallurgy and Materials **16** 304
[12] Miao YC, Zhang XM, Di HS, Wang GD. 2006, J. of Mater. Process. Tech. **174 (1-3)** 7
[13] Li Q, Zhang YK, Liu LG, Zhang P, Zhang Y, Fang Y, Yang QX. 2012, J. of Mater. Science **47(9)** 3953
[14] Sahoo S, Kumar A, Dhindaw BK, Ghosh S. 2013, Materials and Manufacturing process **28** 61
[15] Sahoo S, Ghosh S. 2014, Steel Res. Int. **85** 1
[16] Launder BE, Spalding DB. 1974, Computer Methods in Applied Mechanics and Engineering **3** 269
[17] Wang G-X, Matthys EF. in Melt Spinning, Strip Casting and Slab Casting, E.F. Matthys and W.G. Truckner, eds., TMS, Warrendale, PA, 1996, 205
[18] Guthrie RIL, Isac M, Kim JS, Tavares RP. 2000, Metallurgical and Materials Transactions B **31(5)** 1031

IOP Conf. Series: Materials Science and Engineering 75 (2015) 012014 doi:10.1088/1757-899X/75/1/012014

Effect of temperature on stability behaviour of functionally graded spherical panel

Vishesh R Kar* **and Subrata K Panda**

Department of Mechanical Engineering, National Institute of Technology, Rourkela, India, 769008

*E-mail: visheshkar@gmail.com

Abstract. In the present article, stability of functionally graded spherical panel under thermal environment is examined. The effective material properties are evaluated through Voigt's micromechanical model and continuous gradation is achieved using power-law distribution of the volume fraction of constituents. In addition, material properties are taken as temperature dependent. Finite element solutions are obtained through commercially available tool ANSYS using an eight node serendipity element. The linear eigenvalue buckling problem is solved using Block Lanczos method. Comparison study is made with the available published literatures. Finally, the effect of different geometry and material parameters such as thickness ratio, curvature ratio and power-law index on the critical buckling temperature of functionally graded spherical shell panel under thermal environment is demonstrated.

1. Introduction

Researches on advanced composites are now being increased due to their compatibility and durability in severe environment conditions. Functionally graded material (FGM) is known for its tailor-made properties which are achieved through the continuous gradation of material phase from one surface to another surface. FGMs are made from the designed combination of metal/alloy and ceramic materials. Metals or metal alloys are known for their mechanical strength and fracture toughness whereas ceramics hold good heat and corrosion resistance. This kind of in-homogenous material eliminates delamination as well as inter-laminar thermal stress concentration which can be generally seen in laminated structures. These novel characteristics make FGM most trustworthy for different sectors like aerospace, defence, nuclear plant, biomedical etc. Due to this, many researchers have shown their interest on modelling and analysis of FGM structures.

Na and Kim [1],[2] examined the buckling behaviour of functionally graded (FG) panel under uniform and non-uniform thermal field across the transverse direction. Woo et al. [3] investigated the post-buckling behaviour of FG plate and cylindrical panel under axial loading conditions. Zhao et al. [4] approached element-free kp-Ritz method in conjunction with first order shear deformation theory (FSDT) to analyse the thermo-mechanical buckling behaviour of FG flat panel. This work has been extended by Zhao and Liew [5] for the cylindrical. Tung and Duc [6] employed classical plate theory and von Karman strain terms to obtain the thermo-mechanical buckling and post-buckling behaviour of FG plate. Lee et al. [7] examined the thermo-mechanical post-buckling behaviour of FG plates based on the element-free kp-Ritz method in the FSDT framework. Thai and Choi [8] proposed an efficient refined theory to investigate the buckling behaviour FG plate. Ghannadpour et al. [9] applied finite strip

Content from this work may be used under the terms of the Creative Commons Attribution 3.0 licence. Any further distribution of this work must maintain attribution to the author(s) and the title of the work, journal citation and DOI.
Published under licence by IOP Publishing Ltd

method to obtain the buckling responses of FG plate under various uniform and non-uniform thermal fields across the thickness direction.

It is noted that very few work has been reported on stability analysis of FGM panel under thermal environment. Here, authors have attempted the thermal buckling analysis of FG spherical shell panel under thermal environment with temperature dependent material properties. Finite element (FE) solutions have been presented for FG spherical panel through commercially available FE tool ANSYS. For the discretisation purpose, an eight node serendipity element is considered. The convergence of present finite element model has been carried out and subsequently validation of the proposed model with the previous published results is performed. The effects of thickness ratio, curvature ratio, volume fraction and temperature on stability behaviour of FG spherical shell panel are presented and discussed.

2. Effective material properties of FGM

In this study, FGM is the composition of gradual variation of metal phase (at bottom surface) to ceramic phase (at top surface) across the thickness direction. The FGM constituents are considered as cubic function of temperature T as in [10].

$$P_{c,m}(T) = P_0(P_{-1}T^{-1} + 1 + P_1T + P_2T^2 + P_3T^3) \tag{1}$$

where, $P_{c,m}$ denotes material properties of metal/ceramic, P_0, P_{-1}, P_1, P_2 and P_3 are the temperature coefficients.

In order to evaluate the effective material properties of FGM (P), Voigt's micromechanical model is used [11].

$$P(T,z) = \left(P_c(T) - P_m(T)\right)V_c(z) + P_m(T) \tag{2}$$

where, $V_c(z)$ is the ceramic volume fraction which is the function of thickness coordinate z. This can be achieved through power-law distribution [12] and expressed as

$$V_c(z) = \left(\frac{z}{h} + \frac{1}{2}\right)^n \qquad (0 \le n < \infty) \tag{3}$$

where, n is the power-law index which exhibits the material profile across the thickness. Different material profiles are shown in Figure 1 for volume fractions corresponds to different values of power-law index along with the dimensionless thickness ($Z=z/h$).

By substituting Eq. (3) in Eq. (2), the effective material properties of FGM can be expressed for Young's modulus $E(T,z)$ and thermal expansion coefficient $\alpha(T,z)$ as

$$E(T,z) = \left\{E_c(T,z) - E_m(T,z)\right\}\left(\frac{z}{h} + \frac{1}{2}\right)^n + E_m(T,z) \tag{4}$$

$$\alpha(T,z) = \left\{\alpha_c(T,z) - \alpha_m(T,z)\right\}\left(\frac{z}{h} + \frac{1}{2}\right)^n + \alpha_m(T,z) \tag{5}$$

Here, Poisson's ratio ν is taken constant throughout the thickness of the FG panel. The temperature dependent material properties of ceramic and metal are shown in table 1.

Table 1. Temperature-dependent material properties of ceramic and metal [10]

Material	Properties	P_0	P_{-1}	P_1	P_2	P_3	TID (at 300 °K)
Si_3N_4	E(Pa)	3.48e+11	0	-3.07e-04	2.16e-07	-8.95e-11	3.22e+11
	α (K^{-1})	5.87e-06	0	9.10e-04	0	0	7.47e-06
Ti6Al4V	E (Pa)	1.23e+11	0	-4.59e-04	0	0	1.06e+11
	α (K^{-1})	7.58e-06	0	6.64e-04	-3.15e-06	0	6.94e-06

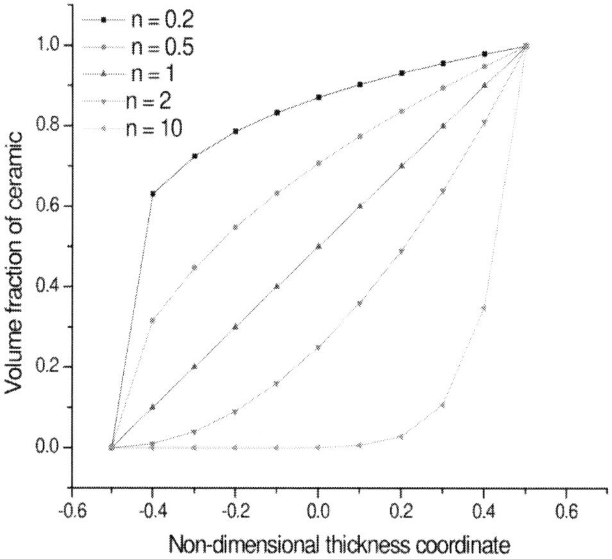

Figure 1. Different material distributions along the non-dimensional thickness coordinate.

3. Finite element modelling

In this study, an FG spherical panel of radius R with uniform thickness h and sides a and b is considered as shown in Figure 2. Here, FG spherical shell panel is modelled and analyse in ANSYS through ANSYS parametric design language (APDL) code. An eight nodded serendipity shell element (SHELL281), defined in the ANSYS library, is utilised to discretise the FG spherical panel. This shell element has total six degrees of freedom per node i.e., translations and rotations in the x, y and z directions [13].

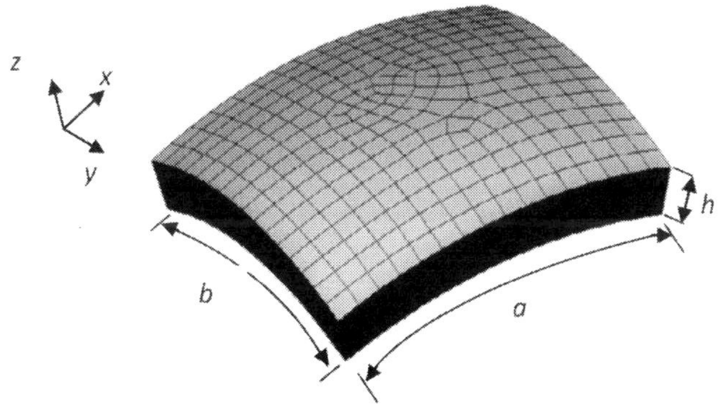

Figure 2. A discretised FG spherical shell panel model.

4. Results and discussions

In this section, the stability behaviour of FG spherical panel is performed under uniform temperature field. Here, FG panel is raised uniformly from the reference temperature (T_0 =300 °K) to the final temperatures (T= 400, 500, 600 and 700 °K). The FG spherical shell panel is discritised and solved using finite element steps in ANSYS APDL platform. Block Lanczos method is used to obtain the eigenvalue bucking responses.

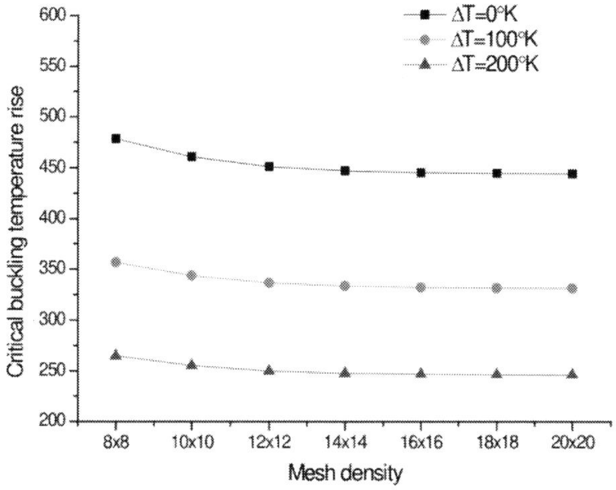

Figure 3. Convergence behaviour of simply-supported FG (*Ti-6Al-4V/Si$_3$N$_4$*) spherical panel with R/a=5, a/h=100 and n=0.2 under thermal environment.

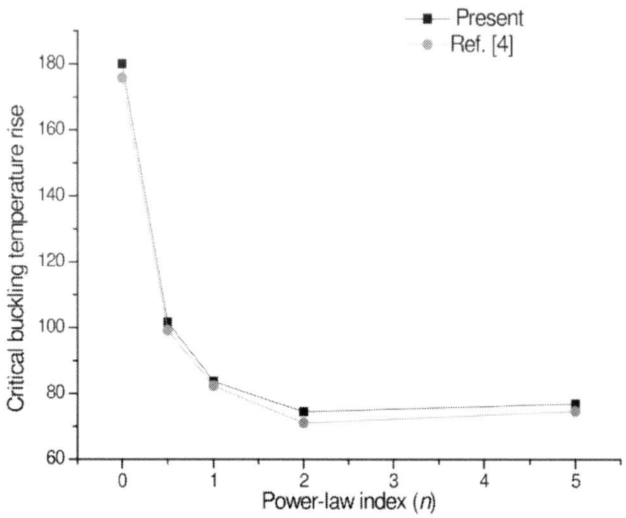

Figure 4. Critical buckling temperature rise of clamped (Al/Al_2O_3) FG flat panel (*a*/*h*=50).

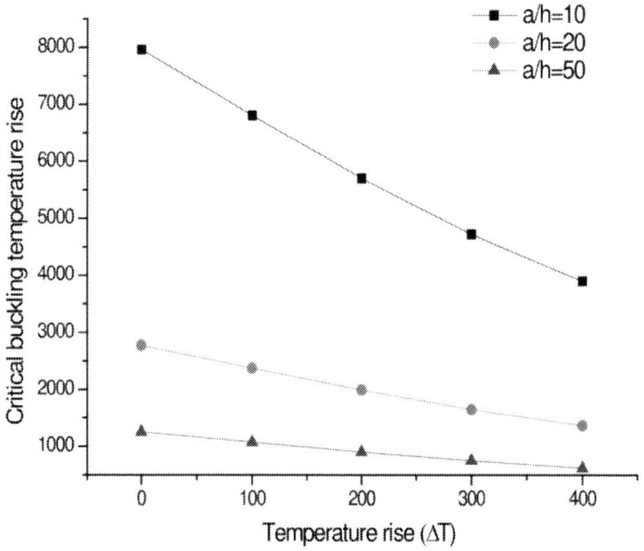

Figure 5. Effect of thickness ratio on critical buckling temperature rise of simply-supported FG spherical panel under thermal environment.

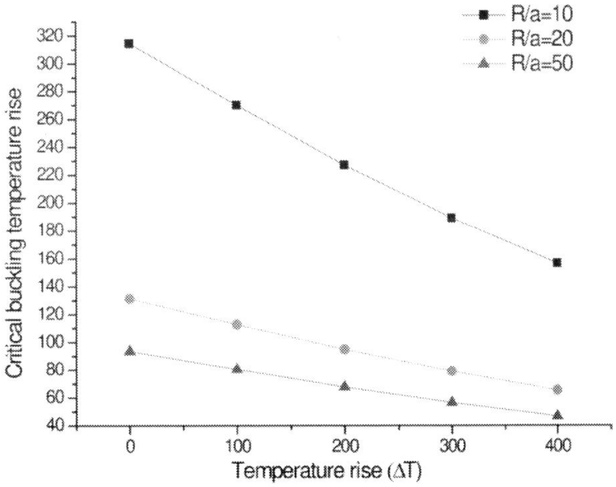

Figure 6. Effect of curvature ratio on critical buckling temperature rise of simply-supported FG spherical panel under thermal environment.

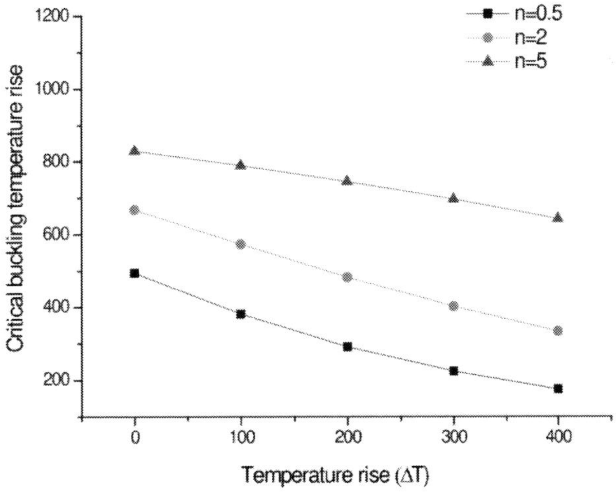

Figure 7. Effect of power-law index on critical buckling temperature rise of simply-supported FG spherical panel under thermal environment.

4.1. Convergence and comparison

The convergence study is performed through the proper mesh refinement of the present FG panel. Buckling responses of simply-supported FG spherical shell panel (a=b=1, R/a=5, n=0.2 and a/h =100) under three different temperature field (ΔT=0, 100 and 200 ^{0}K) are obtained for different meshes as shown in Figure 3. Material properties are taken same as given in Table 1. It is clear from the figure that a (18×18) mesh is sufficient enough to obtain the buckling response throughout in the analysis.

In order to check the efficacy of the present model, a comparison study has been made with the previous published results of [4]. Geometry and material parameters are taken as in [4]. In this problem, a square FG (Al/Al_2O_3) plate (a/h=50) is analysed for five different power-law indices (n=0, 0.5, 1, 2, 5) and presented in Figure 4. It is observed from the figure that the difference between the present results and the reference results are very nominal.

4.2. *Numerical illustrations*

Some numerical problems are carried out to show the robustness of present FG model. Titanium alloy (*Ti–6Al–4V*) as metal and silicon nitrated (Si_3N_4) as ceramic are considered at the bottom and the top surfaces, respectively. The influences of thickness ratios (a/h), curvature ratios (R/a), power-law indices (n) and temperature on the buckling behaviour of FG spherical panel are discussed in the following paragraphs.

Figure 5 exhibits the effect of thickness ratio on the critical buckling temperature rise of simply-supported FG spherical shell panel (a/b=1, R/a=5, n=2) at different elevated temperatures. It is noted that that the critical buckling temperature decreases with the increase in thickness ratio i.e., thin shell panels have less buckling load.

Figure 6 represents the variation of critical buckling parameter of simply-supported FG spherical panel (a/b=1, n=2, a/h=100) along with the different temperature values for three different values of curvature ratio (R/a=10, 20, 50). It is clearly observed that as the curvature ratio increases the critical buckling temperature rise decreases i.e., curved panel is having relatively higher buckling temperature.

Figure 7 exhibits the influence of power-law index on the critical buckling temperature rise of simply-supported FG spherical shell panel (a/b=1, R/a=5, a/h=100) along with the elevated temperature values. It is found that the critical buckling temperature rise is increasing with the increase in power-law index. It is also interesting to note that as the elevated temperature rise increases the critical temperature reduces in all the cases considered.

5. Conclusions

In this study, the thermal bucking behaviour of FG spherical shell panel under uniform temperature field is investigated. The effective material properties of FGM are evaluated through Voigt's model and power-law distribution. In addition, temperature dependent material properties of FGM constituents are considered. Finite element solution for the buckling behaviour of present FG model is proposed using Block Lanczos method. The present model is compared with the previous reported results. The influences of different material and geometrical parameters on the thermal buckling of FG spherical panel are illustrated.

Acknowledgments

Authors are thankful to AICTE (All India Council for Technical Education) for partial support through the sanction 8023/RID/RPS/56/11/12 Dated: 26/03/2012.

References

[1] Na K S and Kim J H 2004 Three-dimensional thermal buckling analysis of functionally graded materials *Compos. Part B-Eng.* **35** 429–37
[2] Na K S and Kim J H 2006 Three-dimensional thermomechanical buckling analysis for functionally graded composite plates *Compos. Struct.* **73** 413–22

[3] Woo J, Meguid S A, Stranart J C and Liew K M 2005 Thermomechanical postbuckling analysis of moderately thick functionally graded plates and shallow shells *Int. J. Mech. Sci.* **47** 1147–71
[4] Zhao X, Lee Y Y and Liew K M 2009 Mechanical and thermal buckling analysis of functionally graded plates *Compos. Struct.* **90** 161–71
[5] Zhao X and Liew K M 2010 A mesh-free method for analysis of the thermal and mechanical buckling of functionally graded cylindrical shell panels *Comput. Mech.* **45** 297–310
[6] Tung H V and Nguyen D D 2010 Nonlinear analysis of stability for functionally graded plates under mechanical and thermal loads *Compos. Struct.* **92** 1184–91
[7] Lee Y Y, Zhao X and Reddy J N 2010 Post-buckling analysis of functionally graded plates subject to compressive and thermal loads *Comput. Method App. M.* **199** 1645–53
[8] Thai H T and Choi D H 2012 An efficient and simple refined theory for buckling analysis of functionally graded plates *Appl. Math. Model.* **36** 1008–22
[9] Ghannadpour S A M, Ovesy H R and Nassirnia M 2012 Buckling analysis of functionally graded plates under thermal loadings using the finite strip method *Comput. Struct.* **108** 93–9
[10] Reddy J N and Chin C D 1998 Thermomechanical analysis of functionally graded cylinders and plates *J. Therm. Stresses* **21** 593–626
[11] Gibson L J, Ashby M F, Karam G N, Wegst U and Shercliff H R 1995 Mechanical properties of natural materials. II. Microstructures for mechanical efficiency *Proc. R. Soc. A* **450** 141–62
[12] Shen H S 2009 *Functionally Graded Material: Nonlinear Analysis of Plates & Shells* (Boca Raton, FL: CRC Press)
[13] ANSYS 15.0 Reference Manual

Effect of Tempering Temperature and Time on Strength and Hardness of Ductile Cast Iron

R. Kumar [1], R. K. Behera, S. Sen,
Department of Metallurgical and Materials Engineering,
National Institute of Technology, Rourkela, Odisha, India

E-mail:- ravindranit2010@gmail.com

ABSTRACT. The effects of tempering temperature and time on the mechanical properties of ductile cast iron is investigated in the current work. The specimens were austenitized at 900°C for 120 minutes and then quenched in mineral oil at room temperature. Immediately after quenching the specimens were tempered at 400°C and 200°C for 60min, 90min, and 120min. In the tempering temperature range of 200°C-400°C, there is sudden increase in impact strength, ductility and toughness of the materials, as the temperature and time increase. The UTS drops initially, and hardness of materials will depends on amount of phase of martensitic and retained austenitic and graphite nodules. In this work alloying elements also effected the microstructure of the specimen. And due to increase tempering time the amount of martensitic phase will decrease and retained austenitic phase will increase, retained austenitic phase is softer then martensitic so hardness will decrease.

Keywords: Ductile Cast Iron, Tempering, Temperature, Time, Graphite Nodules, Retained Austenitic, Martensitic, ferrite phase.

1. Introduction

The mechanical properties of ductile irons are controlled by the amount of matrix phase and distribution of those phase and there microstructure [1]. They can be improved the mechanical properties through proper heat treatment application. Ductile Iron and steel are superficially similar, due to high carbon and silicon levels in ductile iron result in important differences in their response to heat treatment [2]. The higher carbon levels in ductile iron increase hardenability, permitting heavier sections to be heat treated with lower requirements for expensive alloying or severe quenching media [2].Tempered ductile iron has excellent mechanical properties such as tempering reduces the strength and hardness and increases the ductility, toughness and machinability of quenched or normalized ductile Iron. In addition, tempering quenched castings also reduces residual stresses, decreases the amount of retained austenite, and reduces the probability of cracking, Oil quenching at 80°C-100°C to prefer to avoid cracks [3]. At if tempering temperature range of 400°C- 500°C, there is a sudden increase in impact strength and elongation %, but ultimate tensile strength and yield stress decrease [3]. In the relatively newly developed ductile cast irons (DI) with dual matrix structure (DMS), the structure consists of ferrite and martensitic or ausferrite (bainitic ferrite and austenite) [4]. In some components, suspension parts of automobiles, good toughness

Content from this work may be used under the terms of the Creative Commons Attribution 3.0 licence. Any further distribution of this work must maintain attribution to the author(s) and the title of the work, journal citation and DOI.
Published under licence by IOP Publishing Ltd

and high ductility are required. These requirements are met by this newly developed iron that best exhibits tensile strength and proof strengths, together with hardness. It has been shown that the tensile strength and ductility can satisfactorily be optimized by critical combinations of austenitizing and tempering temperature and time [5]. Tempering is one of the most important heat treatment processes that is applied to quenched steel and cast iron. The objectives of this process include reducing the brittleness of materials, improvement of toughness and ductility, and also, reducing the probability of quench cracking and minimize stresses. Tempered hardness depends on as-quenched hardness level, alloy content and tempering temperature, as well as time. Okabayashi whose results indicate that impact strength of the samples, tempered at 600°C for 20 min, is better than those tempered at 200°C for 60 Minutes [6]. About effect of mechanical properties due to change in tempering temperature with a time variations. But for better mechanical properties amount of alloying is also a very important. Mg and Ce help to produced nodules in ductile cast iron [7]. As like carbon is to increase the volume of graphite produced which is turn reduce tensile strength, elongation and hardness and Si & Ni will increase the strength properties. Matrix structure of this type of ductile cast iron consists of a soft phase austenitic and a hard phase martensitic that are formed by a special heat treatment process [8]. The past results reported are mainly leaned towards Tempered ductile iron & the studies were focused on the effect of tempering temperature and time on mechanical properties. Hence the present investigation is concentrated to study the mechanical properties of tempered ductile iron compared with higher tempering temperature and different time

2 Experiment

2.1: *Specimen Preparation*

In order to investigate the structure-property relationship, ductile iron test blocks with different alloying elements were brought from L&T Kansbahal, India. The chemical composition of test block by weight percentage is presented in Table 1. In order to carry out the experiment specimens were machined from the test blocks. Specimens were then austenitized at 900°C for 120 minutes and then it was quenched in mineral oil at room temperature. Immediately after quenching tempering heat treatment processes was done at 400°C and 200°C for 60 min, 90 min, and 120 min respectively. After heat treatment oxide layer from each of the specimen was removed by conventional filing & emery paper polishing and cloth and diamond polishing method and etched 2% nital.

Table 1: Chemical composition of specimens in wt. %

Elements	Wt. %
C	3.52
Si	2.04
Mn	0.17
S	0.008
Mg	0.042
P	0.024
Cr	0.02
Ni	0.15
Mo	0.001
Cu	0.02
Ce	0.007
Fe	Balance

2.2: *Tensile strength*
Tensile test were carried out according to ASTM (A 370-2002) Test were conducted by using Instron 1195 universal testing machine connected to computer to draw the stress-strain curve accounting to tensile strength, tensile load of 50KN applied to the specimen up to the breaking point. Specimen have specific measurement of each sample. After failure we obtained tensile strength of different sample and stress-strain curve at different time and temperature

2.3: *Hardness testing*
The method utilized for hardness testing was Vickers hardness testing. Vickers hardness was measured by applying a load of 20Kg & dwell time being 10seconds on each heat treated tempered specimen. Two diagonals, d^1 and d^2, are measured, take averaged and the surface area calculated then divided into the load applied.

2.4: Microstructural & nodules and nodularity Investigation
Specimens were first polished with belt polisher then followed by 1/0, 2/0, 3/0, 4/0 grades of emery paper & finally cloth polishing was done with alumina slurry followed by diamond polishing. Then etched 2% nital. Metallographic images were taken with the help of computer integrated optical microscope at 200X magnification. Before etched taken metallographic images thought optical microscope and investigate number of nodules present in the $1mm^2$ area and calculate nodularity thought the computer software.

2.5: Impact energy
Charpy V-notch test described in ASTM E23 was for impact testing. The Charpy specimen was placed horizontally across supports with the notch away from the hammer. The indicator pointer was slide to the left until it indicates the maximum energy range on the upper Charpy- Tension scale. The pendulum arm was raised to the right until it is firmly supported by the latching mechanism. The pendulum was released by pushing up on the release knob. The hammer dropped, striking the specimen, with a swing through dependent on the amount of energy absorbed by the test specimen. The indicator moved and stopped when peak swing through was registered, providing a direct reading of the energy absorbed by the specimen. The indicated value from charpy scale was recorded.

2.6: *Fractographic Investigation*
Fracture surface of the sample after tensile test are analyzed by scanning Electron microscopy (SEM). First cleaned with acetone and properly handle, so that the fracture surface does not get damaged. Thereafter the specimen is cut to appropriate size for analysis using SEM (Scanning Electron Microscope.) Then all the specimens are kept under SEM for analysis of fracture surface. A series of photographs are taken in due course of experiment which helps in determining the type of fracture that has taken place.

3: Result and discussion

The effect of tempering temperature and time on different mechanical properties of ductile cast iron. as like yield stress, ultimate tensile strength, impact strength and elongation percentage are demonstrated in figure 1 and table 2. The mechanical properties vary with the Different matrix phase. when tempering time increase retained austenite decrease and reduces the probability of cracking. and ductility again increase with time the increase in strength initially at low time interval is due to the high amount of martensite and unreacted austenite, but the time increase above 30 Minutes the stage reaction in the intercellular regions for which strength decreases and ductility increases, Elongation of sample increases with increasing time, and yield strength, tensile strength decreases with the increase in tempering time, and impact strength increases in tempering process because of increase in the toughness of the sample. And at higher austenitizing temperatures increase the carbon content of the austenite but the bulk hardness is reduced to retained austenite and a lower resultant martensite content [2]. Hardness decrease as the tempering temperature and time will increase. This is due to transformation of martensitic to retained Austenite phase, and retained Austenite phase in softer then the martensitic phase

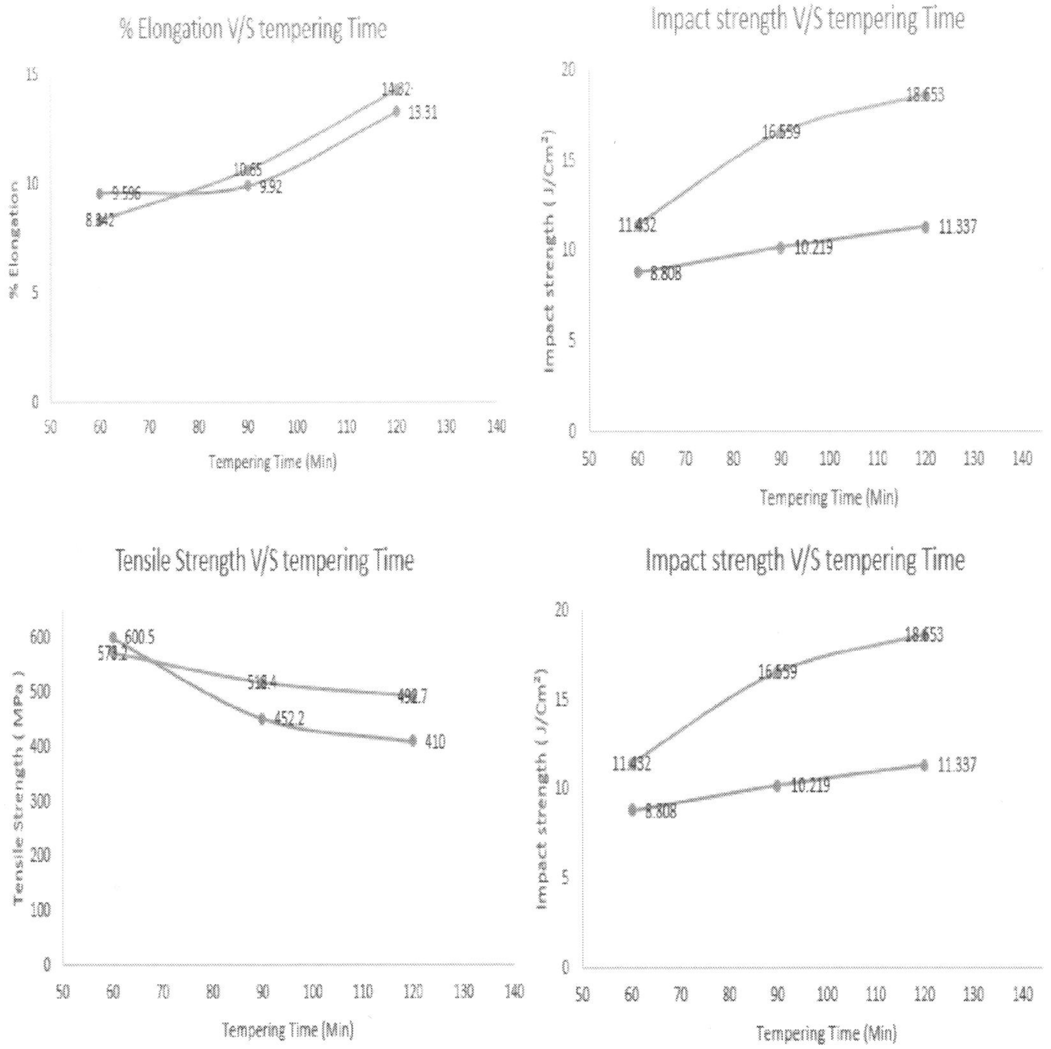

Figure.1 Effect of tempering Time On Elongation %, Impact Strength, Tensile Strength, Yield Strength

The drop in hardness accompanying secondary graphitization produces a corresponding reduction in tensile and fatigue strength as well as because alloy content affects the rate of secondary graphitization, each alloy will have a unique range of useful tempering temperatures [10].

Table 2. Mechanical properties of tempered specimen at 200°C and 400°C at different time

Tempered sample	0.2% Y.S. (MPa)	Tensile strength (MPa)	% Elongation	Impact strength (J/Cm²)	Hardness
200°C At 60 Min	216.6	600.5	9.59	8.80	380HV20
200C At 90 Min	149.3	452	9.92	10.21	337HV20
200°C At 120 Min	145.1	410	13.21	11.33	321HV20
400°C At 60 Min	236.6	573.2	8.34	11.43	326.3HV20
400°C At 90 Min	223.1	518.4	10.65	16.55	278.5HV20
400°C At 120 Min	147.8	492.7	14.32	18.65	250.3HV20

. The microstructures of respective specimens were shown in Fig.2 and this microstructure have different matrix phase, tempered ductile Cast Iron have nodules graphite structure and they have also austenite and martensitic phase[9]. The nodularization amount and graphite shape have a great effect on the workability [11]. In microstructures At 200°C have different amount of phase in table 3. at after austenitized temperature sample immediately quench in mineral oil at room temperature then at quenching martensitic transformation occurs and retained austenite formed, that amount of matrix define the mechanical properties of ductile cast Iron

Table 3. Amount of matrix presence in tempered specimen at different time and temperature

Tempering Temperature &Time	%of Martensitic	%of retained Austenite	% of Graphite	% of Ferrite
200°C at 60 Min	37.06%	45.03%	17.90%	
200°C at 90 Min	35.28%	45.86%	18.85%	
200°C at 120 Min	27.58%	55.47%	16.94%	
400°C at 60 Min	39.64%		16.32%	44.03%
400°C at 90 Min	27.28%		18.28%	54.42%
400°C at 120 Min	19.90%		17.75%	62.34%

Amount of matrix presence in tempered specimen is calculated by microstructure of different specimen we taken 10 different Images of same sample and calculate the threshold value of all Images and calculate area friction from microstructure analysis and take average value of different phases. at 200°C austenite phase not stable that we called retained Austenite and % of retained austenite is because after quenching we suddenly tempered the sample at 200°C therefore martenisite transformation is not happened proper,

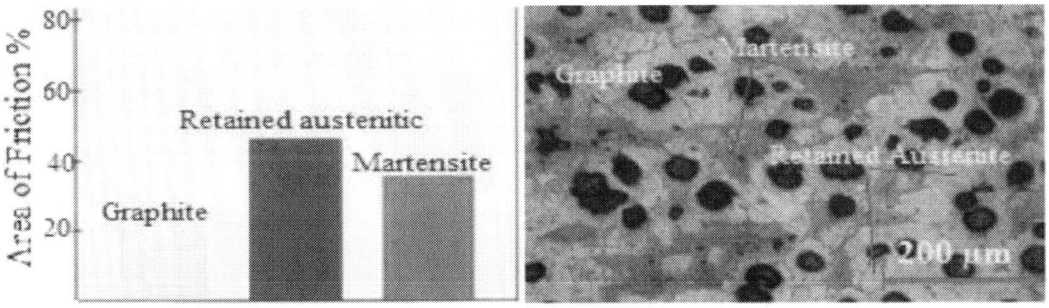

Figure 2.a Microstructure of 200°C tempered specimen at 60 Min

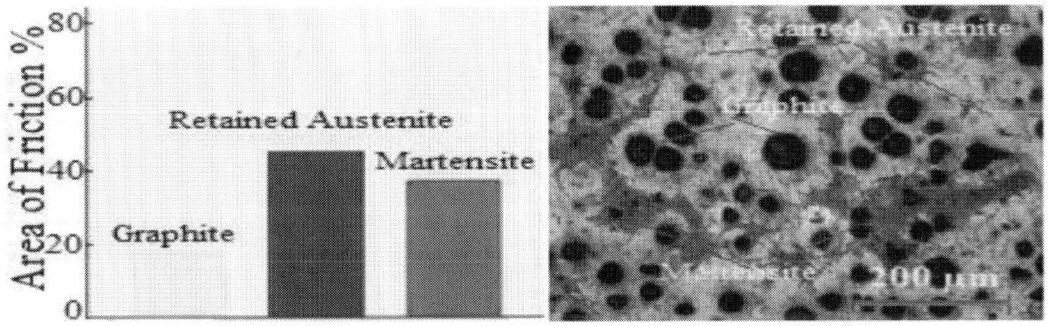

Figure 2.b Microstructure of 200°C tempered specimen at 90 Min

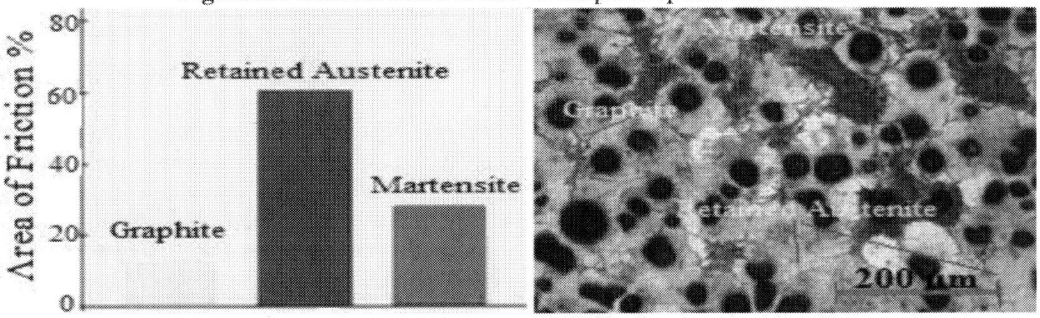

Figure 2.c Microstructure of 200°C tempered specimen at 120 Min

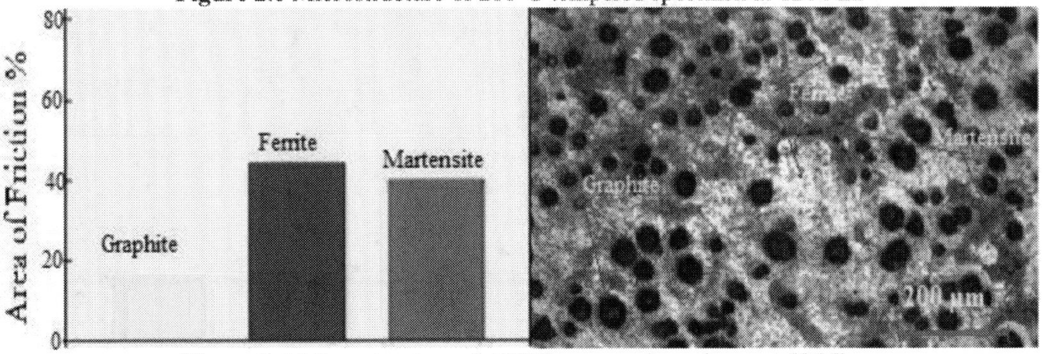

Figure 2.d Microstructure of 400°C tempered specimen at 60 Min

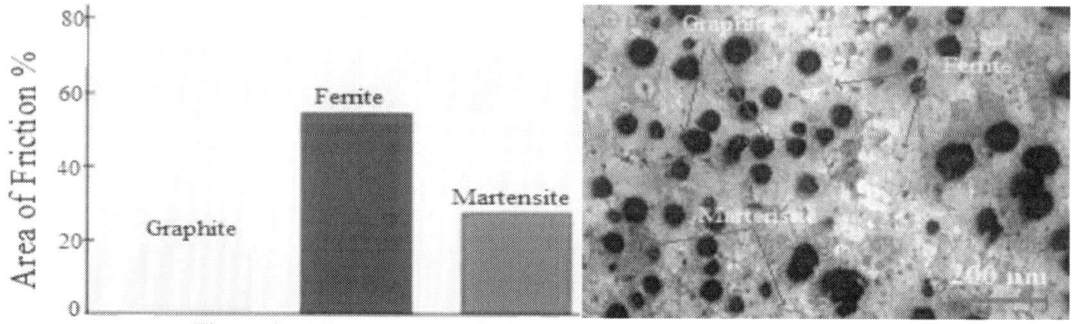

Figure 2.e Microstructure of 400°C tempered specimen at 90 Min

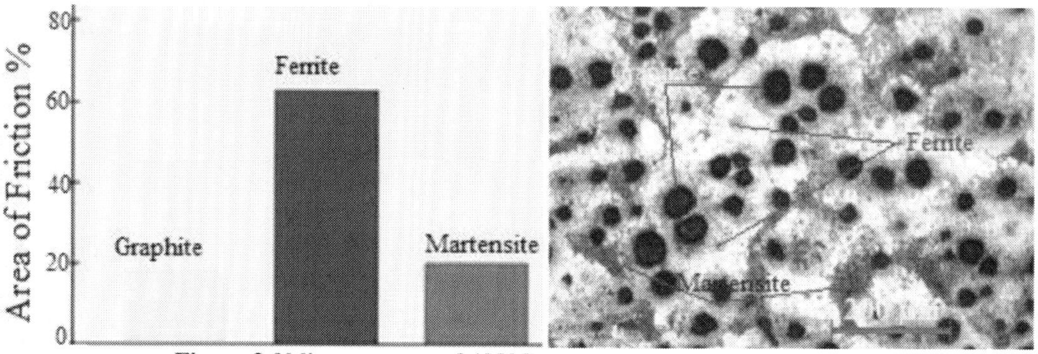

Figure 2.f Microstructure of 400°C tempered specimen at 120 Min

The morphology of the fracture specimens are analyzed by Scanning Electron Microscopy. Figure 3 show the fracture surface of different sample Specimens tempered at 200°C were dominant by brittle failure mode characterized by the presence flat shiny surfaces. Whereas with increase in tempering time shallow dimples were also observed in every specimens & were increased in the specimen tempered for 120 minutes. On the other hand specimens tempered at 400°C for various times showed dimples around the spherical nodules suggesting dominating nature of ductile failure mode. However the specimen tempered for 90 minutes showed mixed mode of failure characterized by presence of flat shiny surfaces along with dimples. On the other hand specimen tempered for 120 minutes showed complete ductile mode of failure.

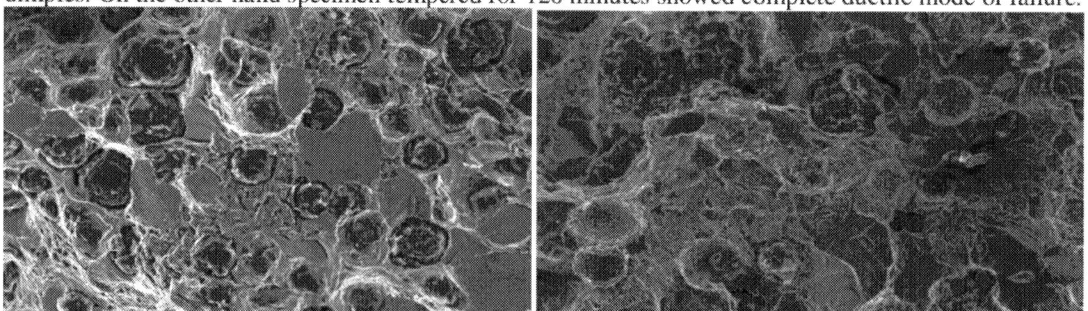

(a) (b)

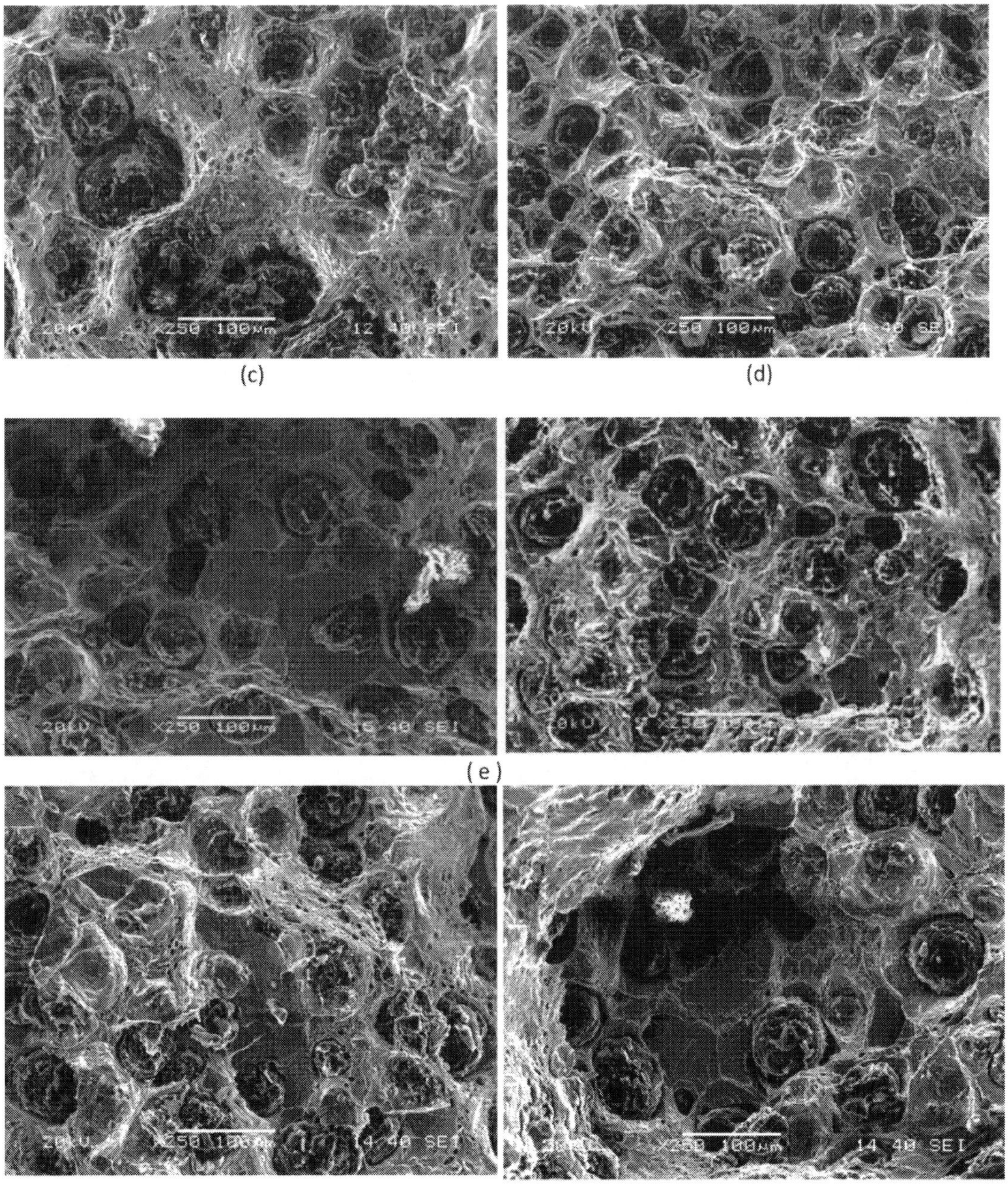

(c) (d)

(e)

(f)

Figure 3: Fractography Image of tempered specimen a) at 200°C in 60 min b) at 200°C in 90 min c) at 200°C in 120 min d) at 400°C in 60 min e) at 400°C in 90 min f) at 400°C in 120 min

4. Conclusion

The result obtained from present work can be summarized in the following points.
1. By increase tempering time martensitic content reduced resulting in increase of ductility and reduction of ultimate tensile strength and yield strength,
2. Within the temperature range of 200°C–400°C, there is a sudden increase in impact strength and elongation percentage, whereas within the same temperature range, the ultimate tensile strength and yield stress decreased.
3. Longer duration of tempering period at 200°C & 400°C increases the elongation percentage for tempering periods up to 120 min, and ductility increases
4. Whereas with increase in tempering time light pits were also observed in every specimens & were increased in the specimen tempered for 120 minutes. On the other hand specimens tempered at 400°C for various times showed pits around the spherical nodules suggesting dominating nature of ductile failure mode
5. As the tempering temperature & time increases the impact energy increases, this means the energy absorption capacity of Spheroidal graphite iron increases as the time & temperature of tempering increases.
6. Stress relieving, a low-temperature treatment, to reduce or relieve internal stresses remaining after quenching.
7. Oil is preferred as a quenching medium to minimize stresses and quench cracking, then the water or brine.
8. The nodularization amount and graphite shape have a great effect on the workability.

5. Reference

[1]. A guide to mechanical properties of ductile iron, Mid –Atlantic Casting Casting Service.
[2]. Ductile Iron Society, http: www.ductile.org/ didata/Section3/3part1.htm
[3]. Ali M. Rashidi, M.Moshrefi-torbati/*Materials latters **45** (2000) 203-207*, Mechanical Engineering Department, Razi University, Kermanshah, Iran
[4]. Hughes ICH, *Mater Des* **6** (1985) pp. 124
[5]. AVNER Sidney H, Introduction to Physical Metallurgy, *Second Edition*, Chapter-11
[6]. K. Okabayashi, M. Kawamoto, A. Ikenaga, M.Tsujikawa, *Trans. Jpn. Foundrymen's Soc*.1982
[7]. Anita bisht, http://ethesis.nitrkl.ac.in/1501/1/motu-thesis.pdf, Rourkela odisha, india, june-2009
[8]. N. Wade, C. Lu, Y. Ueda, *Trans. Jpn. Foundrymen's Soc Ž1985*. 22–26, April.
[9]. M.Nili Ahmadabadi, H.M. Ghasemi, M.Osia, A.R. Ghaderi. *Wear **255** (2003) 410–416*
[10]. Janina M. Radzikowska, The Foundry Research Institute, Krako´w, Poland
[11]. M.A.Shaker, Journal of Materials Processing Technology, 32 (1992) **545**-552, july-24-1991,

Effects of hygrothermal conditions on free vibration behaviour of laminated composite structures

Trupti R Mahapatra[1,*] **and Subrata K Panda**[2]

[1]KIIT University-Bhubaneswar, Odisha, India, 751024
[2]National Institute of Technology-Rourkela, Odisha, India, 769008

*E-mail: trmahapatrafme@kiit.ac.in

Abstract. Free vibration behaviour of laminated composite structure under combined hygrothermal load is investigated in this article. The responses are obtained numerically through a finite element model using hygrothermal dependent composite material properties. The convergence behaviour of the present numerical model has been established and validated by comparing the responses with those available open literatures. The effects of various hygrothermal conditions and geometrical parameters (curvature ratio, side-to-thickness ratio and aspect ratio) are examined in details.

1. Introduction

Laminated structures made of composites are exposed to combined hygrothermal loading both during manufacturing as well as in service. It is well known that, the mechanical properties (Young's modulus and rigidity modulus) of laminated composites are significantly influenced due to hygrothermal and high strain rate loading. The corrugated fiber properties due to uneven environment and fiber volume fractions may alter the mechanical responses substantially. The structural components made up of laminated composites found in aerospace, marine, automotive, agriculture, sports, and biomedical industries are exposed to vibration which experiences fatigue/cyclic loading during their service life. In recent years researchers quest to analyse (analytical, numerical and/or experimental) the complex loading in conjunction with degraded material properties of laminated structures accurately for finished products. It is noted that, free vibration behavior of laminated shell panels under hygrothermal loading have been analysed by using various displacement models say, classical laminated shell theory (CST), the first order shear deformation theory (FSDT), and the higher order shear deformation theory (HSDT). In this regard some of the important contributions are discussed briefly in the following line to address the objective and necessity of the present study. Sai Ram and Sinha [1] investigated the effects of moisture and temperature on the free vibration responses of laminated composite plates using finite element method (FEM). A quadratic isoparametric finite element (FE) formulation is presented by Parhi et al. [2] for the hygrothermal free vibration and transient analysis of multiple delaminated plates and shells. Their formulation is based on the FSDT. Patel et al. [3] analyzed the static and dynamic behavior laminated composite plates using FEM. Their model is based on the modified HSDT and includes hygrothermal dependent material properties. Huang and Zheng [4] analysed nonlinear free and forced vibration of moderately thick laminated plates using HSDT and von-Karman type nonlinear kinematics under transient loading. Shen et al. [5] reported dynamic behaviour of laminated composite plate resting on elastic foundation under

Content from this work may be used under the terms of the Creative Commons Attribution 3.0 licence. Any further distribution of this work must maintain attribution to the author(s) and the title of the work, journal citation and DOI.
Published under licence by IOP Publishing Ltd

hygrothermal load through a micromechanical based HSDT model. Naidu and Sinha [6] investigated the nonlinear vibration response of laminated composite shells in hygrothermal environment using nonlinear FEM steps. They have developed the model based on the FSDT and the Green-Lagrange type nonlinearity. Analytical solution of the hygrothermal stresses in thick multilayered composite plates by taking the hygrothermal dependent material properties was presented by Benkhedda et al [7]. Lo et al. [8] presented the effect temperature dependent material properties and hygrothermal loading on multilayer composites using global–local HSDT mid-plane kinematics in conjunction with four-node quadrilateral plate element. Nonlinear free vibration and transient response of laminated composite spherical and cylindrical shells with imperfections under hygrothermal environment is investigated by Nanda and Pradyumna [9] using FSDT and von-Karman nonlinearity. The severity due to combined variation in temperature and moisture concentrations on the nonlinear vibration behaviour of laminated composite structure is reported by Ashraf [10]. The effects of temperature and moisture content on natural frequencies of woven fiber glass/epoxy delaminated composite plate is analysed through combined numerical (FEM) and experimental steps by Panda et al. [11]. They modelled the composite plate based on the FSDT kinematics by considering the degraded hygrothermal properties. Panda and Mahapatra [12] studied the nonlinear thermal free vibration behaviour of laminated composite spherical shell panel using the HSDT and Green-Lagrange nonlinearity.

We note that limited studies related to linear/nonlinear vibration behaviour of laminated composite flat/curved panels under hygrothermal environment have been reported in the open literature. Based on the authors' knowledge, studies related to free vibration behaviour of laminated composite spherical shell panel under hygrothermal loading by using HSDT mid-plane kinematics are also scarce. The present work aims to analyse the free vibration behaviour of laminated composite shear deformable spherical shell panels under hygrothermal loading using a mathematical model based on the HSDT. The temperature and/or moisture dependent material properties are considered in order to obtain a realistic response. The sets of governing equations are obtained using Hamilton's principle and discretised using FEM steps. The effect of various geometrical parameters on the free vibration behaviour of laminated composite spherical panels under combined hygrothermal loading is investigated through subsequent numerical illustrations.

2. Numerical Methodology

Figure 1 shows the geometry and lay-up of a laminated spherical shell panel of length a, width b and thickness h with composed of 'N' number of equally thick anisotropic layers. R1 and R2 are the radii of curvature of the shell panel at the mid surface (Z=0). In order to derive the hygro-thermo elastic mathematical model of laminated composite spherical shell panel, a HSDT based displacement field is adopted which is assumed to be in following form:

$$\begin{aligned}
\bar{u}(x,y,z,t) &= u(x,y,t) + z\phi_1(x,y,t) + z^2\psi_1(x,y,t) + z^3\theta_1(x,y,t) \\
\bar{v}(x,y,z,t) &= v(x,y,t) + z\phi_2(x,y,t) + z^2\psi_2(x,y,t) + z^3\theta_2(x,y,t) \qquad \left[-\frac{h}{2} \le z \le \frac{h}{2}\right] \qquad (1) \\
\bar{w}(x,y,t) &= w(x,y,t)
\end{aligned}$$

where, $\bar{u}$, $\bar{v}$ and $\bar{w}$ are the displacements at any point of the shell panel along the (X, Y, Z) coordinates corresponding to any point of time t. (u, v, w) are the mid plane displacements and ϕ_1 and ϕ_2 are the shear rotations about the y and x-axes, respectively. ψ_1, ψ_2, θ_1, θ_2 represent the higher order terms of the Taylor series expansion defined in the mid-plane and account for the parabolic variation of shear stress.

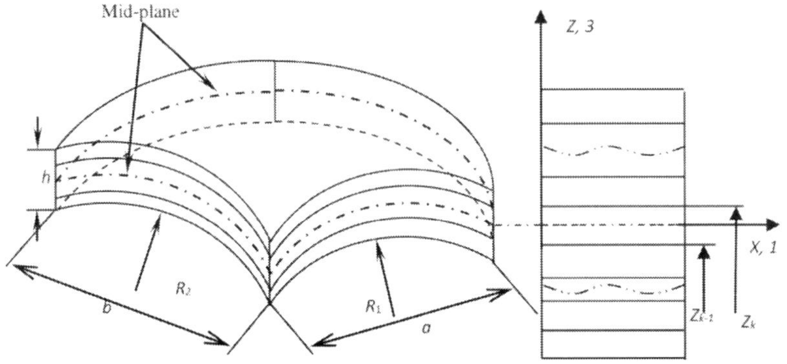

Figure 1. Lay-up of laminated composite spherical shell.

2.1 Strain Displacement Relation

The strain-displacement relation for a shell panel can be defined as:

$$\{\varepsilon\}^{T}=\{\varepsilon_{xx}\ \ \varepsilon_{yy}\ \ \gamma_{xy}\ \ \gamma_{xz}\ \ \gamma_{yz}\}^{T}$$
$$=\left\{\left(\frac{\partial\bar{u}}{\partial x}+\frac{\bar{w}}{R_1}\right)\ \left(\frac{\partial\bar{v}}{\partial y}+\frac{\bar{w}}{R_2}\right)\ \left(\frac{\partial\bar{u}}{\partial y}+\frac{\partial\bar{v}}{\partial x}\right)\ \frac{\partial\bar{u}}{\partial z}+\left(\frac{\partial\bar{w}}{\partial x}-\frac{\bar{u}}{R_1}\right)\ \frac{\partial\bar{v}}{\partial z}+\left(\frac{\partial\bar{w}}{\partial y}-\frac{\bar{v}}{R_2}\right)\right\}^{T} \tag{2}$$

Using Equations (1) and (2), the strain vector $\{\varepsilon\}$ may be expressed in terms of unknowns (for the structural deformation) as $\{\varepsilon\}^{T}=[\mathrm{H}]\{\bar{\varepsilon}\}^{T}$ (3)

where, $\{\bar{\varepsilon}\}^{T}=\{\varepsilon_1^0\ \ \varepsilon_2^0\ \ \varepsilon_6^0\ \ \varepsilon_5^0\ \ \varepsilon_4^0\ \ k_1^1\ \ k_2^1\ \ k_6^1\ \ k_5^1\ \ k_4^1\ \ k_1^2\ \ k_2^2\ \ k_6^2\ \ k_5^2\ \ k_4^2\ \ k_1^3\ \ k_2^3\ \ k_6^3\ \ k_5^3\ \ k_4^3\}$, the terms in $\{\bar{\varepsilon}\}$ having superscripts '0', '1', '2' and '3' are bending, curvature and higher order terms and $[\mathrm{H}]$ is the function of thickness coordinate.

2.2 Hygro-thermo elastic equations

The hygro-thermo elastic constitutive matrix equation of generalized stress tensor considering plane stress condition for any general k[th] isotropic composite lamina and any fibre orientation angle θ, can be expressed as

$$\{\sigma_{ij}\}^{k}=\left[\overline{Q_{ij}}\right]^{k}\{\varepsilon_{ij}-\alpha_{ij}\Delta T-\beta_{ij}\Delta C\}^{k} \tag{4}$$

where, $\{\sigma_{ij}\}^{k}=\{\sigma_1\ \ \sigma_2\ \ \sigma_6\ \ \sigma_5\ \ \sigma_4\}^{T}$ is the stress vector for the kth layer, $\{\varepsilon_{ij}\}^{k}=\{\varepsilon_1\ \ \varepsilon_2\ \ \varepsilon_6\ \ \varepsilon_5\ \ \varepsilon_4\}^{T}$ the strain vector for the kth layer, $\left[\overline{Q_{ij}}\right]^{k}$ the transferred reduced elastic constant for the kth layer, $\{\alpha_{ij}\}^{k}=\{\alpha_1\ \ \alpha_2\ \ 2\alpha_{12}\}^{T}$ the thermal expansion/contraction coefficient vector, $\{\beta_{ij}\}^{k}=\{\beta_1\ \ \beta_2\ \ 2\beta_{12}\}^{T}$ the moisture expansion/contraction coefficient vector, $\Delta\mathrm{T}$ the uniform change in temperature and ΔC the uniform change in moisture concentration.

2.3 Work done, kinetic energy and strain energy

The in-plane hygro thermal forces are obtained using the steps in [14] and given by

$$\begin{Bmatrix} \{N\} \\ \{M\} \\ \{P\} \end{Bmatrix} = \left[\sum_{k=1}^{N} \int_{z_{k-1}}^{z_k} \left\{ \left[\bar{Q}\right]^k \left\{ \{\alpha\}^k (1, z, z^3)\Delta T + \{\beta\}^k (1, z, z^3)\Delta C \right\} \right\} dz \right] \tag{5}$$

where, $\{N\}$, $\{M\}$ and $\{P\}$, are the resultant vectors of compressive in-plane hygrothermal forces, moments and the higher order terms due to combined temperature and moisture variation.

The work done (W) due to in-plane hygrothermal load is given by

$$W = \frac{1}{2} \int_A \{\varepsilon_G\}^T [D_G] \{\varepsilon_G\} dA \tag{6}$$

The kinetic energy of a vibrating laminated plate is expressed in following step

$$T = \frac{1}{2} \int_A \left\{ \sum_{k=1}^{NL} \int_{z_{k-1}}^{z_k} \rho^{(k)} \left\{\dot{\delta}\right\}^T [f]^T [f] \left\{\dot{\delta}\right\} dz \right\} dA = \frac{1}{2} \int_A \left\{ \left\{\dot{\delta}\right\}^T [m] \left\{\dot{\delta}\right\} \right\} dA \tag{7}$$

where, $[\rho]$, $\{\delta\}$ and $\{\dot{\delta}\}$ are the density, global displacement vector and first order derivative of the global displacement vector with respect to time.

$[m]$ is the inertia matrix written as $[m] = \sum_{k=1}^{NL} \int_{z_{k-1}}^{z_k} \rho^{(k)} [f]^T [f] dz$ and $\{\delta\} = \{\bar{u} \quad \bar{v} \quad \bar{w}\}^T = \{f\}\{\delta\}$

where, $\{f\}$ is the function of thickness coordinate.

The total strain energy of the laminated shell panel is expressed as

$$U = \frac{1}{2} \int_V \{\varepsilon\}_i^T \{\sigma\}_i dV \tag{8}$$

Substituting the values of stress and strain components from Equations (3) and (4) the strain energy expression can be re-written as

$$U = \frac{1}{2} \iint_{xy} \{\bar{\varepsilon}\}_i^T [D] \{\bar{\varepsilon}\}_i dxdy \tag{9}$$

where, $[D] = \sum_{k=1}^{NL} \int_{Z_{k-1}}^{Z_k} [H] \left[\bar{Q}\right]^k [H] dZ$ and '*NL*' is number of layer.

3. Finite element implementation

The FEM has proved to be a powerful and widely appreciable method for the vibration analysis of laminated composite structures whose analytical solutions are difficult to obtain. A nine-noded isoparametric element having nine degrees of freedom per node is employed in the present model.

The displacement vector $\{\delta\}$ can be expressed after implementing the FEM as

$$\{\delta\} = \sum_{i=1}^{NN} N_i \delta_i \tag{10}$$

where, N_i is the nodal interpolation function for i^{th} node and NN is the numbers of node per element.

Employing FEM the strain vector in Eq. (3), can be rewritten in terms of global displacement vector

$$\{\varepsilon\} = [B]\{\delta\} \tag{11}$$

where, $[B]$ is the product of operator matrix and interpolation functions of structural displacement vector.

4. System governing Equation

The system governing equation of free vibrated laminated orthotropic spherical shell panel under hygrothermal loading can be obtained using Hamilton's principle as follows.

$$\delta\int_{t_1}^{t_2} L\mathrm{d}t = 0 \tag{12}$$

where, $L = [T\text{-}(U+W)]$

Now employing the Equation (10) into Equation (6) and (9), the elemental form is obtained as in [12] and subsequently Equation (12) can be expressed as in [15] as

$$\left([\mathrm{K}]-\omega^2[\mathrm{M}]\right)\delta = 0 \tag{13}$$

where, $\{\delta\}^{\mathrm{T}} = [u \quad v \quad w \quad \phi_1 \quad \phi_2 \quad \psi_1 \quad \psi_2 \quad \theta_1 \quad \theta_2]$ is the displacement vector, $[\mathrm{M}]$ and $[\mathrm{K}]$ are the global mass and stiffness matrices, respectively.

Now, the numerical solutions are obtained using a direct iterative method.

5. Results and Discussion

A homemade finite element computer code is developed in MATLAB 7.10.0 environment based on the present mathematical formulation for the laminated composite spherical shell panel. The graphite/epoxy composite lamina properties are considered for computation purpose throughout the analysis. The composite properties are also taken as temperature and moisture dependent ([3]) and presented in Table 1. (ν_{12}=0.3, G_{12} = G_{13}, G_{23} =0.5G_{12}, α_1=-0.3x10^{-6}/K, α_2= 28.1x10^{-6}/K, β_1=0, β_2= 0.44).

Table 1 (a) Elastic moduli of graphite/epoxy lamina at different temperatures

	Temperature, T (K)					
Elastic Moduli (Gpa)	300	325	350	375	400	425
E_{11}	130	130	130	130	130	130
E_{22}	9.5	8.5	8.0	7.5	7.0	6.75
G_{12}	6.0	6.0	5.5	5.0	4.75	4.5

Table 1(b) Elastic moduli of graphite/epoxy lamina at different moisture concentration

	Moisture concentration, C (%)				
Elastic Moduli (Gpa)	0	0.25	0.5	0.75	1
E_{11}	130	130	130	130	130
E_{22}	9.5	9.25	9	8.75	8.5
G_{12}	6.0	6.0	6.0	6.0	6.0

The following sets of support conditions are considered in the present analysis.

(a) All edges simply supported condition (SSSS):

$v = w = \Phi_2 = \Psi_2 = \theta_2 = 0$ at $x = 0, a$ and

$u = w = \Phi_1 = \Psi_1 = \theta_1 = 0$ at $y = 0, b$

(b) All edges clamped condition (CCCC):

$u = v = w = \Phi_1 = \Phi_2 = \Psi_1 = \Psi_2 = \theta_1 = \theta_2 = 0$ for both $x = 0, a$ and $y = 0, b$.

IOP Conf. Series: Materials Science and Engineering **75** (2015) 012016 doi:10.1088/1757-899X/75/1/012016

The fundamental frequency in nondimensional form is obtained using the equation $\bar{\omega} = \omega a^2 / h\sqrt{(\rho / E_2)}$.

5.1. Convergence and Comparison Study

In order to establish the convergence behaviour the free vibration under unlike environmental conditions, responses computed using the present model is plotted with respect to mesh divisions in figure 1. The geometrical, material properties, lamination scheme and support conditions are taken same as [6] and [9]. The results show very good convergence rate with respect to mesh refinement. Based on the convergence study, it is concluded that a (5×5) mesh is adequate to obtain desired responses throughout the analysis. For validation study, simply supported square (a=b=0.5 m), laminated composite spherical shells (R1=R2, h=5mm, a/h=100) are considered. The material properties are taken same as [2]. The responses obtained using the present model along with the reference values are presented in Table 2. It is clearly observed that the present results are in good agreement with the reference values. The difference in results is because the fact that the reference uses FSDT based mathematical model where as the present model is developed based on HSDT mid plane kinematics, which is more realistic in nature.

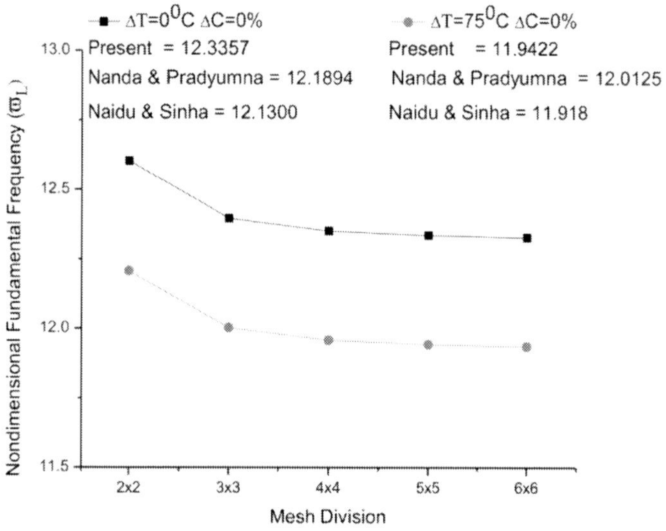

Figure 1. Convergence study of simply supported, square, laminated composite spherical panel ($R_1=R_2$, $R/a=10$, $a/h=10$, $[0/0/30/-30]_2$).

Table 2. Comparison of fundamental frequency (Hz) of simply supported laminated composite spherical shells

		R/a=5		
Source	Lamination	C=0.25	C=0.75	C=1.0
Present	[0/90]2	217.1000	196.7728	185.6448
	[45/-45]2	385.3415	370.9074	358.5730
[2]	[0/90]2	201.9100	201.7200	201.6400
	[45/-45]2	346.3500	344.1800	343.1800
Difference (%)		6.9970	-2.5140	-8.6160
		10.1190	7.2050	4.2920
		R/a=10		
Source	Lamination	T=325	T=375	T=400
Present	[0/90]2	140.1045	125.125	122.7846
	[45/-45]2	269.3122	252.7032	248.4685
[2]	[0/90]2	129.0800	128.9000	128.8400
	[45/-45]2	299.31	296.4900	295.5100
Difference (%)		7.8690	-3.0170	-4.9320
		-11.139	-17.3270	-18.9330

5.2. Parametric Study

In order to address the effect of various geometrical parameter like curvature ratio (R/a = 5, 10, 50, 100), thickness ratio (a/h = 5, 10, 50, 100) and aspect ratio (a/b = 1, 1.5, 2, 2.5) on the nondimensional fundamental frequency of laminated composite spherical shell panel under hygrothermal environment is computed and discussed in detail in the following examples.

5.2.1. Example 1. In this example, a simply supported square (a/b=1) eight layer anti-symmetric ($[0°/90°]_4$) cross-ply laminated composite spherical panel (a/h=20) is used for the computational purpose under different hygrothermal load. The responses are presented in figure 2. It is observed that the nondimensional fundamental frequency parameter increases as temperature and moisture concentrations increases for each case. It can also be seen that frequency parameter decreases as the curvature ratio increases.

5.2.1. Example 2. Figure 3 shows the effect of thickness ratio (a/h) and hygrothermal conditions on the nondimensional linear frequency responses of angle-ply ($[\pm 45]_2$), square (a/b=1) laminated composite spherical (R/a=50) panel under CFCF support condition. It is observed that the nondimensional fundamental frequency responses are higher for thin shell panels in comparison to the thick panels.

5.2.3. Example 3. Square hybrid ([±30 /0/75]) laminated spherical panels (R/a=50, a/h=100), under CSCS support condition and different environmental conditions are considered. The computed results are presented in figure 4. It can be seen that the nondimensional fundamental frequency parameter increases with hygrothermal load and aspect ratios.

IOP Conf. Series: Materials Science and Engineering **75** (2015) 012016 doi:10.1088/1757-899X/75/1/012016

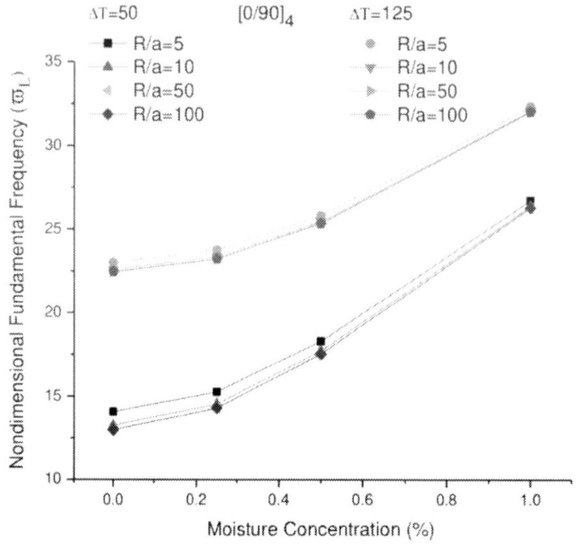

Figure 2. Variation of nondimensional fundamental frequency parameters with curvature ratio under hygrothermal loading

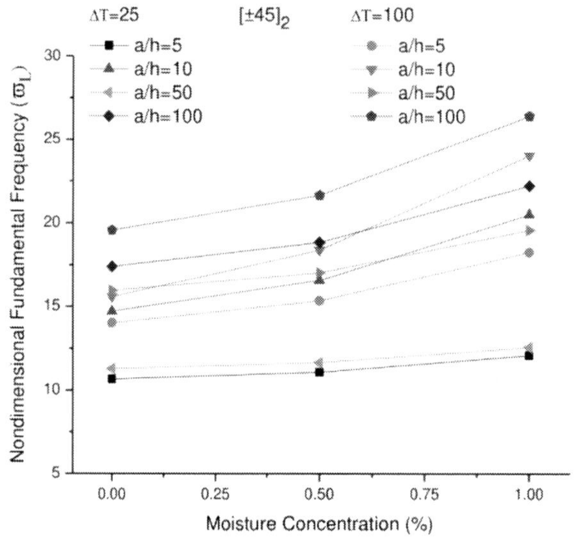

Figure 3. Variation of nondimensional fundamental frequency parameters with thickness ratio under hygrothermal loading

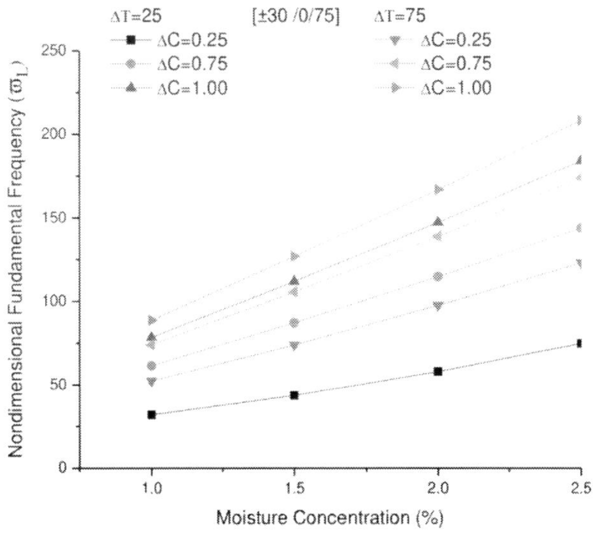

Figure 4. Variation of nondimensional fundamental frequency parameters with aspect ratio under hygrothermal loading

6. Conclusion

Effect of hygrothermal load and geometrical parameter on the free vibration behaviour of laminated composite spherical shell panels have been investigated in the present study. A mathematical model based on the HSDT mid-plane kinematics is developed and discretised using FEM. The convergence and validation of the present model has been established. The significance of the present HSDT based model for the analysis of laminated structures under combined hygrothermal loading condition is observed. The subsequent numerical illustrations indicate that considering the effect of hygrothermal conditions is inevitable for laminated composite structures, particularly when they are exposed to a severe environmental condition. Irrespective of the hygrothermal load and support conditions, the nondimensional fundamental frequency responses decrease with curvature ratios whereas they increase with increase in thickness ratios and aspect ratios under all sorts of hygrothermal conditions.

Acknowledgement

Authors are thankful to AICTE (All India Council for Technical Education) for partial support through the sanction 8023/RID/RPS/56/11/12 Dated: 26/03/2012.

Reference

[1] Sai Ram K S and Sinha P K 1992 *Journal of Sound and Vibration* **158(1)** 133–148
[2] Parhi P K, Bhattacharyya S K and Sinha P K 2001 *Journal of Sound and vibration* **248(2)** 195-214
[3] Patel, B P, Ganapathi M and Makhecha D P 2002 *Composite Structures* **56** 25–34
[4] Huang X L and Zheng J J 2003 *Engineering Structures* **25** 1107–1119
[5] Shen H S, Zheng J J and Huang X L 2004 *Journal of Reinforced Plastics and Composites* 23 (1095) DOI: 10.1177/0731684404037038
[6] Naidu N V S and Sinha P K 2007 *Composite Structure* **77** 475-483
[7] Benkhedda A, Tounsi A and Adda Bedia E 2008 *Composite Structures* **82** 623–35

[8] Lo S H, Zhen W, Cheung Y K and Wanji C 2010 *Composite Structures* **92** 633–646
[9] Nanda N and Pradyumna S 2011 *Journal of Composite Material* **45(20)** 2103-2112
[10] Ashraf M Z 2012 *Composite Structures* **94** 3685–3696
[11] Panda H S, Sahu S K and Parhi P K 2013 *Composite Structures* **96** 502–513
[12] Panda S K and Mahapatra T R 2014 *Meccanica* **49** 191–213
[13] J N Reddy (2004), *Mechanics of laminated Composite Plates and Shells*, (Second Edition), CRC Press
[14] Cook R D, Malkus D S and Plesha M E 2000 *Concepts and applications of finite element analysis* Third Edition John Willy and Sons (Asia) Pvt. Ltd., Singapore
[15] Sundaramoorthy R, David W and Murray 1973 *Incremental finite element Matrices* (ASCE J. Structural Division **99 (ST12)** 2423-2438

Experimental and simulation study of flexural behaviour of woven Glass/Epoxy laminated composite plate

Sushree S Sahoo*, **Vijay K Singh and Subrata K Panda**

National Institute of Technology-Rourkela, Odisha, India, 769008

*E-mail: sushree.sahoo101@gmail.com

Abstract. Flexural behaviour of cross ply laminated woven Glass/Epoxy composite plate has been investigated in this article. Flexural responses are examined by a three point bend test and tensile test carried out on INSTRON 5967 and Universal Testing Machine INSTRON 1195 respectively. The finite element model is developed in ANSYS parametric design language code and discretised using an eight nodded structural shell element. Convergence behaviour of the simulation result has been performed and validated by comparing the results with experimental values. The effects of various parameters such as side-to-thickness ratio, modular ratio on flexural behaviour of woven Glass/Epoxy laminated composite plate are discussed in details.

1. Introduction

Laminated composite materials over the last many years have been the dominant emerging materials, penetrating and conquering markets of aerospace, marine and civil infrastructure relentlessly owing to their advantages like light weight, high strength/stiffness, superior fatigue response characteristics, facility to vary fibre orientation, material and stacking pattern, resistance to electrochemical corrosion, and other superior material properties of composites. Based on the design requirement, the laminated structural materials must have enough strength against complex loading, deformation capability and excessive vibration to increase the service life since these materials are prone to fail under such circumstances in real life situation. Hence, the need of the hour is to understand the load deformation behaviour of laminated structure and predict the exact structural responses for various loading for final design of desired component. In order to do so, the finite element method has been ascertained as a versatile tool for the analysis of complex structural behaviour, especially the laminated composite structures as they are statically indeterminate due to their geometry. It is well known that, the deflection behaviour of composites has a great importance on their design and analysis because they are flexible in comparison to the other available conventional materials.

In order to foresee the realistic responses of laminated structures several attempt have been made in past by the researcher. Reddy et al. [1] carried out numerous finite element analysis for various parameters to study the effect of transverse shear deformation on deflection and stresses of laminated composite plates subjected to uniformly distributed load in ANSYS environment. Catangiu et al. [2] used a bending fatigue testing machine to obtain results of bending fatigue tests of composites plates of orthogonal woven Glass/Epoxy composite made up in hand lay-up method preceded by a preliminary degradation. Gupta et al. [3] investigated the static response of laminated composite plate subjected to uniformly distributed loading using finite element approach based on first-order shear deformation theory considering geometric nonlinearity and damage evolution. Ahmed et al. [4] analysed static and

Content from this work may be used under the terms of the Creative Commons Attribution 3.0 licence. Any further distribution of this work must maintain attribution to the author(s) and the title of the work, journal citation and DOI.
Published under licence by IOP Publishing Ltd

dynamic response of Graphite /Epoxy laminated composite plates under transverse loading based on First Order Shear Deformation Theory. Parhi et al. [5] evaluated laminated composite plates with randomly located multiple delaminations subjected to transverse static and impact by applying finite element analysis procedure using first order shear deformation via eight-node isoparametric quadratic elements. Kumar et al. [6] investigated the transverse deflection and stresses of laminated composite skew plate with elliptical hole when the plate is subjected to transverse pressure loading via a finite element method based on three dimensional theory of elasticity. Aagaah et al. [7] analysed deformations of a rectangular multi layered laminated composite plate due to mechanical loads by deriving its linear dynamic equations based on a third order shear deformation plate theory in conjunction with the Von Karman strains. Dash and Singh [8] presented transverse bending analysis of shear deformable laminated composite plates in Green–Lagrange sense using finite element approach based on the higher order shear deformation theory considering ten degree of freedom per node. Pandya and Kant [9] applied finite element formulation to obtain flexure of an orthotropic plate based on higher-order displacement model and a three-dimensional state of stress and strain. Lal et al. [10] examined nonlinear transverse central deflection of laminated composite plates subjected to transverse uniform lateral pressure and thermal loading and the effect of sensitivity of randomness in system parameters on central deflection applying higher order shear deformation theory in the von-Karman sense. Kommineni and Kant [11] studied linear and geometrically non-linear finite element analysis of fibre reinforced composite and sandwich laminates applying a refined higher order shear deformation theory in von Karman sense considering nine degree of freedom per node. Raju and Kumar [12] developed an analytical procedure to examine the bending features of laminated composite plates based on higher order shear displacement model with zig-zag function. Sahoo and Singh [13] proposed a trigonometric zigzag theory for the static analysis of laminated composite and sandwich plates employing a displacement based C^0 finite element model. Goswami [14] incorporated refined higher-order shear deformation theory and 3-dimensional state of stress and strain for thick and thin laminated composite plates to a C^0 plate bending element formulation. Liu et al. [15] presented static and free vibration analyses of isotropic composite plates through a linearly conforming radial point interpolation method whose mesh-free shape functions were constructed by combining the radial and polynomial basis functions. Park et al. [16] presented quasi-conforming formulation of four node stress resultant shell element for the linear static and dynamic analysis of composite plates and shells which uses interrelated displacement-rotation interpolations applicable for moderately thick and thin composite shells. Derras et al. [17] established a new refined shear deformation theory for the nonlinear cylindrical bending behaviour of functionally graded plates which are subjected to pressure loading, and their geometric nonlinearity is introduced in the strain-displacement equations based on Von-Karman assumptions.

The flexural behaviour of laminated structure has attracted many researchers not only because it is interesting but also challenging. We note that the experimental study of laminated structure for transverse bending and subsequent simulation study using commercial finite element package and/or computational code are not much in number. Here in this study the bending behaviour of woven laminated composite plate has been investigated experimentally and compared with commercial finite element software package ANSYS parametric design language (APDL) code to study the flexural behaviour. The model has been discretised using Shell 281 element and the responses are compared with the experimental values obtained from three point bend test. The flexural behaviour of laminated composite panels are obtained for different side-to-thickness ratio (a/h) and modular ratios (E_1/E_2) and discussed in details.

2. Numerical Methodology

A laminated doubly curved shell element is considered with thickness 'h'. The radii of curvature in x and y direction are R_x and R_y respectively as shown in the figure 1. Each of the thin lamina can be oriented at an arbitrary angle 'ϕ' with reference to the x-axis. The displacement field is assumed to be in following form:

$$\left.\begin{aligned} U(x,y,z)&=u_0(x,y)+z\theta_x(x,y)\\ V(x,y,z)&=v_0(x,y)+z\theta_y(x,y)\\ W(x,y,z)&=w_0(x,y)+z\theta_z(x,y)\end{aligned}\right\} \quad (1)$$

Where U , V and W are the displacements at any point within the shell in x, y and z directions respectively. u_0, v_0 and w_0 are the associated mid-plane displacements and θ_x, θ_y and θ_z are the rotations about the respective axes at z = 0.

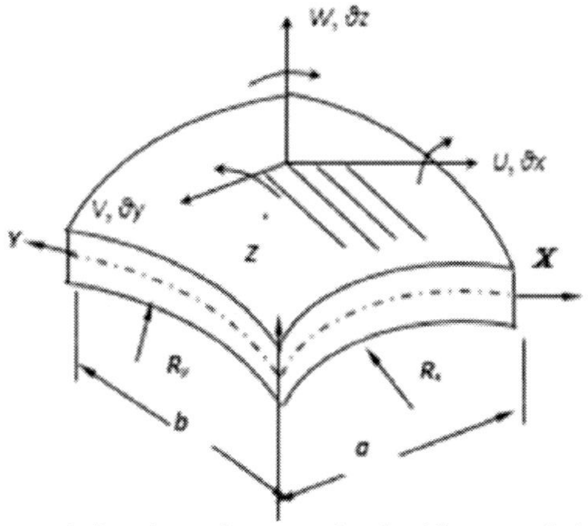

Figure 1. Laminated composite doubly curved shell.

The stress strain relations for the kth lamina oriented at an arbitrary angle 'ϕ' about any arbitrary axes are given by:

$$\{\sigma\}=\left[\overline{Q}_{ij}\right]\{\varepsilon_j\} \quad (2)$$

Stress vector can be rewritten in force form as:

$$\{F\}=[D]\{\varepsilon\} \quad (3)$$

The elements of the stiffness matrix $[D]$ are defined as:

$$[D]=\left[A_{ij},B_{ij},D_{ij}\right]=\sum_{k=1}^{n}\int_{z_{k-1}}^{z_k}\left(\overline{Q}_{ij}\right)_k\left(1,z,z^2\right)dz \quad (4)$$

2.1. *Finite element formulation*

An eight-nodded isoparametric element with six degrees of freedom viz., u_0, v_0, w_0, θ_x, θ_y and θ_z at each node is used. The displacement vector 'd' at any point on the mid surface is given by:

$$d = \sum_{i=1}^{8} N_i(x, y) d_i \tag{5}$$

Where d_i and N_i are the displacement vector and the interpolating function, respectively, associated with the node 'i'. Total strain is written in the linear form:

$$\{\varepsilon\} = [B_L]\{d_i\} \tag{6}$$

Where $[B_L]$ is a strain displacement relation matrix. The virtual work equation for free vibration in Lagrangian coordinate system may be written as:

$$\int_A \{\delta d\}^T [\rho]\{d\} dA + \int_A \{\delta\varepsilon\}^T \{F\} dA = 0 \tag{7}$$

Where $\{d\}$ is the generalized displacements, 'δ' denotes variation and $[\rho]$ is the mass matrix.

Substituting Equations (3), (5) and (6) into Equation (7), it can be written in the finite element form for an element as:

$$[M]\{\ddot{d}_i\} + [K]\{d_i\} = 0 \tag{8}$$

Where $[K]$ $[M]$ are the stiffness and mass matrix respectively can be expressed as:

$$[K] = \int_A [B_L]^T [D][B_L] dA$$
$$[M] = \int_A [N]^T [\rho][N] dA \tag{9}$$

For the flexural analysis:

$$([K])\{d\} = \{F\} \tag{10}$$

For the present analysis, the specimen is applied with point load at the center and two opposite edges of the specimen are clamped. The boundary conditions are as follows:-

$$u_0 = v_0 = w_0 = \theta_x = \theta_y = \theta_z = 0 \quad \text{at } x=0 \text{ and } a \tag{11}$$

2. Results and Discussion

2.1. Experimentation

The properties of Glass/Epoxy laminated composite plate can be defined completely by its Young's modulus, shear modulus and Poisson's ratio in principal material directions. The desired properties are determined experimentally by performing unidirectional tensile test on specimens (longitudinal and transverse directions and at an angle of 45° inclined to the longitudinal direction) based on ASTM standard: D 3039/D 3039M. A thin flat strip of specimen having a constant rectangular cross section was prepared in all cases with 20cm, 2.25cm and 0.5 cm as its length, breadth and height respectively. For measuring the Young's modulus, the specimen is loaded in UTM INSTRON 1195 with a recommended rate of loading of 1 mm/minute as shown in figure 2(a) and figure 2(b). Poisson's ratio for the present analysis is taken 0.17 as in Jensen et al. [18]. The shear modulus was determined using the formula as in Jones [19] as:

$$G_{lt} = \frac{1}{\frac{4}{E_{45}} - \frac{1}{E_l} - \frac{1}{E_t} - \frac{2\upsilon_{12}}{E_l}} \tag{11}$$

The material properties are presented in table.1.

Table 1. Material properties of Glass/Epoxy composite plate.

No. of layers	10
Stacking sequence	$(0/90)_5$
Young's modulus x direction (E_l)	5.802GPa
Young's modulus y direction (E_t)	4.966GPa
Young's modulus z direction (E_z)	4.966GPa
Shear modulus (G_{lt})	1.898GPa
Shear modulus (G_{tz})	0.949GPa
Shear modulus (G_{lz})	1.898GPa
Poisson's ratio (υ_{lt})	0.17
Poisson's ratio (υ_{lz})	0.17
Poisson's ratio (υ_{tz})	0.17
Density (ρ)	1646.00 Kgm^{-3}

In order to determine flexural behavior experimentally, three point bend test was conducted on Glass/Epoxy laminated composite plate specimen using UTM INSTRON 5967 with environmental chamber with a commended rate of loading of 2 mm/minute as shown in figure 3(a) and figure 3(b). The dimension of the specimen was taken according to ASTM standard: D790 that is, for depth of 5mm the length and breadth of the specimen was taken to be 9 cm and 2.25 cm, respectively.

Figure 2(a). UTM INSTRON 1195.

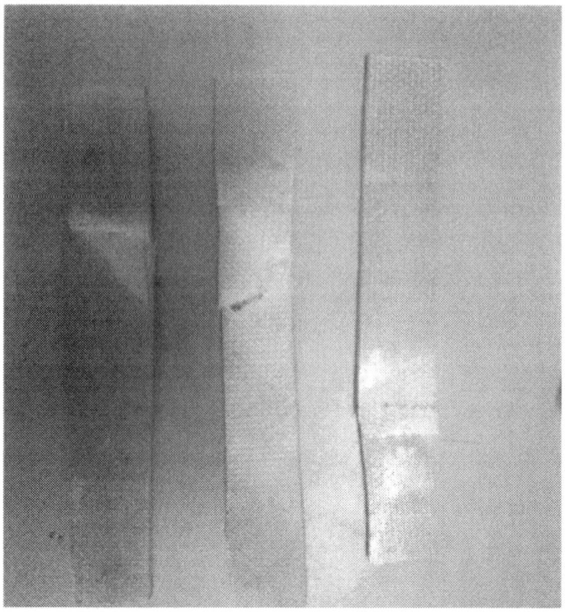

Figure 2(b). 10-layer cross ply $(0^0/90^0)_5$ Glass/Epoxy laminated composite specimen after tensile test.

Figure 3(a). Three point bend test set up (INSTRON 5967).

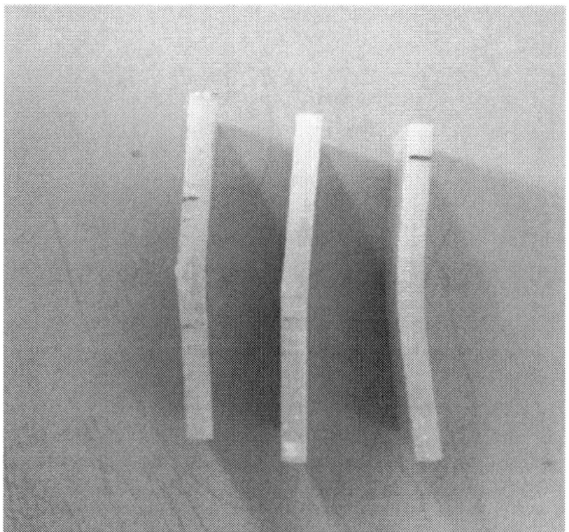

Figure 3(b). Ten layer cross $(0^0/90^0)_5$ Glass/Epoxy laminated composite specimen after three point bend test.

2.2. Finite element modelling :

In this section, the flexural behaviour of laminated composite panel has been inspected using APDL code developed in ANSYS environment. An eight-nodded, six degrees of freedom structural shell element (shell281) has been developed for the current analysis which allows modeling thin to moderately thick plate and shell structures as shown in figure 4(a) and the deformation curve is presented in figure 4(b). The numeric values are compared with the obtained experimental results.

Figure 4 (a). Finite element model of 10-layer cross ply Glass/Epoxy laminated composite plate.

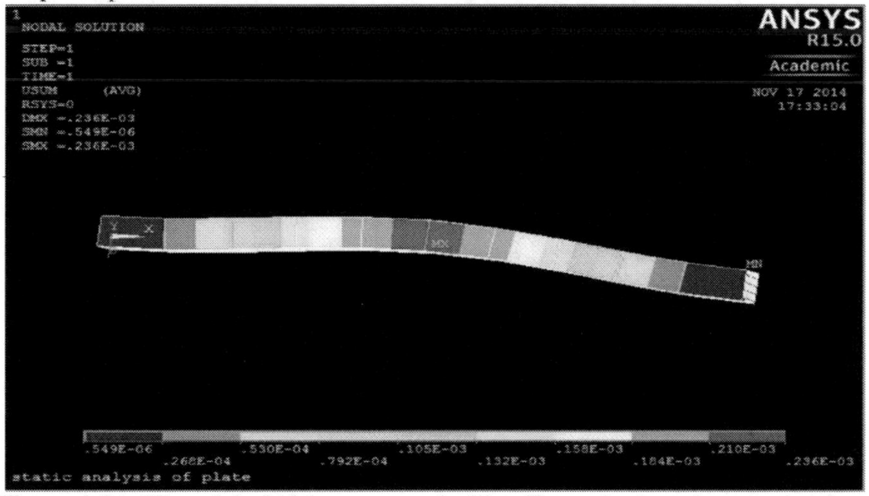

Figure 4 (b). FE model of 10-layer cross ply Glass/Epoxy laminated composite plate.

2.3. Convergence and Comparison Study

The convergence study with respect to mesh divisions of glass epoxy laminated flat panel for flexural behaviour is plotted in figure 5. The central deflections of plate when applied to a various point loads at the centre are compared with the experiment results. It is quite obvious from Table 4 that the present results are showing very small difference with the experimental result and the present model will be able to solve any general type of geometry effortlessly. Based on the convergence, it is noted that a (8×8) mesh is appropriate to obtain the flexural responses.

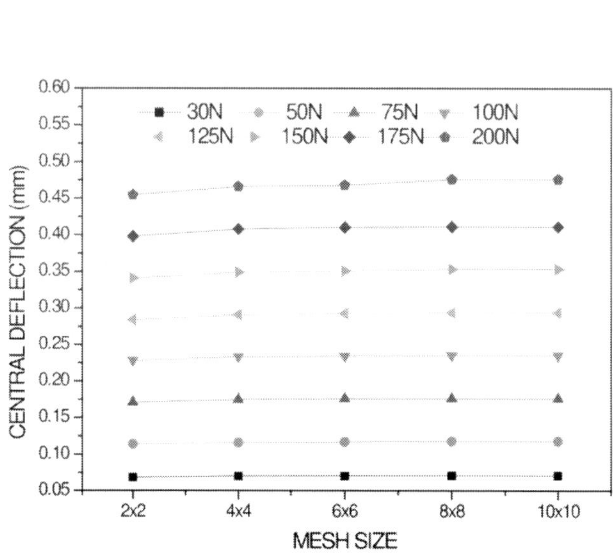

Figure 5. Convergence study of laminated $(0/90)_5$ Glass/Epoxy laminated composite plate.

Table 2. Comparison of Central deflection values of Glass/Epoxy composite plate

Load(N)	Central deflection(mm)		
	Experimental	Present	Difference (%)
30.00297	0.08011	0.0706	11.88
50.00085	0.12681	0.118	6.95
75.0372	0.18236	0.176	3.49
100.09009	0.23906	0.235	1.69
125.03559	0.29566	0.294	0.57
150.01338	0.35448	0.353	0.42
175.03528	0.4159	0.411	1.18
200.00079	0.48054	0.476	0.94

3. Parametric Study

In this section, some new results are computed for different parameters. The central deflections are plotted for side to thickness ratios *(a/h)* and modular ratio *(E_1/E_2)* and their effects are discussed in the upcoming subsections.

3.1. Effect of side to thickness ratio

In general, with the increase in side to thickness ratio, there is a decrease in stiffness of the material due to which the central deflection tends to increase. From figure 6 it can be concluded that the present model is in clear agreement with the theory as when the side to thickness ratio increases from 30 to 60, the central deflection is seen increasing.

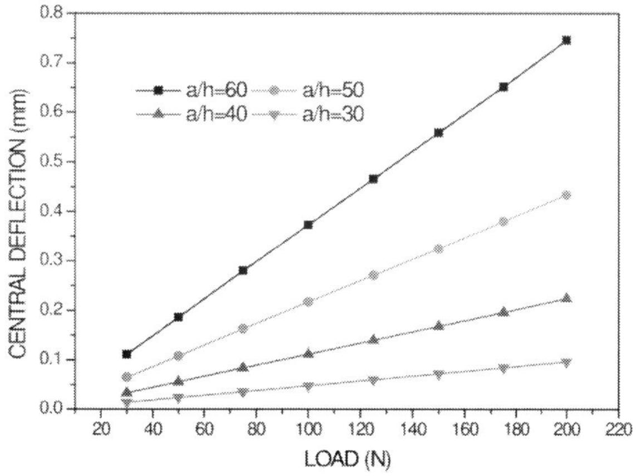

Figure 6. Effect of side to thickness on deflection of Glass/Epoxy composite plate at different loads.

3.2. Effect of modular ratio

Mathematically, it is well known that as the modular ratio increases, the reduced transformed stiffness matrix increases and hence the central deflection decreases. The responses computed using the present model follow the same as depicted in figure 7. As the modular ratio increases from 5 to 20, the central deflection of the plate decreases.

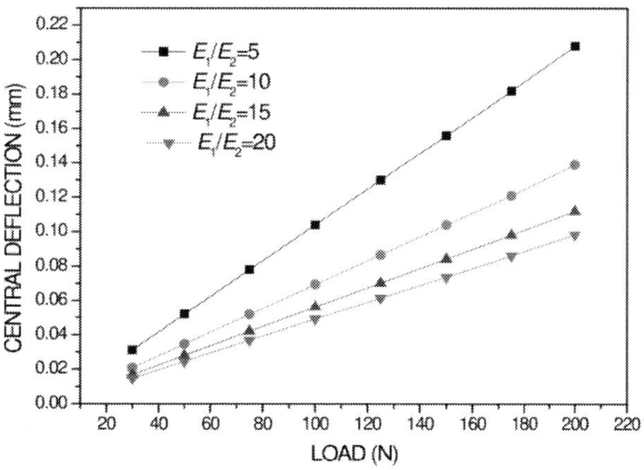

Figure 7. Effect of modular ratio on deflection of Glass/Epoxy composite plate at different loads.

4. Acknowledgement

This work is under the project sanctioned by the department of science and technology (DST) through grant SERB/F/1765/2013-2014 Dated: 21/06/2013. Authors are thankful to DST, Govt. of India for its consistent support.

5. Conclusion

Flexural behaviour of Glass/Epoxy $(0/90)_5$ laminated composite plate has been investigated in this present study. Flexural responses are examined by a three point bend test carried out on INSTRON 5967 and material properties were found out by performing tensile test on UTM INSTRON 1195, respectively. The experimental results were compared with commercial Finite element package ANSYS using APDL. The model has been discretised using an eight nodded structural shell element (shell 281). The convergence study shows that the present results are converging well with mesh refinement and the difference between the numerical and experimental values are within the expected line. The parametric study indicates that the central deflection of flat panel increases with side to thickness ratio and decreases with increase in modular ratio.

Reference

[1] Reddy B S, Reddy A R, Kumar J S and Reddy K V K 2012 *International Journal of Engineering, Science and Technology* **4** 177-190

[2] Catangiu A, Dumitrescu A T and Ungureanu D 2011 *Materials and mechanics* **6** 47-51

[3] Gupta A K, Patel B P and Nath Y 2013 *Acta Mechanica* **224** 1285–1298

[4] Ahmed J K, Agarwal V C, Pal P and Srivastav V 2013 *International Journal of Innovative Technology and Exploring Engineering* **3** 56-60

[5] Parhi P K, Bhattacharyya S K and Singh P K 2001 *Buletin of Material Science* **24** 143-149

[6] Kumar M S R N, Sarcar M M M and Murthy V B K 2009 *Indian Journal of Engineering and Materials Science* **16** 37-43

[7] Aagaah M R, Mahinfalah M and Jazar G N 2003 *Composite Structures* **62** 27-39

[8] Dash P and Singh B N 2010 *Communication Nonlinear Science Numerical Simulation* **15** 3170-3181

[9] Pandya B N and Kant T 1988 *Computers & Structures* **28** 119-133

[10] Lal A, Singh B N and Kumar R 2011 *Archive of Applied Mechanics* **81** 727–743

[11] Kommineni J R and Kant T 1992 *Computers and Structures* **45** 511-520

[12] Raju T D and Kumar J S 2011 *ARPN Journal of Engineering and Applied Sciences* **6** 106-110

[13] Sahoo R and Singh B N 2014 *Aerospace Science and Technology* **35** 15-28

[14] Goswami S 2006 *Composite Structures* **72** 375–382

[15] Liu G R, Zhao X, Dai K Y, Zhong Z H, Li G Y and Han X 2008 *Composites Science and Technology* **68** 354–366

[16] Parka T, Kimb K and Han S 2006 *Composites: Part B* **37** 237–248

[17] Derras M, Kaci A, Draiche K and Tounsi A 2013 *Journal of Theoretical and Applied Mechanics* **51** 339-348

[18] Jensen D W, Crawley E F and Dugnani J 1979 *Journal of Composite Materials* **13** 195

[19] Jones R M 1975 *Mechanics of Composite Materials* (Philadelphia: Taylor and Francis) p 97

[20] *Ansys Mechanical APDL 15.0 User Manual*

Fabrication and characterization of carbon nanofiber (CNF) based epoxy composites

S Bal[1] and S Saha[2]

[1]Department of Metallurgical and Materials Engineering, National Institute of Technology, Rourkela-769008, Odisha, India.
[2]Department of Physics and Astronomy, National Institute of Technology, Rourkela-769008, Odisha, India.

Email: balss@nitrkl.ac.in

Abstract. The behaviour of epoxy based nanocomposites containing 1.5, 2 and 2.5 wt.% of carbon nanofibers (CNFs) have been analysed through mechanical and viscoelastic characterizations. The nanocomposites thus fabricated were allowed to cure at two different temperature conditions- one was at room temperature (27 °C) and other was at low temperature (4 °C). Among all the room temperature cured (RTC) and low temperature cured (LTC) composites, maximum improvement in flexural modulus (83% & 128%), storage modulus (34% & 21%) and glass transition temperature (34% & 35%) were found in case of 2 wt.% CNF loaded samples (CNF2). In addition, LTC composites at 2 wt.% were showing enhanced flexural and thermal properties in comparison to the RTC composites. However at relatively higher concentration less increment in flexural modulus was noticed for both RTC and LTC composites. Microscopical investigation supported the above results by showing the interfacial arrangement between carbon nanofibers and epoxy matrix.

Keywords. Carbon nanofiber, epoxy composite, flexural modulus, glass transition temperature.

1. Introduction

For about a decade and more, carbon nanofibers (CNFs) are acting as extraordinary filler constituents offering overall improvement in mechanical and physical properties of the polymer matrix material. Due to light weight, high specific modulus ~ 500 GPa and high specific strength ~3 GPa of individual nanofibers, CNFs-reinforced polymer becomes eye-catching structural materials [1]. Further electrical conductivity ~ 10^3 S/cm, thermal conductivity ~ 1900 Wm^{-1} K^{-1} and huge specific surface area (SSA) of up to 1000 m^2/g together with their low manufacturing cost make them suitable for improving the performance of polymeric materials in numerous applications like in automotive, aerospace, electronics or biotechnology [2]. In polymer, epoxy resin is most often used as the polymer matrix due to its series of interesting properties like high specific strength, good stiffness, chemical resistance, dimensional stability and also strong adhesion to the embedded reinforcement [3-6]. Enhancement in mechanical properties of CNF based epoxy composites have been well illustrated in the literature [7-12] by the achievement of strong interfacial adhesion between the composite phases. Previous research showed that CNFs are prone to agglomerate because of their surface properties and the process of

Content from this work may be used under the terms of the Creative Commons Attribution 3.0 licence. Any further distribution of this work must maintain attribution to the author(s) and the title of the work, journal citation and DOI.
Published under licence by IOP Publishing Ltd

fabrication [13-17]. So the significant improvement in CNF dispersion in the composite is very much necessary as it facilitates the stress transmission from matrix to the reinforcing agent.
In the present study, the entanglement and agglomeration of CNFs were removed by using the solvent to the CNFs and by curing of the fabricated specimen at low temperature. Our group already worked on this procedure [18] and fabricated CNF/epoxy composites by the incorporation of low wt.% of CNFs (0.5, 0.75, 1 wt.%) where qualitative dispersion has been reported. The originality of present work is to study the influence of reinforcing strategies of CNF/epoxy composites of higher wt.% (up to a maximum 2.5%) of CNFs in the epoxy matrix.

2. Materials and Methods

2.1. Materials

Epoxy polymer matrix in the current study was prepared by mixing epoxy resin (araldite LY-556 based on Bisphenol A) and hardener HY-951 (aliphatic primary amine) in wt. ratio 100/12. Epoxy and hardener were procured from CIBA-GEIGY, INDIA. This epoxy resin (5.3-5.4 equiv/kg) was of low processing viscosity and good overall mechanical properties. Carbon nanofibers (CNFs) used for experimental study were procured from Nanostructured & Amorphous Materials Inc. (Nano Amour), USA. They are 10-50 μm long and 200-500 nm in diameter. The purity of the nanofibers was >95%.

2.2. Preparation of composites

The Carbon nanofibers (1.5 wt.%, 2 wt.% and 2.5 wt.%) were first dispersed inacetone by sonication for 4 hours. The obtained powder was then again dispersed in acetone and sonicated. This process assisted to de-bundle the nanofibers at this higher concentration which was optimized to be the best dispersion, confirmed through optical microscope. Finally the powder of nanofibers were mixed to epoxy resin and sonicated at controlled power levels for 6 hours. Then the CNF and epoxy resin mixture were kept in vacuum oven for 12 hours to eliminate the moisture and air bubbles. Thereafter with manual stirring hardener was added to the mixture inside a water bathto avoid further reaction. Another degassing process was conducted to the mixture in vacuum oven for a few minutes. The mixture was injected into two molds one of which was kept in room temperature atmosphere (27 °C) and other was in low temperature (4°C). These prepared samples were then post cured at 90 °C for 6 hours.
The fabricated specimens in this work are named as follows:

Room temperature cured specimens= RTC

Low temperature cured specimens= LTC

1.5 wt.% CNF based epoxy composites= CNF1.5

2 wt.% CNF based epoxy composites= CNF2

2.5 wt.% CNF based epoxy composites= CNF2.5

3. Characterization

As the content of CNFs, mixed in the form of reinforcement to the epoxy is not so large, hence it is expected that the volume occupied by the fillers is small. The fibers were oriented in a random way throughout the matrix which results the nanocomposites isotropic in behaviour. Hence, a test standard for polymeric materials was used to evaluate the mechanical properties of nanocomposite materials.

3.1. Flexural test

Flexural tests were accompanied according to ASTM D790–02 "Standard Test Methods for Flexural Properties of Un-reinforced and Reinforced Plastics and Electrical Insulating Materials" on an universal testing machine (Instron 5967). The mode of test was three point bending and the rate at which the load was imposed was 2.0 mm/min. The span between two supports, width and thickness of the specimen was shown in the figure 1. In order to reproduce the results five specimens were tested in each group.

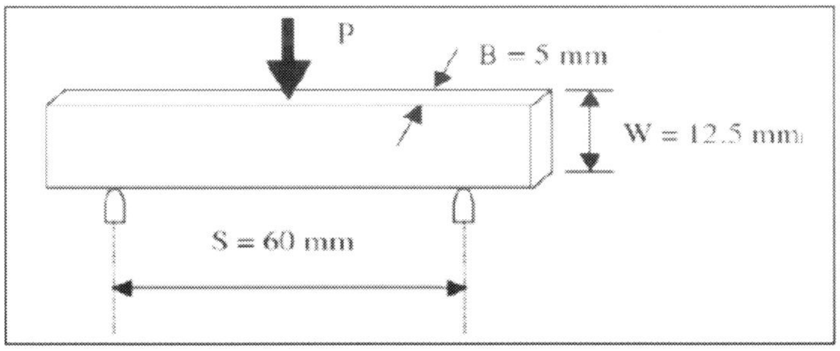

Figure 1. Flexural testing specimen configuration showing thickness = B, width =W, span = S and applied force = P.

3.2. Dynamic mechanical analysis (DMA)
Viscoelastic properties of composites were analyzed using Gabo Qualimeter (Eplexor 150N, Germany) in dual cantilever mode. Treatment conditions included a frequency 1 Hz with a strain of 0.1%. A temperature range from 20-260 °C was applied at a heating rate of 2 °C min in air.

3.3. Microscopical investigation
Field emission scanning electron microscope (Nova NANOSEM 450) was used to observe the fracture surface topography of all the composites specimens.

4. Results and Discussion

4.1. Flexural test results
Figure 2 (a) and (b) represented the elastic moduli of parent epoxy and nanocomposites for CNF1.5, CNF2 and CNF2.5 for both RTC and LTC specimens respectively. Earlier we have reported [18] the behaviour of mechanical properties of CNF/epoxy composites by incorporating low wt.% of CNFs (0.5, 0.75, 1 wt.%). In the present studies higher CNF contents were taken and flexural modulus values were evaluated.

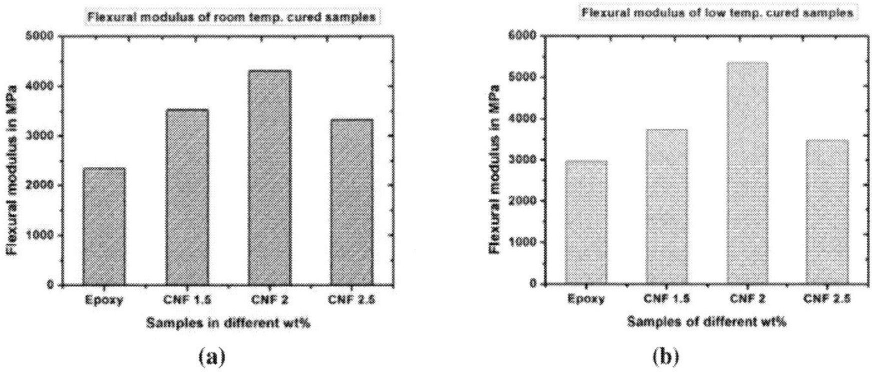

Figure 2. Flexural modulus of (a) RTC samples and (b) LTC samples.

For RTC samples the evaluated flexural modulus of neat epoxy was 2344 MPa. Modulus was increased by 50% and 83% for CNF1.5 and CNF2 respectively. Similar trend was observed in the LTC specimens. The modulus values for the CNF1.5 and CNF2 were found to be 3736 MPa and 5356 MPa i.e. increment in modulus by 25% and 79% respectively in comparison to the neat epoxy (2976 MPa). The modulus was almost double for CNF2 in case of LTC specimens compare to the neat epoxy of RTC specimens. This result is well agreed with earlier reported result [19] where the mechanical properties of CNF/epoxy composites were increased due to the addition of CNFs below 3 wt.%. The increment in modulus value was attributed to the qualitative dispersion of CNFs with the epoxy which results formation of strong interface between the filler and the host [20]. The improved interface is in turn facilitates load transfer from matrix to the nanofiller enhancing overall mechanical properties of the composites. However at higher loading of CNFs (CNF2.5) the trend was different. The modulus value obtained was lower than the other nanocomposites, though it was still higher than that of the parent epoxy for both the cases. For the LTC specimens CNF2.5 showed only 16% increment than that of the neat epoxy composite. Due to high aspect ratio and vander Waals' interaction of CNFs agglomeration may be created at higher concentration. As a consequence the whole nanocomposites may act as a stress concentrator and hence the mechanical properties did not increase to greater extent. These results were in accordance with the study by Gershon et al. [21] and Palencia et al. [11] who established the existence of a limit of dispersion of CNFs in the epoxy matrix which was evidenced by the mechanical properties.

On the otherhand, it was worth noting that LTC specimens showed greater modulus to their RTC counterparts. The result indicates that the low temperature facilitates the curing procedure in an effective manner by delaying the settling time and hence mechanical properties were improved [18].

Flexural stress-strain curves of neat epoxy and CNF filled epoxy were recorded during the three-point bending test of the materials studied shown in figure 3 (a) and (b). All the specimens failed immediately after the flexure stress reached the maximum value. An increased strain to failure was observed in case of some LTC nanocomposites (figure 3(b)) in comparison to RTC nanocomposites (figure 3 (a)).

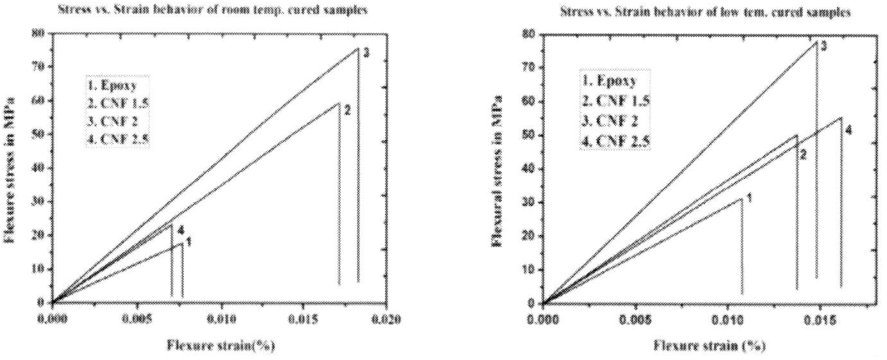

(a) (b)

Figure 3. Stress vs. strain behaviour of (a) RTC and (b) LTC composites.

4.2. DMA test results

DMA was performed to evaluate the viscoelastic properties of the composites such as storage and loss modulus, glass transition temperature etc. Figure 4 (a) and (b) show the effect of temperature on the storage modulus of the LTC and RTC composites respectively. It was found that compared to the neat epoxy composites the value of storage modulus was increased upon increasing CNF contents up to 2 wt.% for both LTC and RTC specimens.

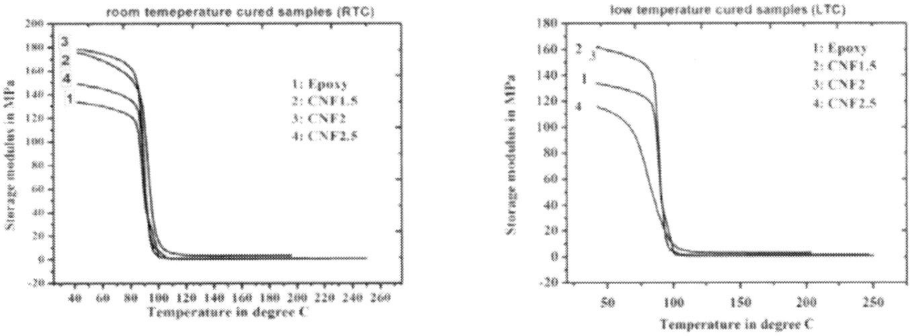

(a) (b)

Figure 4. Storage modulus vs. temperature plot of (a) LTC (b) RTC composite samples.

The value of the storage modulus of neat epoxy composites was found to be around 133 MPa for both the cases. In case of RTC specimens for CNF1.5, CNF2, and CNF2.5 the percentage of increment in modulus were 32%, 34% and 12% respectively. CNF2 showed the maximum enhancement in modulus (179 MPa) among the RTC specimens. On the other hand in LTC specimen CNF1.5 and CNF2 demonstrated almost equal modulus values (162 MPa) which is 21% greater than the neat epoxy. The increase in modulus may be due to the uniform dispersion of the CNFs as well as confinement of CNFs to the epoxy polymer [22]. However CNF2.5 displayed lowest value of storage modulus (115 MPa) among all. The modulus is even lesser (13% decreased) than the respective neat epoxy. Lower modulus value clearly indicates that reinforcement effect of CNFs was decreased when concentration of CNFs went towards higher side attributing the agglomeration of CNFs and weak interfacial bonding between CNFs and epoxy matrix [23].

Tan delta is defined as the ratio of loss to storage modulus of a composite where delta represents the phase angle. The phase angle is very much sensitive to the structural change of the materials and decreases with increasing filler contents [24]. The peak of tan delta is used to determine the glass transition temperature (T_g) of a material from the onsets of the sharp drop in tandelta vs.

Table 1. Glass transition temperature (T_g) of RTC & LTC samples.

RTC samples	T_g in °C	LTC samples	T_g in °C
Epoxy	65	Epoxy	67
CNF 1.5	81.8	CNF 1.5	83.5
CNF 2	87	CNF 2	90.7
CNF 2.5	85.4	CNF 2.5	74

temperature graphs. The evaluated T_g value for different composite samples from the graph were presented here in tabulated form. In case of RTC the value of T_g for the neat epoxy was found to be 65°C and that of LTC specimens was 67 °C. It was observed that as the CNF content goes higher, T_g values for both the RTC and LTC nanocomposites were increased. This was increased by 26%, 34% & 31% for RTC and 24%, 35% and 10% for LTC for CNF1.5, CNF2 and CNF2.5 respectively compare to their neat epoxy composites. These results indicated significant enhancement in glass transition temperature compare to the parent epoxy. Maximum enhancement in T_g was recorded for CNF2 (90.7°C) among all the samples. The increase in T_g of the nanocomposites was due to the addition of CNFs which restricts the mobility of polymer chains, made the nanocomposites brittle in nature and hence improves the thermal stability of the composites [25-27]. However at higher wt.% (for 2.5) the T_g for the LTC samples showed lower value than the others (still higher than neat epoxy). The detrimental results were may be due to presence of aggregated phases of nanofibers at higher concentration which eventually prevented from achieving uniform dispersion. As a result the interface between CNFs and epoxy became much weaker which in turn made the composite less stiff compared to others. There were some other reason behind the reduction in T_g according to which changes in chemical equivalent between matrix and curing agent may decrease the cross linking density of the composites and hence T_g was decreased [7]. Another observation from present study revealed that the LTC samples were showing greater T_g values than RTC specimens except higher loading. The same trend was noticed in mechanical results also. The possible reason is that at low temperature the nanofibers were well dispersed and cured by delaying the settling time which enhanced the overall mechanical and thermal properties of the composites.

4.3. Microscopical results

Figure 5-7 shows the micrograph of fracture surfaces of RTC and LTC specimens. It was clearly observed in figure 5(a) i.e. in the case of CNF1.5 (RTC) nanofibers were intimately attached to the matrix. As a result of this bonding matrix can transfer the applied load on the composites to the high modulus CNFs. In figure 5(b) i.e. in the case of CNF2 (RTC), nice integration of CNFs with the epoxy resin was found. This is in turn denoted good interfacial strength of the composites even if the fracture occurred. In the case of LTC CNF1.5 (figure 6(a)) adequate wetting of CNFs inside the matrix were seen. It seems like fractographs were much denser where CNFs were attached to the epoxy and

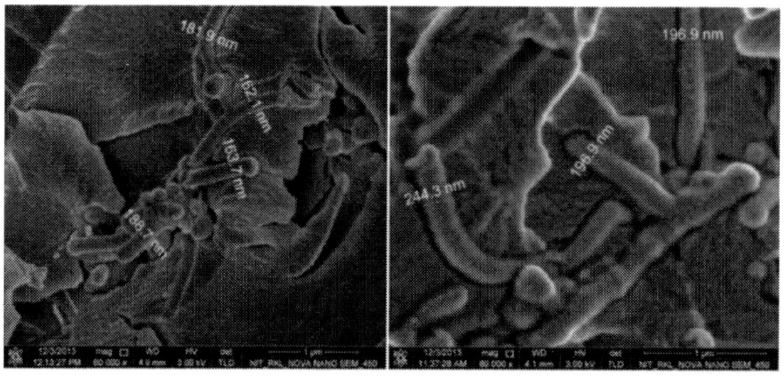

(a) **(b)**

Figure 5. Fracture surfaces of the RTC specimens (a) CNF1.5and (b) CNF2.

(figure 6(a) and (b)) very less number of CNFs were removed after the application of load. This indicated wrapping of polymers around the nanofibers suggesting stronger bonding compare to the RTC specimens. That is why significant enhancement in mechanical results was seen in case of LTC samples. Another worth noting from figure 6(a) was that most of the CNFs were aligned in the same direction which was along the direction of application of load. Since load is transferred effectively in the axial direction hence in LTC, the CNF1.5 specimen showed greater flexural and storage modulus than that of the RTC samples. Further figure 6(b) showed the bridging and spanning of nanofibers between the composite cracks that offers resistance to the crack propagation. It has already been reported that bridging of nanofillers inhibits the crack initiation [26, 28]. Hence, the presence of CNFs in the form of a bridge between the crack regions can be the reason for increased mechanical behaviour in LTC samples.

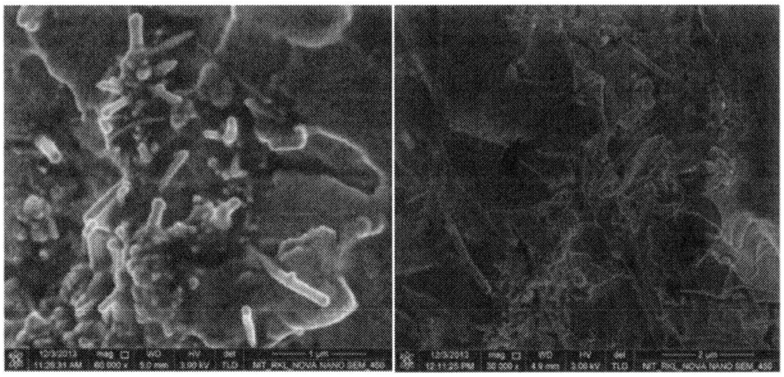

(a) (b)

Figure 6. Fracture surface of LTC specimens showing (a) wetting of nanofibers in CNF1.5 and (b) bridge formation in CNF2.

Now according to the observation under mechanical test it was found that at higher concentration of CNFs (for CNF2.5) the overall mechanical properties were decreased for both the RTC and LTC specimens. The prediction of the detrimental effect was may be due to the presence of aggregated phases of CNFs. It was observed that the prediction shows well agreement after the microscopical investigation was performed. The investigation revealed the formation of aggregates of CNFs at CNF2.5 for both the cases shown in figure 7(a) and (b).This aggregation are very much responsible for creating voids and internal damages inside the composites [8] resulting lower mechanical properties.

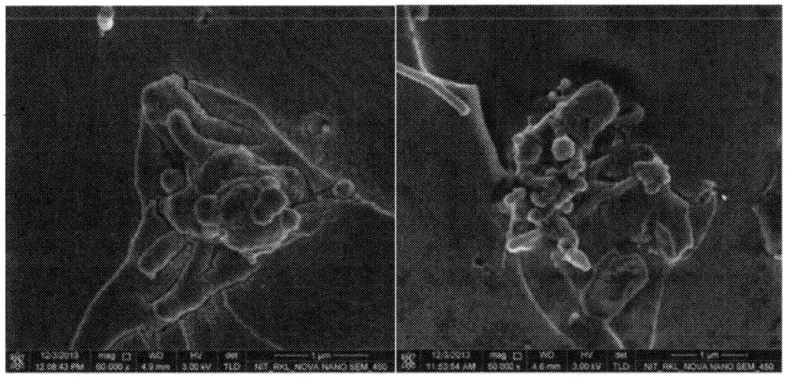

(a) (b)

Figure 7. Agglomeration of CNFs in (a) CNF2.5 (RTC) and (b) CNF2.5 (LTC).

According to Bal [18], low improvement (49%) in flexural modulus value was found in case of 1 wt.% RTC samples in comparison to the RTC 0.75% (i.e. 60%) than neat epoxy (1682 MPa). Comparatively, in the present investigation, remarkable improvement was found in case of RTC CNF1.5 (50%) as well as in RTC CNF2 (83%) specimens. So the curing procedure followed in this study facilitates better dispersion for relatively higher concentration. Further investigation is however required for complete removal of agglomeration of CNFs for more than 2 wt.%.

5. Conclusion

This study investigated the influence of adding higher amount of carbon nanofibers on the mechanical and thermal properties of RTC and LTC epoxy composites. Flexural modulus was improved to 50% and 25% for CNF1.5 and 83% and 79% for CNF2 respectively. Storage modulus and glass transition temperature demonstrated improved results for these specimens attributing to good interfacial adhesion. The micrographs of fracture surface revealed the interfacial attachment of nanofibers and matrix in composites that supports the results obtained from mechanical and viscoelastic studies. On the contrary, CNF2.5 failed to show further improvement due to the appearance of aggregated phases in the composites evidenced by micrographs.

Acknowledgment

The authors would like to thank Naval Research Board (DRDO), Govt. of India for the financial support to carry the research work.

References

[1] Maruyama B and Alam 2002 *SAMPE J.* **38(3)** 59–70
[2] Thostenson X E T, Li C and Chou T W 2005 *Compos. Sci. Technol.* **65**491
[3] Puglia D, Valentini L and Kenny J M *J. Appl. Polym. Sci.* **88** 452–458
[4] Li P, Zhao TJ, Zhou J H, Sui Z J, Dai Y C and Yuan W K 2005 *Carbon* **43** 2701–2710
[5] Ren Y, Li F, Cheng H M and Liao K 2003 *Carbon* **41** 2177–2179
[6] Zhang C S, Ni Q Q, Fu S Y, Kurashiki K 2007 *Compos. Sci. Technol.* **67** 2973–2980
[7] Miyagawa H, Rich M J and Drzal L 2006 *Thermochim Acta* **442** 67–73
[8] Lee H *et al* 2007 *Compos. Part* B **38(10)** 58–65
[9] Xu L R et al 2004 *J. Compos. Mater.* **38(18)** 1563–82
[10] Choi Y K, Sugimoto K, Song S, Gotoh Y, Ohkoshi Y and Endo M 2005 *Carbon* **43** 2199–208
[11] Gibson R F, Ayorinde E O and Wen YF 2007 *Compos. Sci. Technol.* **67** 1–28
[12] Pervin F, Zhou Y, Rangari V K and Jeelani S H 2005 *Mater. Sci. Eng.* A **405** 246–253
[13] Feng Q P, Yang J P, Fu S Y and Mai Y W 2010 *Carbon* **48** 2057–2062
[14] Gojny F H, Wichmann M H G, Kopke U, Fiedler B and Schulte K 2004 *Compos. Sci. Technol.* **64** 2363-2371
[15] Breton Y *et al* 2004 *Carbon* **42** 1027-1030
[16] Fidelus J D, Wiesel E, Gojny F H, Schulte K and Wagner H D 2005 *Compos. Part* A **336** 1555-1561
[17] Fiedler B, Gojny F H, Wichmann M H G, Nolte M C M and Schulte K 2006 *Compos. Sci. Technol.* **66** 3115–3125
[18] Bal S 2010 *Mater. Des.* **31** 2406–2413
[19] Zhou Y X, Pervin F, Jeelani S and Mallik PK 2008 *J. Mater. Process Technol.* **198** 445–53
[20] Salam M B A, Hosur M V, Zainuddin S and Jeelani S 2013 *Open journal of composite materials* **3** 1-9
[21] Gershon A L, Cole D P, Kota A K and Bruck H A 2010 *J. Mater. Sci.* **45** 6353–6364
[22] Zhu J, Wei S, Ryu J, Budhathoki M, Liangd G and Guo 2010 *J. Mater. Chem.* **DOI: 10.1039/c0jm00063a**
[23]Jagtap S B and Ratna D 2013 *eXPRESS Polym. Lett.* **7(4)** 329–339 **DOI: 10.3144/expresspolymlett.2013.30**
[24] Xiao K Q, Zhang L C and Zarudi I 2007 *Compos. Sci. Technol.* **67** 177-182
[25] Abdalla M, Dean D, Adibempe D, Nyairo E, Rob- inson P and Thompson G 2007 *Polymer* **48** 5662-5670
[26] Ma P C, Kim J K and Tang B Z 2007 *Compos. Sci. Technol.* **67** 2965-2972
[27] Tian S, Wang X 2008 *J. Mater. Sci.* **43** 4979-4987
[28] Cooper C A *et al* 2002 *Appl. Phys. Lett.* 2002 **81(20)** 3873–3875

IOP Conf. Series: Materials Science and Engineering **75** (2015) 012019 doi:10.1088/1757-899X/75/1/012019

Influence of semisolid heat treatment on aging behavior of Al-Cu alloy

K Sala and M. Mallik
Department of Metallurgical and Materials Engineering, National Institute of Technology, Durgapur – 713209, W.B., India

E-mail: manabmallik@gmail.com

Abstract: The present study focuses on the aging response of semisolid heat treated Al–5 wt.% Cu alloy. Semisolid heat treatment has been employed to modify cast dendritic structure. Globular structure with grain size of 90±25 μm was obtained after semisolid heat treatment. Subsequently T6 heat treatment has been carried out on both as cast and semisolid heat treated conditions. The semisolid heat treated alloy is found to peak-age in 4 h at 170 °C after solution treatment at 500 °C whereas, as cast sample was found to peak age in 18 h. Semisolid heat treatment leads to higher Cu dissolution in Al-α and these accelerated precipitation kinetics during ageing.

1. Introduction

A large number of semisolid metal (SSM) processing techniques have been developed for production of nondendritic structures in cast alloys [1]. The SSM technique such as rheocasting and thixocasting was first discovered by Spencer et al. [2] in early seventies. After that number of researchers carried out development work on different nonferrous metals and alloys by different semi solid processing. Isothermal heat treatment in the semisolid temperature is one of them. During isothermal heat treatment microstructures consisting of globular [3, 4, 5] grains of the primary phase are produced. In the as-cast condition, the microstructure of Al-4.5 wt.% Cu alloy contain dendrites of α–aluminium and eutectic which are located between dendrite arms and grain boundaries [5]. Generally, as-cast Al-4.5% Cu alloys are not utilized due to its poor mechanical properties. Mechanical properties of Al-4.5% Cu alloys can be improved either by microstructure modification through SSM techniques [6] or by ageing. Prasad et al. [6] have reported that microstructure modification by rheocasting has led to improvement of mechanical properties of the Al-Cu alloys. Therefore, in the present study combined effects of microstructure modification and ageing behavior of Al-5 wt.% Cu alloy on mechanical properties have been reported.

2. Experimental

Aluminum alloy plate with compositions (wt. %) of Cu-5, Si-0.008, Mg-0.086, Mn- 0.062, Ti-0.038, Fe-0.083 and Al-balance has been prepared by conventional casting route. Commercially pure aluminium (CP-Al) and copper were used as the basic raw materials. The temperatures of 590 °C, which approximately correspond to 5% liquid content was selected for the semisolid isothermal heat treatment. Small samples with dimensions of 10 mm × 10 mm ×10 mm were heated in a resistance furnace to the semisolid regime and quenched rapidly in water. As cast and semisolid heat treated (SSH) samples were metallographically prepared for microstructural characterization. Optical

Content from this work may be used under the terms of the Creative Commons Attribution 3.0 licence. Any further distribution of this work must maintain attribution to the author(s) and the title of the work, journal citation and DOI.
Published under licence by IOP Publishing Ltd

IOP Conf. Series: Materials Science and Engineering **75** (2015) 012019 doi:10.1088/1757-899X/75/1/012019

microscopy and grain size analysis were carried out using Leica image analyzer (Model: DM2500M, Leica Microsystems, Wetzlar, Germany) interfaced with Image analysis software. In addition, the microstructural investigation of the alloys was done using a scanning electron microscope (SEM) (Model: JSM5800, JEOL, Japan) coupled with energy dispersive X-ray (EDX) microanalyser (Model: ISIS300, Oxford Instruments Ltd., UK) for compositional analysis. As cast and SSH samples were solutionised at 500 °C for 1 h and quenched in water. Subsequently, samples were subjected to ageing treatment at 170 °C for 1–24 h in a furnace. To assess the strength, Vickers hardness (5 kg load) measurements were carried out on the as cast, SSH and aged samples. High resolution transmission electron microscope (HRTEM, JEM 2100, JEOL, Japan) operated at an acceleration voltage of 200 kV, primarily to examine the precipitate.

3. Results and Discussion

The microstructures of as cast and SSH Al-5 wt.% Cu alloy are shown in Fig. 1 (a&b). Figure 1(a) exhibiting a typical dendritic structure of Al-5 wt.% Cu alloy, with α –Al dendrites which were surrounded by eutectic composition. Figures 1 (b) shows the microstructure of SSH Al-5 wt.% Cu alloy at 590 °C for 60 min. It was found that the dendritic structures present in cast structure were transformed to globular microstructure.

Scanning electron micrograph of SSH sample has been shown in Figure 2 which indicates that a large number of CuAl2 particles with average size of less than 1 μm have been homogeneously distributed throughout the α-Al grains. EDX analysis show that average Cu content in Al-α in as cast and SSH conditions were 2.51 and 4.73 wt.% Cu, respectively. This result shows that more Cu is dissolved in Al-α in the SSH structure

Figure 3 shows the variations of hardness as a function of aging time of as cast and SSH Al-5 wt.% Cu alloy. From this figure, average peak hardness values were measured to be about 103±2 and 124±2 Hv, for as cast and SSH samples, respectively. Time required for achieving these peak harnesses was about 18h and 4h, respectively, exhibiting a shorter aging time with microstructure modification. This is due to higher diffusion rate and hence faster precipitation. Isothermal heat treatment in semisolid region leads to higher Cu content in solution in SSH structure, delaying supersaturation and consequent precipitation of CuAl2 in the ageing process. These results lead to achieve maximum ageing response in shorter time in the SSH sample. The microstructure of needle like precipitates formed in the peak aged condition of SSH sample has been shown in Figure 4. During ageing the precipitation sequence of Al–Cu alloys is: GP zones → θ″→ θ′→ θ. The precipitates (Figure 4) formed after ageing of SSH sample are considered to be intermediate phases of θ″or θ′, which have good coherency with the matrix and contribute to the increased strength [7].

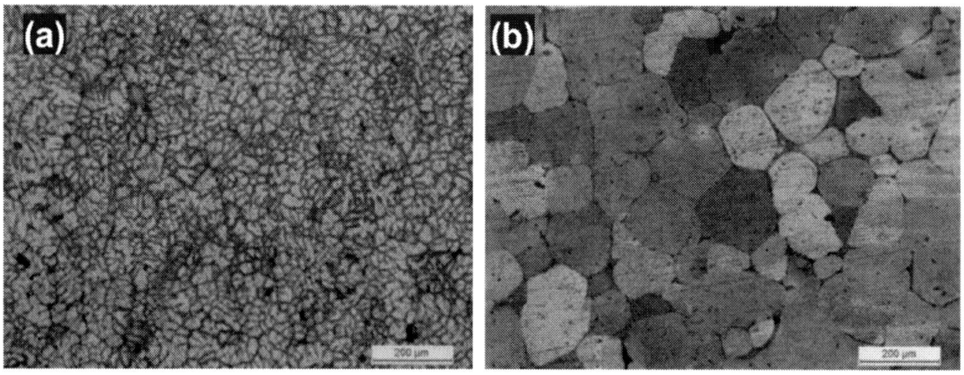

Figure 1. Microstructures of the Al 5 wt. % Cu alloy after different conditions: (a) As cast and (b) semisolid heat treatment at 590 °C for 60 min.

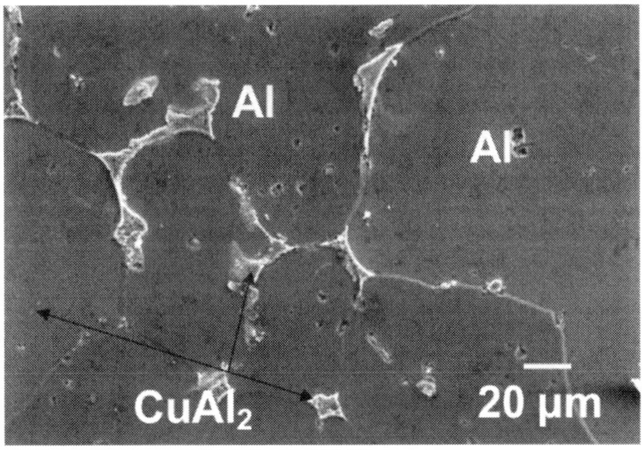

Figure 2. SEM micrograph of Al-5 wt. % Cu alloy after SSH at 590 °C for 60 min.

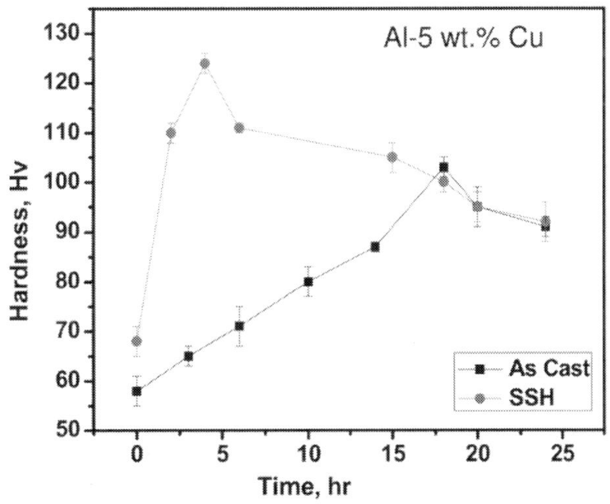

Figure 3. Variation of hardness values with the aging time for as cast SSH samples.

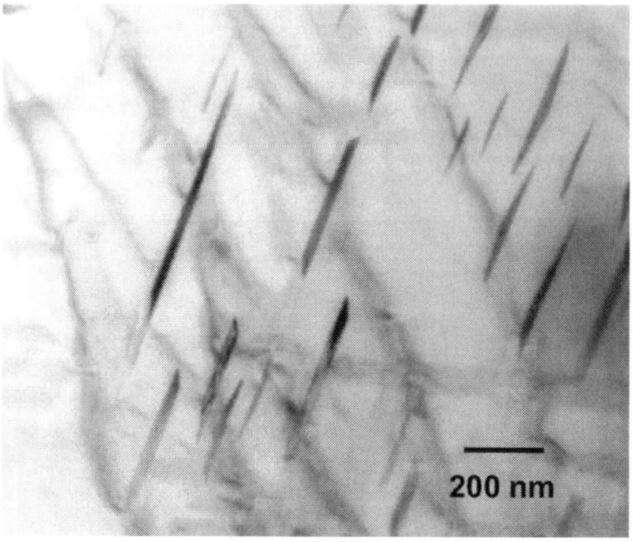

Figure 4. TEM micrograph showing precipitate present on Al-α matrix in the SSH structure after ageing at 170 °C for 4h.

The above results clearly indicate that isothermal heat treatment in semisolid region leads to faster precipitation kinetics of Al-5 wt. % Cu alloy.

4. Conclusions

The present studies reveal the effect of semisolid heat treatment on ageing behavior of Al–Cu alloy. The experimental results support the following conclusions:

1. Isothermal heat treatment at semisolid region produce globular α–Al grains. EDX analysis suggests supersaturation of the α-phase with respect to Cu. Very fine CuAl2 precipitates are homogeneously distributed throughout the Al grains.
2. Semisolid heat treatment accelerated precipitation kinetics (4h) in comparison to cast microstructure (18h).

Acknowledgement

The authors wish to acknowledge Dr. A. K. Chakraborti for his valuable suggestions.

References

[1] Mohammed M N, Omar M Z, Salleh M S, Ahawau K S and Kapranos P 2013 *The Scientific World J.* ID752175 1
[2] Spencer D B, Mehrabian R and Flemings M C 1972 *Met. Trans.* 3 1925
[3] Chen G, Jiang J, Du Z, Cao Q, Li H and Zhang X 2013 *J. Mat. Sci. Tech.* 29 979
[4] Bhattacharya A, Chakrabarti A K and Mondal D K 2011 *Indian Foundry Jr.* 57 23
[5] Chakraborti A K and Mandal D K 2013 *IIM Metal News* 16 31
[6] Zoqui E J J 2003 *J. Mater. Process. Technol.*145-144 195
[7] Prasad P R, Ray S, Gaindhar J L and Kapoor M L 1988 *J. Mat. Sc.* 23 823
[8] Shanmugasundaram T, Murty B S and Sarma V S 2006 *Scripta Materialia* 54 2013

IOP Conf. Series: Materials Science and Engineering **75** (2015) 012020 doi:10.1088/1757-899X/75/1/012020

Mechanical and structure studies of $Zr_{50}Cu_{50}$ glass matrix composites during nano-indentation-a molecular dynamics study

Ashwani Kumar[1] and Natraj Yedla
Metallurgical and Materials Engineering Department, NIT-769008, Odisha, India

E-mail: [1]ashwanik2010@gmail.com

Abstract. In this paper we report molecular dynamics simulations of nano-indentation on $Zr_{50}Cu_{50}$ metallic glass matrix composite (14% crystalline volume fraction) at various strain rates. The objective of this paper is to investigate the effect of strain rate on the deformation behaviour and understand the deformation mechanism during deformation. Structural analysis during deformation is done by centro-symmetry parameter (CSP) studies. The load-displacement plots are drawn for the loading portion of indentation to analyze the deformation behaviour. It is found that strain rate has significant effect on the nature of the load-displacement plot. With increasing strain rate serrations decreased and flat load-displacement regime is observed with progress of indentation (~10 Å) at strain rate of 1×10^{11} s^{-1}. This could be due to atoms getting less time to get rearranged themselves so as to bear further load. Also, the structure studies by CSP indicated that, at low strain rates (2×10^{10} s^{-1} and 5×10^{10} s^{-1}) there is significant plastic deformation of the crystallite as compared to that at higher strain rate value of 1×10^{11} s^{-1} at a particular indentation depth. This indicates that there is load transfer from the glassy matrix to the crystallite much earlier at low strain rates. However, at indentation depths of 20 Å at all the strain rates amorphization of the crystallite is observed.

1. Introduction

Bulk metallic glasses (BMG) are used for structural applications because of distinctive mechanical properties compared to crystalline materials like higher strength (1.8 GPa), higher elastic strain (~2 %) good corrosion resistance, higher hardness [1-3]. But they exhibit limited plasticity in compression and negligible in tension (<0.5 %) [1, 4 and 5]. So, they become prone to catastrophic failure which bounds their use in limited functional area. There are several types of BMGs such as Cu-based, Zr-based, Ti based, Fe-based [6]. The most widely studied are Cu-based metallic glass as they have good glass forming ability, high thermal stability, high corrosion resistance and high strength [6]. The deformation mechanism in BMGs is different from that of crystalline materials due to the absence of dislocations. The deformation studies revealed that at low strain rate, BMGs fail in discrete manner by shear banding which results in brittle failure while this deformation becomes continuous at high strain rates and not in discrete manner. This is due to the formation of a single shear band at low strain rate which is not able to bear sudden high strain values so multiple shear bands are formed at high strain rates [7]. To improve the ductility, BMG composites were synthesized. It was found that there was significant increase in the plastic strain due to the obstruction of the rapid propagation of shear bands in the presence of crystallites [5, 8]. Several deformation studies were done on BMG reinforced with

Content from this work may be used under the terms of the Creative Commons Attribution 3.0 licence. Any further distribution of this work must maintain attribution to the author(s) and the title of the work, journal citation and DOI.
Published under licence by IOP Publishing Ltd

crystalline particulates [2, 9]. It was found that, this reinforcement changed their mechanical properties like fracture toughness and ductility by formation of multiple shear bands. Addition of Nb to $Zr_{57}Nb_5Al_{10}Cu_{15.4}Ni_{12.6}$ BMG to form composite increased its toughness [2]. Wang et al. [10] did experiments to show that the addition of Nb to Zr-Ti-Cu-Ni-Be metallic glass system resulted in increase of ductility due to formation of crystalline phases. The deformation mechanism in glasses is different to that of crystalline materials due to the disordered structure. There are several proposed deformation mechanism theories, such as STZ (shear transformation zone), shear band theory and free volume theory. As proposed by Argon [11] STZ is the fundamental unit of plastic deformation of metallic glasses similar to dislocation in crystalline materials. Further, STZ's are local cluster of atoms of few nanometres that rearrange by the application of shear stress. This involves displacement of atoms, their inelastic reconfiguration and redistribution of free volume in these atomic clusters. A different theory for deformation behaviour in metallic glasses proposed by Spaepen [12] was based on free volume creation by stress and it's destruction by diffusion.

In order to reveal the deformation behaviour of metallic glasses at the atomic level, several atomistic simulations have been done by shear, indentation, tensile and compression [13-16]. Most of the simulation studies are done by tensile or compression as it is easier to model. However, nano-indentation studies are seldom reported. The nano-indentation techniques employed were useful to detect sub-nanometre disturbances connected with "plasticity" phenomenon such as dislocation nucleation in crystalline materials and shear band localization and propagation in metallic glasses [17]. In the MD deformation indentation studies of Qiu et al., [18] on $Cu_{50}Zr_{50}$ metallic glass substrate at different loading speed and temperature it was shown that, with increase of loading rate, the force required for indentation also increases but decreases with increase in temperature. Further, Liu et al. [19] did nano-indentation simulation study of Zr-based metallic glass sample ($Zr_{55}Cu_{30}Ni_5Al_{10}$ and $Zr_{69.5}Cu_{12}Ni_{11}Al_{7.5}$). In these studies serrated plastic flow was observed in the load-displacement plots and found to be dependent on loading rate. Pang et al. [20] did experiment on binary MG thin film $Cu_{49.3}Zr_{50.7}$ grown on Si substrate. It was shown that plasticity was delayed at high loading rate. Wu et al. [21] did nano-indentation experimental study of $Cu_{52}Zr_{37}Ti_8In_3$ BMG and found that Young's modulus and hardness both decrease with increase in depth of indentation. However, seldom nano-indentation studies have been reported on BMG composites.

The mechanical properties of amorphous metals or metallic glasses are affected by addition of nano-crystallites and depend on crystal particle size, crystal volume fraction, strain rate of indentation, temperature of working and the amorphous-crystal phase interaction strength [22]. In the present investigation, the deformation behaviour of $Zr_{50}Cu_{50}$ based glass matrix composites (GMC) and structural changes during deformation have been studied using nano-indentation technique at different strain rates (2×10^{10} s^{-1}, 5×10^{10} s^{-1} and 1×10^{11} s^{-1}) at room temperature. Since modelling of binary amorphous alloy is easier than alloys containing more than two alloys and also it reveals the properties of amorphous alloys so $Zr_{50}Cu_{50}$ system was selected.

2. Molecular dynamics (MD) simulation procedure

$Zr_{50}Cu_{50}$ model metallic glass matrix composite containing 14 % crystallite by volume was prepared by simulated melting and quenching. The simulation box size is of 100 Å × 100 Å × 100 Å dimension comprising of 87,808 atoms. Here, crystallites are arranged in FCC crystal structure. The equilibration of model was done at minimum energy configuration at a temperature of 300 K. Sample was relaxed under NPT (constant pressure and temperature) ensemble. The initial sample was heated from 300 K to 2300 K for 20 ps, allowing solid to melt and then held at 2300K for 100 ps. After that the sample was quenched from 2300K to 300K for 200 ps corresponding to cooling rate of 10^{13} K/s. The pressure was maintained at 0 bar during whole process. Nano-indentation was carried out on the sample with an indenter having a spherical tip (diameter 40Å) up to a depth of 20Å along Y-direction. The MD simulation of nano-indentation was employed under NVT (constant temperature and constant volume) ensemble. The bottom part of sample was kept fixed to prevent the substrate from moving in the direction of indentation during the whole simulation. The size of substrate was 100 Å × 20 Å × 100 Å.

Periodic boundary condition was applied in non-loading directions i.e., along X-axis and Z-axis. MD simulation of nano-indentation was carried out on glass matrix composite (GMC) to study its effect on the deformation behaviour. The strain rate used was 2×10^{10} s^{-1}, 5×10^{10} s^{-1}, and 1×10^{11} s^{-1}. The simulation of nano-indentation comprised of indentation and equilibrium stages. In equilibrium stage the indenter was kept at a distance of 3 Å from sample in order to avoid long range attractive force between them. In the second stage the indenter moved downward to 20 Å inside the sample at above mentioned strain rates. The motion of each atom was governed by Newton's second law of motion. Newton's law of motion was integrated by velocity-Verlet algorithm with a time-step of 0.002 ps. Load-displacement curves were drawn for sample at various strain rates to study the nature of the curve. The EAM (embedded atom method) potential was used to describe the interaction between atoms. The total energy of an atom E_i is given by Mendelev et al. [23] as

$$E_i = F_\alpha\left(\sum_{j\neq i}\rho\beta(r_{ij})\right) + \frac{1}{2}\sum_{j\neq i}\phi_{\alpha\beta}(r_{ij})$$

Where, embedding energy F_α is a function of atom's electron density, is a pair potential interaction, and are the element types of atoms i and j. It is a widely used potential for studying the properties of materials [16, 18, 24-26]. All the simulations are carried out on LAMMPS platform which is a widely used open source code [27].

3. Results and Discussions

Figure 1 shows the atomic positions snap shot of $Zr_{50}Cu_{50}$ metallic glass composite in full (figure 1a) and sliced view (figure 1b). Spherical crystal (blue coloured) of 32 Å radius is embedded at the centre occupying 14 % volume.

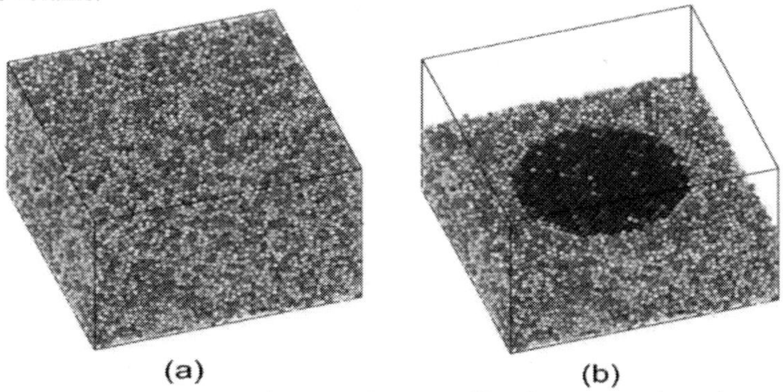

(a) (b)

Figure 1. Atomic position snap shots of the $Zr_{50}Cu_{50}$ metallic glass composite: a) complete simulation box (red coloured atoms: Zr; green coloured atoms: Cu; b) sliced box revealing the central crystallite position (the atoms are coloured based on the centro-symmetry parameter values (blue colour: perfect crystal)).

Figure 2 shows the load applied on the indenter and the depth of indentation in the $Zr_{50}Cu_{50}$ metallic glass composite sample at different strain rates of 2×10^{10} s^{-1}, 5×10^{10} s^{-1}, 1×10^{11} s^{-1}. All the curves show linear elastic behaviour and non-linear plastic behaviour. Elastic behaviour is shown up to first load drop in the curve. The vertical force on the indenter increases elastically from zero according to Hertz theory before the beginning of plastic deformation [28]. The expression for the load is given by the following equation.

$$P = \frac{\pi a^2 E^*}{4R} L_y$$

Where, P is load, R is the reduced elastic constant and a is half of the projected contacted length. L_y is the length of the simulation box in the y-direction. E^* is the reduced Young's modulus in plain strain conditions. The curve changes its slope after the first load drop indicating plastic deformation occurring with further loading. The first load drop or yield point appears at lower indenter displacement with increase in strain rate indicating plastic deformation at early stage of deformation. These plastic deformations are indicated by "pop ins" or serrations [17] in the curve which are more pronounced (of larger size) at lower strain rates of 2×10^{10} s^{-1} and 5×10^{10} s^{-1} and decreases with increase of strain rate to 1×10^{11} s^{-1}. This is so because at low strain rates atoms get enough time to rearrange and bear further load, so there are more fluctuations in the load value leading to more pronounced serrated curve. Number of rearranged atoms is also more which is an indication of high degree of plastic deformation. Whereas, at high strain rate atoms get very less time to rearrange and so load drops are less and so number of serrations are less. Number of rearranged atoms is also less causing less degree of plastic deformation. This phenomenon is coherent with the simulation result presented by Jiang et al. [13]. The load drops are irregularly spaced in all curves which may be due

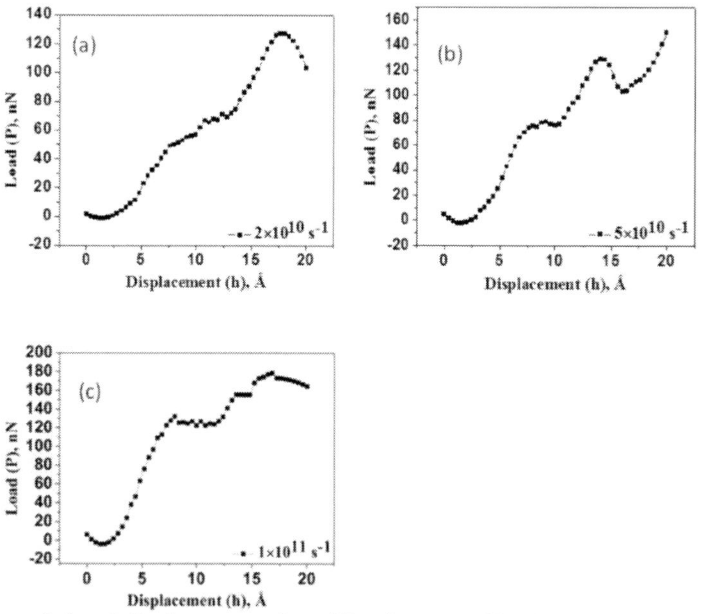

Figure 2. Load-displacement plot of $Zr_{50}Cu_{50}$ metallic glass composite at different strain rates: a) 2×10^{10} s^{-1}; b) 5×10^{10} s^{-1}; c) 1×10^{11} s^{-1}.

to short range order structure of GMC. We can observe that the amplitude of serrations decrease with increasing strain rate as shown in figure 2(a) to (c). This is so because atoms do not get enough time to rearrange and bear further load at high strain rates but as amorphization of atoms is delayed with increase in strain rate so the maximum load value increases gradually with rise of strain rate value as shown in Table 1. So, with increase in indenter displacement even after the yield point, the load on the indenter also increases.

Table 1. Mechanical properties of $Zr_{50}Cu_{50}$ metallic glass composite containing at different strain rates

Strain rate (s^{-1})	Yield point (nN)	Maximum load value (nN)
2×10^{10}	49	127
5×10^{10}	73	149
1×10^{11}	132	178

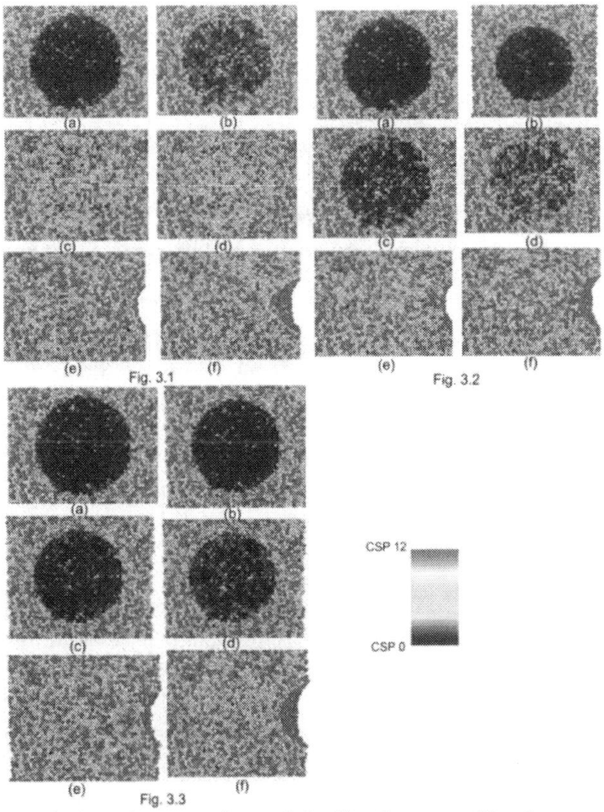

Figure 3. Atomic snap shots of the $Zr_{50}Cu_{50}$ metallic glass composite during nano-indentation at different depth of indentation a) initial; b) 0.16 Å; c) 0.36 Å; d) 0.52 Å; e) 12 Å; f) 20 Å and at different strain rates:$2 \times 10^{10} s^{-1}$ (Fig. 3.1); $5 \times 10^{10} s^{-1}$ (Fig. 3.2); $1 \times 10^{11} s^{-1}$ (Fig. 3.3).

In order to reveal the structural details during deformation, CSP studies are done. figure 3 shows the atomic snap shots of $Zr_{50}Cu_{50}$ metallic glass composite at different depths of indentation during nano-indentation along y-direction and at different strain rates i.e., 2×10^{10} s^{-1}; 5×10^{10} s^{-1} and 1×10^{11} s^{-1}. The centro-symmetry parameter (CSP) is used for analysis of the structural changes during indentation. The colour coding is done according to the intensity of defects. CSP value of zero indicates no defects while CSP value of twelve indicates surface and stacking fault defects. The red and green atoms surrounding the blue circular region (crystallite) is $Zr_{50}Cu_{50}$ metallic glass. The crystallite initially is free of defects (figure 3.1a, figure 3.2a, and figure 3.3a), but with progress of indentation plastic deformation of the crystallite occurs at all strain rates along with that of glassy matrix. It can also be seen that the crystallite becomes amorphous at the indenter displacement of 20 Å at all strain rates. It is well known that the deformation mechanism in glass is different from that of crystals. At indenter displacement of 0.52 Å (figure 3.1d) in the strain rate of 2×10^{10} s^{-1} (figure 3.1) more defects (colour index) are introduced in the crystallite as compared with the other two strain rates (figure 3.2d and figure 3.3d) resulting in flat load-displacement plot in that regime. This clearly indicates that at higher strain rates, deformation of the glass occurs rather than crystallite. In the sense there is no load transfer from the glassy phase to the crystallite phase at higher strain rates.

4. Conclusions

1. Strain rate has significant effect on the load-displacement plot. With increasing strain rate yielding load increases.
2. With increasing strain rate serrations decrease, indicating that atoms get sufficient time for atomic rearrangements at lower strain rates and contribution for plastic deformation.
3. Structural studies by centro-symmetry parameter showed that at strain rate of 2×10^{10} s^{-1}, plastic deformation of the crystallite occurs even at smaller indenter displacements as compared to the 1×10^{11} s^{-1} strain rate.

References

[1] Sun GY, Chen G and Chen GL 2007 Comparison of microstructures and properties of Zr-based bulk metallic glass composites with dendritic and spherical bcc phase precipitates *Intermetallics* **15** 632-34

[2] Conner RD, Choi-Yim H, and Johnson WL 1999 Mechanical properties of $Zr_{57}Nb_5Al_{10}Cu_{15.4}Ni_{12.6}$ metallic glass matrix particulate composite *J. of Mat. Res.* **14** 3292.

[3] Chen KW, Jian SR, Wei PJ, Jang JSC and Lin JF 2010 A study of the relationship between semi-circular shear bands and pop-ins induced by indentation in bulk metallic glasses *Intermetallics* **18** 1572-78

[4] Zheng GP and Shen Y 2011 Simulation of shear banding and crack propagation in bulk metallic glass matrix composites *J. of Alloy and Comp.* **509S** S136-S140

[5] Hofmann DC, Suh JY, Wiest A, Duan G, Lind ML, Demetriou MD and Johnson WL 2008 Designing metallic glass matrix composites with high toughness and tensile ductility, *Nature* **451** 1085

[6] Inoue A and Takeuchi A 2011 Recent development and application products of bulk glassy alloys *Acta Materialia* **59** 2243-67.

[7] Schuh CA and Nieh TG 2003 A nanoindentation study of serrated flow in bulk metallic glasses *Acta Materialia* **51** 87-99

[8] Leonhard A, Xing LQ, Heilmaier M, Gebert A, Eckert J and Schultz L 1998 Effect of crystalline precipitations on the mechanical behaviour of bulk glass forming zr- based alloys *Nanostructured Materials* **10** 805-17

[9] Xue YF, Cai HN, Wang L, Wang FC and Zhang HF 2007 Dynamic compressive deformation and failure behavior of Zr-based metallic glass reinforced porous tungsten composite *Materials Science and Engineering A* **445-46** 275-80

[10] Wang WH, Bian Z, Wen P, Zhang Y,Pan MX and Zhao DQ 2002 Role of addition in formation and properties of Zr-based bulk metallic glasses *Intermetallics* **10** 1249–57

[11] Argon AS 1979 Plastic deformation in metallic glasses *Acta mater.* **27** 47-58

[12] Spaepen F 1977 A microscopic mechanism for steady state inhomogeneous flow in metallic glasses *Acta Met.* **25** 407-415
[13] Jiang SY, Jiang MQ, Dai LH and Yao YG 2008 Atomistic Origin of Rate-Dependent Serrated Plastic Flow in Metallic Glasses *Nanosclae res. Lett.* 3 524-29
[14] Delogu F 2008 Molecular dynamics of shear transformation zones in metallic glasses *Intermetallics* **16** 658-61
[15] Ogata S, Shimizu F, Wakeda M, Li J and Shibutani Y 2006 Atomistic simulation of shear localization in Cu–Zr bulk metallic glass *Intermetallics* **14** 1033-37
[16] Park KW, Fleury E, Seok HK and Kim YC 2011 Deformation behaviors under tension and compression: Atomic simulation of $Cu_{65}Zr_{35}$ metallic glass, *Intermetallics* **19** 1168-73
[17] Packard CE, Franke O, Homer ER and Schuh CA 2010 Nanoscale strength distribution in amorphous versus crystalline metal *J. Mater. Res.* **25** 2251-63
[18] Qiu C, Zhu P, Fang F, Yuan D and Shen X 2014 Study of nanoindentation behavior of amorphous alloy using molecular dynamics *Applied Surface Science* **305** 101-10
[19] Liu L and Chan KC 2005 Plastic deformation of Zr-based bulk metallic glasses under nanoindentation *Mat. Lett.* **59** 3090-94
[20] Pang JJ, Tan MJ, Liew KM and Shearwood C 2012 Nanoindentation study of size effect and loading rate effect on mechanical properties of a thin film metallic glass $Cu_{49.3}Zr_{50.7}$ *Physica B condensed matter* **407** 340-46
[21] Wu J, Pan Y and Pi J 2014 Nanoindentation study of $Cu_{52}Zr_{37}Ti_8In_3$ bulk metallic glass *App. Phy. A* **115** 305-12
[22] Matsumoto R and Nakagaki M 2006 Mechanical properties of amorphous metals with dispersed nanocrystalline particles molecular dynamics study on crystal volume fraction and size effect *JSME Int. J.* **49** 513
[23] Mendelev MI, Kramer MJ, Ott RT, Sordelet DJ, Yagodin D and Popel P 2009 Development of suitable interatomic potentials for simulation of liquid and amorphous Cu–Zr alloys *Philosophical Magazine* **89** 967-87
[24] Tsuru T, Shibutani Y and Kaji Y 2010 Nanoscale contact plasticity of crystalline metal: Experiment and analytical investigation via atomistic and discrete dislocation models *Acta Materialia* **58** 3096-3102
[25] Xie L,Brault P, Thomann AL and Bedra L 2013 Molecular dynamic simulation of binary Zr_xCu_{100-x} metallic glass thin film growth *Applied Surface Science* **274** 164-70
[26] Albe K, Ritter Y and Sopu D 2013 Enhancing the plasticity of metallic glasses: Shear band formation-nanocomposites and nanoglasses investigated by molecular dynamics simulations *Mechanics of Material* **67** 94-103
[27] Plimpton S 1995 Fast Parallel Algorithms for Short-Range Molecular Dynamics *J. Comp. Phys.* **117** 1-19
[28] Shi Y and Falk ML 2005 Stress-induced structural transformation and shear banding during simulated nanoindentation of a metallic glass *Acta mat.* **55** 4317-24

Metallurgical Analysis of Cracks Formed on Coal Fired Boiler Tube

Rajat Kishor[1], Tushal Kyada, Rajesh K Goyal and T S Kathayat
Research and Development, Welspun Corp Limited, Anjar-370110, India

E-mail: rajat_kishor@welspun.com; rajatkishor@hotmail.com

Abstract. Metallurgical failure analysis was carried out for cracks observed on the outer surface of a boiler tube made of ASME SA 210 GR A1 grade steel. The cracks on the surface of the tube were observed after 6 months from the installation in service. A careful visual inspection, chemical analysis, hardness measurement, detailed microstructural analysis using optical and scanning electron microscopy coupled with energy dispersive X-ray spectroscopy were carried out to ascertain the cause for failure. Visual inspection of the failed tube revealed the presence of oxide scales and ash deposits on the surface of the tube exposed to fire. Many cracks extending longitudinally were observed on the surface of the tube. Bulging of the tube was also observed. The results of chemical analysis, hardness values and optical micrographs did not exhibit any abnormality at the region of failure. However, detailed SEM with EDS analysis confirmed the presence of various oxide scales. These scales initiated corrosion at both the inner and outer surfaces of the tube. In addition, excessive hoop stress also developed at the region of failure. It is concluded that the failure of the boiler tube took place owing to the combined effect of the corrosion caused by the oxide scales as well as the excessive hoop stress.

Keywords: Boiler tube, Crack, Ash corrosion, Hoop stress, EDS, SEM

1. Introduction

Failure of boiler tubes is common in industries where fossil fuel is used. There are several factors that cause this type of failure. The use of high sodium, sulphur and ash containing fuel, high working temperature and pressure exceeding the design limit and lack of periodic maintenance are the major factors that affect the performance of materials employed for boilers [1, 2]. The hostile environment is the source of corrosion and stress which leads to the failure of boiler tube. The failure of boiler tubes appears in the form of bending, bulging, cracking, wearing or rupture and these hampers the performance of boilers or makes it obsolete in applications [3].

In most of the cases ash corrosion is found to be the prime factor for failure of boiler tubes. This ash corrosion occurs basically due to use of low quality feed coal as fuel and poor cleaning procedures. In these types of failure ash deposition takes place on the fire side of the boiler which provides excessive heat input and leads towards failure while in other cases, accumulation of corrosion products takes place which induces localized corrosion on waterside of the boiler [4]. In the present investigation a systematic case study of the boiler tube failed in Welspun India plant located at Anjar is presented.

[1] To whom any correspondence should be addressed.

Content from this work may be used under the terms of the Creative Commons Attribution 3.0 licence. Any further distribution of this work must maintain attribution to the author(s) and the title of the work, journal citation and DOI.
Published under licence by IOP Publishing Ltd

2. Experimental procedure

A boiler tube of ASME SA 210 GR A1 grade was taken for the present investigation. The detailed parameters used during its application are summarized in Table 1. In order to find the route cause several analyses were carried out. This includes visual inspection, chemical composition analysis, hardness testing, microstructural characterization using optical microscopy and scanning electron microscopy (SEM) coupled with energy dispersive X-ray spectroscopy (EDS).

Table 1. Operational parameters of the failed boiler tube

Parameters	Range
Dimension of Tube	50.8 mm OD & 6.35 mm WT
External temperature	700-800 °C
Internal temperature	250-295 °C
Pressure	50-80 kg/cm^2
TDS of water used in boiler	400-700 ppm
pH of water used in boiler	8.5
Quality of coal on GCV scale	4000
Duration of failure	6 months

Visual analysis was the first step and for it photographs of the cracked region of the boiler were taken. Thickness of the tube at cracked region and away from the failed region was measured using digital micrometer. The chemical analysis of the failed tube was performed using optical emission spectrometer. Hardness measurement at the failed region and away from the failed position was carried out using Rockwell hardness. A small specimen was taken from the failed region and it was then prepared for microstructural observation following conventional polishing and etching. The microstructures were taken using optical microscope at several positions. The morphology and chemical analysis of the deposits on the tube were checked by SEM and EDS.

3. Results and discussion

3.1 *Visual inspection*

Careful visual inspection of the failed tube revealed the presence of oxide scales and ash deposits on the surface of the tube exposed to fire. Many cracks extending longitudinally were observed on the surface of the tube as shown in Figure 1(a) and 1(b). In addition, many small blown up portions were observed on the outer surface of the tube as shown in Figure 1(c). Bulging of the tube was also observed as shown in Figure 1d. Measurement revealed that the diameter and the wall thickness of the tube at the region of failure changed from 50.8 mm to 57.8 mm and 6.35 mm to 3.5 mm, respectively. Therefore, the diameter of the tube increased and the wall thickness of the tube deceased at the region of failure. Crack section was cut in transverse direction for further analysis as shown in Figure 1(e). The presence of corrosion products at the inner surface of the tube was observed as shown in Figure 1(f).

IOP Conf. Series: Materials Science and Engineering **75** (2015) 012021 doi:10.1088/1757-899X/75/1/012021

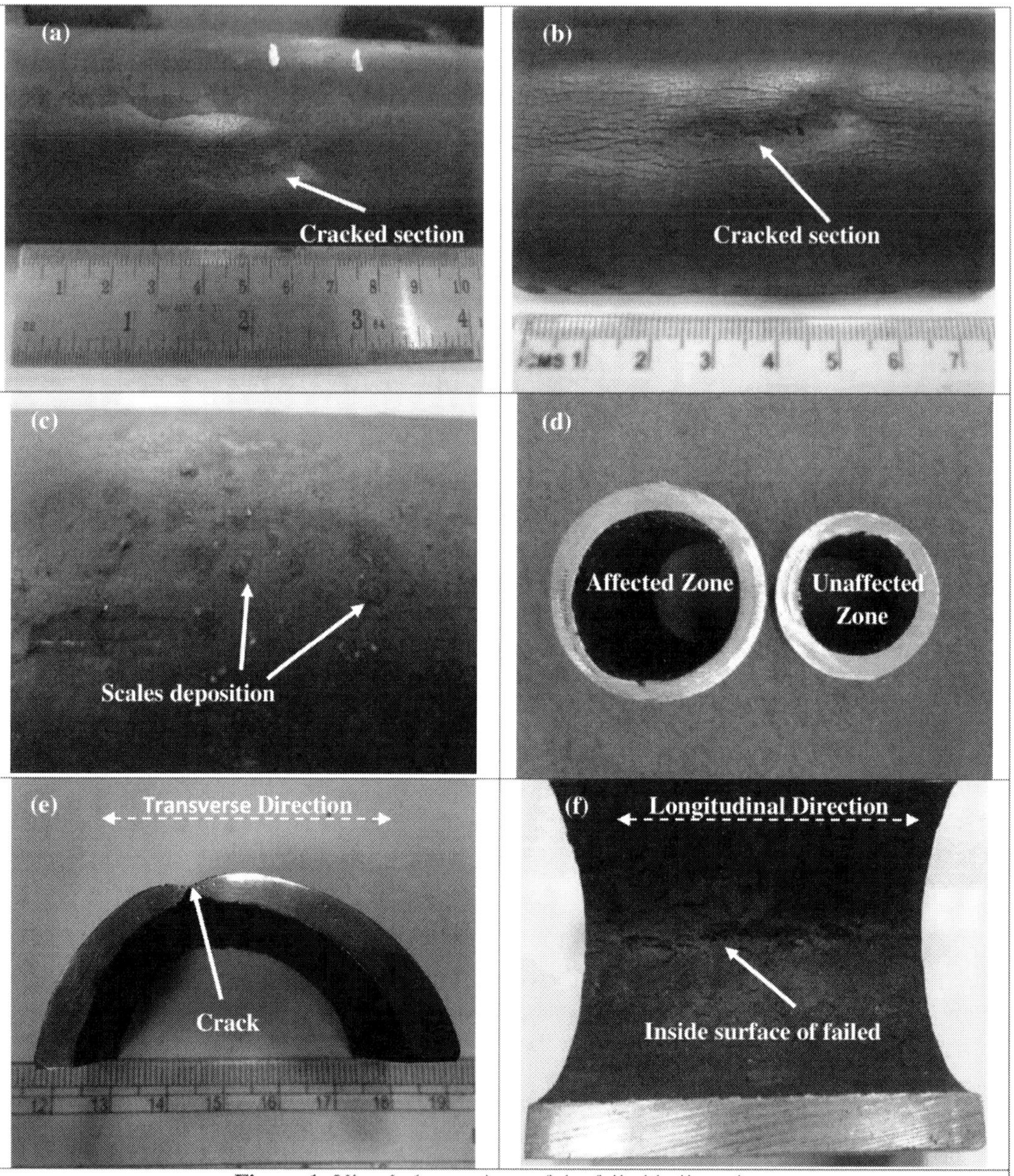

Figure 1: Visual observations of the failed boiler tube

3.2 Chemical analysis and hardness test

Chemical analysis performed on the affected and not affected regions of the tube revealed that there was no variation in composition at both the regions as shown in Table 2. The chemical analysis was further supported by hardness tests performed on these regions. Hardness measurement was carried out in polished transverse cross sections near crack area and unaffected area as shown in Figure 1(e). The hardness value of the not affected region of the pipe was 73 HRB whereas it was 70 HRB at the failed region. Therefore, it is inferred that the tube conforms the material requirement in terms of its chemical composition and hardness.

Table 2: Chemical composition of standard tube and failed tube

Sample ID	Elements (in wt. %)										
	C	Mn	P	S	Si	Cr	Mo	Ni	V	Al	Cu
Specified Composition of Standard Tube	0.15	0.68	0.015	0.012	0.24	0.023	0.004	0.010	0.007	0.033	0.008
Failed Tube	0.154	0.65	0.0084	0.0045	0.19	0.014	<0.0005	0.0053	0.0016	0.034	0.012

3.3 Microstructural analysis

3.3.1. Analysis using optic al microscopy.

In order to examine the microstructural changes in the failed tube with respect to the standard tube, optical microscopic examinations were carried out at the cross section of cracked lip portion and at a distance away from the crack region (Figure 2) on fireside area. It was observed that there are clean polygonal ferrite-pearlite grains on fireside area and was similar to those grains as they were originally specified for the material. Microstructure taken at cross sectional position of cracked lip also indicates a typical pearlite/ferrite grains. Hence, it can be stated that that the cracked area might have not undergone any microstructural changes due to localized temperature variation during its service.

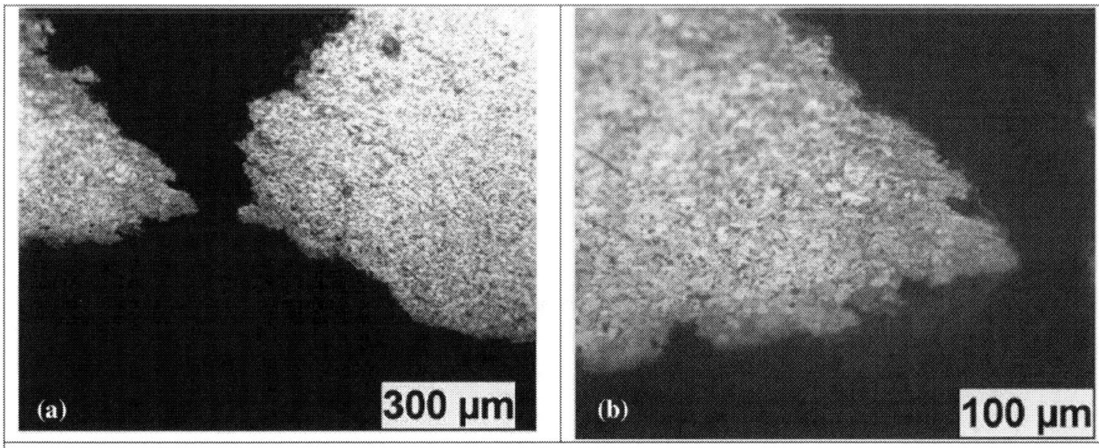

Microstructures at crcak tip (a)100X (b) 200X

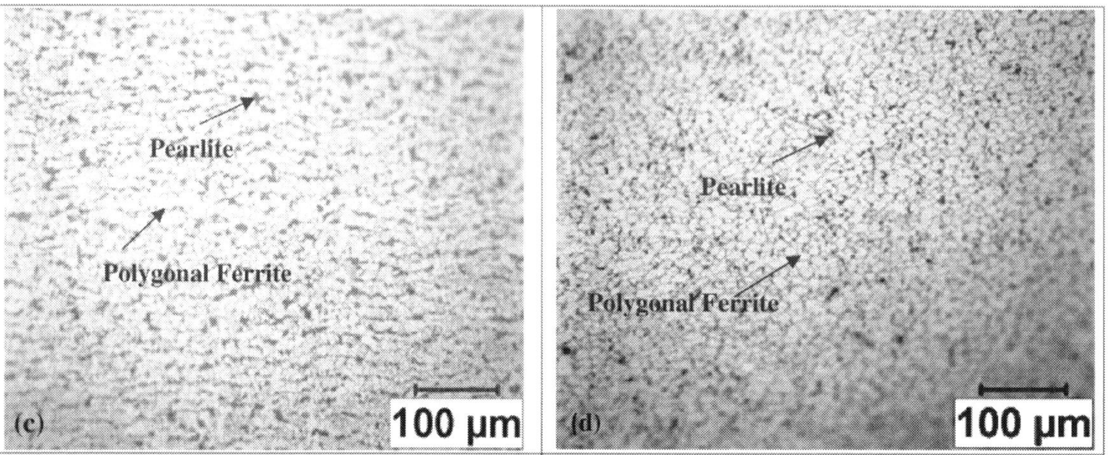

Microstructure at away from the crack region @ 200X

Microstructure near the crack region @200X

Figure 2: Microstructural details of failed tube at several locations.

3.3.2 Analysis using SEM with EDS

To analyze the type of oxide scales or deposits on the fireside as well as waterside of the failed tube, scanning electron microscopy along with EDS (Figure 3) was performed. With the help of SEM images and EDS analysis detailed mechanism of failure was determined. Through SEM images thick black oxide scales can be seen on the waterside surface and some deposits can also be found in fireside surface of the failed tube. It was confirmed by EDS analysis that thick scale on waterside surface is composed of magnetite (Fe_3O_4) while the deposits on fireside surface contain hematite and wastage (Fe_2O_3+FeO), SiO_2.

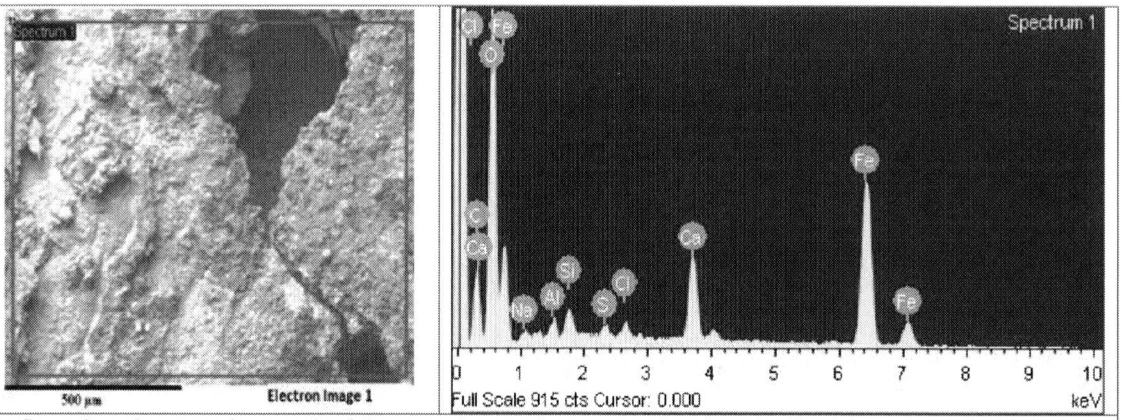

Scanning electron micrograph of the corrosion product accumulated on the fireside surface of failed tube

Scanning electron micrograph of the corrosion product accumulated on the waterside surface of failed tube

Figure 3: Scanning electron micrograph of the corrosion product accumulated on the fireside and waterside surface of failed tube coupled with EDS

Visual analysis reveals the presence of ash deposition on the fireside surface of the failed tube. SEM-EDS analysis of the region closer to the crack lip confirms the presence of iron, oxygen, and carbon as major constituents while aluminum, silicon, sodium, sulphur and chlorine were also present in significant quantity. The existence of these elements resulted from the coal used as fuel and therefore, fly-ash and slag, which are main ingredients of combustion products, are also composed of them. Characteristics such as melting point and viscosity of slag and ash are dependent on the composition of these elements. Fly-ash particles collide with tubes by their inertial force and can adhere or deposit on tube. The presence of sodium, sulphur, iron and oxygen also suggests the possible formation of complex sulphates [5]. Deposition of Na_2SO_4 in the molten or nearly molten state would contribute to catching colliding particles. Similarly, sulfates of sodium and fly-ash deposit on to the fireside will develop the second layer further. As the second layer increases in thickness and its surface temperature rises beyond the saturation temperature of Na_2SO_4, these sulfates would stop condensing on to the wall. After that, only fly- ash would be deposited and would grow in a layer. Serious fouling will be developed if the fly-ash carried by the gas is in a molten state. There are two main mechanisms, sulfidation attack due to alkali sulfates of sodium, and corrosion caused by complex compounds of the sulfates $Na_3Fe\ (SO_4)_3$, which are formed by alkali sulfates and Fe_2O_3 contained in fly-ash. Na_2SO_4 and Fe_2O_3 deposited on to wall react with SO_3 contained in combustion gas and form complex compounds, as described by the following equation:

$$3Na_2SO_4 + Fe_2O_3 + 3SO_3 \rightleftharpoons 2Na_3Fe\ (SO_4)_3 \quad \ldots\ldots.. \ (1)$$

Thus complex compounds of $Na_3Fe(SO_4)_3$, [6] advance the above corrosion process cyclically. The complex compounds are not formed in the high-temperature region beyond 704 °C. That is why the corrosion rate decreases abruptly when the temperature reaches this level.

Along with this, stress also generates in the tube which in turn causes the rupture of tube. In order to evident this phenomenon hoop stress (σh) is calculated by using the formula PD/2t where 'P' is the operational pressure, 'D' is the mean diameter of the tube and 't' is the mean thickness of the tube at the time of failure. It was found that the calculated hoop stress is ~ 9000 psi and the recommended hoop stress for boiler tubes is 4000 psi. At the same operating pressure 'P', increase in hoop stress (σh) is attributed to wall thinning, due to ash corrosion, which gradually leads to increase in tube diameter (D) in the failed region. This observed over stress causes the rupture of boiler tubes.

4. Conclusions

- Ash corrosion is considered to be the main mechanisms in operation from the fireside for early failure of boiler tube. Localized water corrosion mechanism operating from the waterside of the tube is normal and slow. This type of corrosion failure occurs mainly due to the accumulation of ash deposits on the fireside as well as improper maintenance during cleaning procedures of the boiler.
- The hoop stress developed on the failed tube is almost twice than recommended stress value. Hence, the tube might have exceeded the upper limit of allowable stress in the operating temperature of 700-800 °C, resulting in rupture of the tube.

5. Acknowledgement

Authors are thankful to the management of Welspun Corp Limited for giving support and permission to prepare and publish this paper.

6. References

[1] Melville J and Foster G G 1975 *Int. J. of Pressure Vessels and Piping* **3** p 1-25

[2] Harding N S and O'Connor D C *2007 Fuel Processing Technology* **88** p 1082-1093

[3] Khajavi M R, Abdolmaleki A R, Adibi N and Mirfendereski S 2007 Engg. Failure Analysis 14 p 731-738

[4] Siddhartha Bhatt M 2006 *Energy Conversion and Management* **47** p 1247-1263

[5] Ananda Rao M and Sankara Narayanan T S N *Engg. Failure Analysis* **26** p 325-331

[6] Marcel M and Dimitris P 1998 *Mossbauer Spectroscopy in material science* (Netherlands: Kluwer Academic) p 51-54

Modeling of Dephosphorization using Bloated Droplet Theory in Basic Oxygen Steelmaking

Ankit[1], **T. K. Kundu**
Department of Metallurgical and Materials Engineering,
IIT Kharagpur, Kharagpur, W.B, India- 721302

[1]E-mail: ankitv89@gmail.com

Abstract.A considerable fraction of refining of hot metal in basic oxygen steelmaking is carried out in emulsion layer by interaction between the metal droplets and slag. The top oxygen blowing ejects metal droplets into the slag which then undergo refining reactions and return to the metal bath. During this period, the carbon in metal droplets reacts with available oxygen to form carbon monoxide. Above a certain threshold decarburization rate, the evolved carbon monoxide inside the droplet causes the droplet to bloat which increases its surface area. The bloating affects the residence time and the rate of certain interfacial reactions. Here, efforts have been made to study dephosphorization in metal droplet in emulsion by coupling dephosphorization kinetics with the 'bloated droplet theory'. As observed in industrial furnaces, the calculated phosphorous concentration in droplet was found to be lower than that of metal bath. Dephosphorization model developed using mixed transport control model was found to fit industrial data. The results obtained were in satisfactory agreement with the available data and a step ahead in improving the understanding of dephosphorization in steelmaking.

1. Introduction

In basic oxygen steelmaking, metal droplets are ejected into the slag due to impingement of supersonic oxygen jet on the hot metal bath which leads to the formation of an emulsion consisting of ejected metal droplets, gas phase and slag. Models have been developed to incorporate this phenomenon for comprehensive understanding of steelmaking process. Researchers have argued that considerable fraction of refining occurs in the emulsion zone as the formation of droplets increase the interfacial area for reactions. Molloseau and Fruehan [1] have studied the bloating or swelling of metal droplets when reacting with slags which further increase the surface area of the reaction. Brooks and co-workers [2] were able to explain this through a model which was subsequently used for developing of a comprehensive model by Dogan [3-5]. However, these works considered only the decarburization reaction but the effect of bloating on the kinetics of oxidation of phosphorous, manganese and silicon is a matter of further study. In current work, efforts have been made to model the dephosphorization in emulsion using bloated droplet theory. Various dephosphorization model is coupled with bloated droplets theory and the results were compared with the experimental data.

2. Model Development

The basic oxygen steelmaking furnace has been divided into two zones, as shown in Fig.1, namely - the emulsion zone and the metal bath. However, the present study is limited to dephosphorization in the emulsion zone only. The model is divided into three subroutines as shown in Fig. 2., namely, i) The

Content from this work may be used under the terms of the Creative Commons Attribution 3.0 licence. Any further distribution of this work must maintain attribution to the author(s) and the title of the work, journal citation and DOI.
Published under licence by IOP Publishing Ltd

bloated droplet model which calculates the rate of droplet generation, the interfacial area and the residence time, ii) the slag model that calculates the physical properties of the slag, and iii) the dephosphorization model to couple the earlier two models to calculate the phosphorous concentration of metal droplets in emulsion. The important parameters considered for this study are the amount of metal in the emulsion zone at a given time, the residence time of the droplet, the size distribution of the droplets and the total interfacial area between the metal and slag phases.

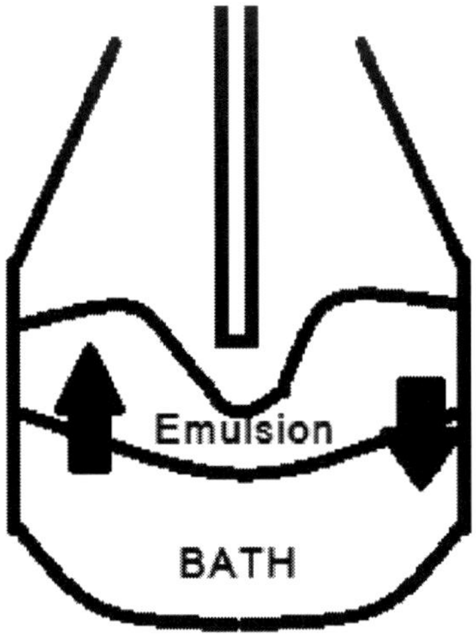

Figure 1. Schematic diagram of BOF Model

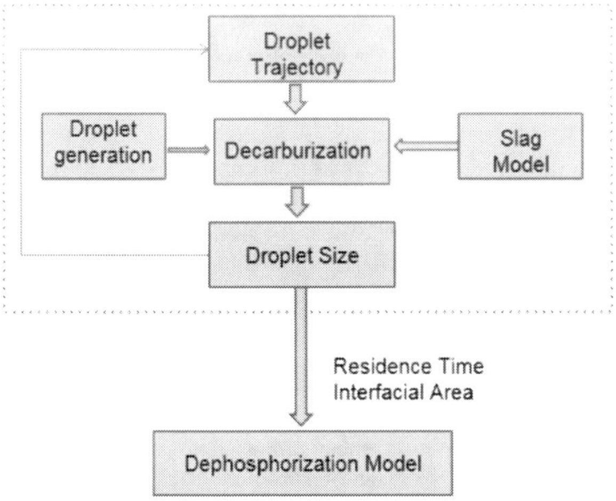

Figure 2. Workflow of the model

2.1. Droplet Model

The mechanism of droplet generation by the impinging oxygen jet has been studied earlier. Subagyo et al [6], in their work had related the droplet generation rate with blowing number (N_B) and gas flow (F_G) rate. The Blowing number (N_B) depends on the critical gas velocity (U_G), the density of the gas (ρg), the surface tension of the metal (σ), the density of the metal bath (ρ_M) and acceleration due to gravity (g) as given in equation 1. The value of N_B should be greater than 3 for signification number of droplets formation [7].

$$N_B = \frac{\rho_G U_G^2}{2\sqrt{\sigma g \rho_M}} \quad (1)$$

Subagyo et al [6] further proposed an empirical correlation between droplet generation rate (R_B, Kg/sec.), the blowing number (N_B) and the volumetric oxygen input flow rate (F_G, Nm^3/s) as expressed in equation 2.

$$R_B = \frac{(N_B)^{3.2} \times F_G}{[2.6\times 10^6 + 2.0\times 10^{-4}\ (N_B)^{12}]^{0.2}} \quad (2)$$

To model the trajectory of the metal droplet in emulsion the initial velocity (u_i) and diameter of the droplet (D) are required. The model for trajectory has been proposed by Subagyo et al [7] and is referred to as "Ballistic Droplet Theory". The initial velocity is calculated by energy balance between the kinetic energy of droplets ($KE_{droplets}$) and impinging jet (KE_{Jet}). The kinetic energy of droplets is given by equation 3 and that of impinging jet by equation 4.

$$KE_{droplets} = \frac{1}{2} R_B \, u_i^2 \qquad (3)$$

$$KE_{jet} = \frac{1}{2} \rho_g R_G U_G^2 \qquad (4)$$

Both of these kinetic energies are co-related by empirical expression as given in equation 5.

$$KE_{droplets} = 0.00143(N_B)^{0.7} KE_{jet} \qquad (5)$$

As assumed by earlier models [2-7], the diameter of metal droplet in this study also was taken to be 2 mm. This theory was further extended by Brooks [1] to include decarburization in the droplet which was closer to the actual phenomenon occurring in the steel converter. They co-related the density of the droplet with the decarburization rate as shown in equation 6 where, r_c is the decarburization rate and r_c* is the critical decarburization rate. The modified density is then used to calculate the bloated or swelled diameter of the droplet. The droplet is bloated only when r_c exceeds r_c* otherwise, r_c is taken equal to r_c*[2].

$$\rho_{d} = \rho_o \frac{r_c^*}{r_c} \qquad (6)$$

The equation for r_c* is given by 2.86 x 10^{-4} **x** (FeO), where (FeO) is the wt % of FeO in slag [2]. The residence time is also dependent on the density of slag-gas mixture, the gas fraction, the initial size of the droplet and initial velocity of the droplets. All these were used to calculate the residence time of the droplet in the emulsion [2]. Extensive discussion on these parameters can be found in the work of Dogan [4]. Steady state is assumed for calculating the mass of metal droplets in slag. The mass of droplets in slag is given as product of residence time and generation rate of the droplets [6].The area is taken as the average area of a droplet during its motion in the emulsion. The total area in emulsion is the sum of individual areas of the droplets at a given instant of time. Droplets are also generated by bottom blowing. However, the size of droplet generated by bottom blowing is too small and they have very low residence time [8].

2.2. Slag Model

The slag model was used to calculate the density and viscosity of the slag as a function of temperature and composition. The methodologies outlined in the Slag Atlas [9-10] was followed to calculate these quantities. The slag composition was taken from Millman et al [11] and fed into the model to calculate the viscosity and density.

2.3. The dephosphorization Model

Dephosphorization of metal bath is favoured by high basicity, low temperature, and high FeO in slag [12]. Since, the reaction takes place at high temperatures, the chemical reactions at interface are not considered as rate limiting reactions, instead the mass transfer in slag or metal or both in slag and metal (mixed) is taken as rate controlling step. Mori et al [13] and Wei et al [14] have considered the mass transfer as mixed controlled while Monaghan et al [15] have concluded that the mass transfer in slag phase is the rate controlling step. The equations 7 and 8 were proposed by Monaghan for first order dephosphorization with mass transport in slag and metal as rate controlling step respectively. The subscript 'b' is for the bulk concentration, 'o' is for initial concentration and 'e' is for equilibrium concentration. Here the mass transfer coefficient in slag and metal are k_s and k_m respectively, W_m is the mass of the metal in emulsion, W_s is the mass of the slag, ρ_s is the density of the slag. In these cases the initial phosphorus concentration in slag was taken as zero.

$$ln\left[\frac{[pct\ P]_b-[pct\ P]_e}{[pct\ P]_o-[pct\ P]_e}\right]\left(\frac{[pct\ P]_e}{[pct\ P]_o}\right)\left(\frac{W_S}{\rho_M A}\right)=-k_s t \quad (7)$$

$$ln\left[\frac{[pct\ P]_b-[pct\ P]_e}{[pct\ P]_o-[pct\ P]_e}\right]\left(\frac{[pct\ P]_o-[pct\ P]_e}{[pct\ P]_o}\right)\left(\frac{W_m}{\rho_M A}\right)=-k_M t \quad (8)$$

In the current model, mixed mass transport equation proposed by Manning and Fruehan [16] was used. The initial phosphorous concentration of slag was not taken to be zero instead the %P in slag was taken from Millman et al [11] based on P_2O_5 evolution with time. Therefore, equation 9 was used,

$$[pct\ P]_{emulsion}=\frac{(Z4*exp(-Z1*Z3*t)+Z2)}{Z1} \quad (9)$$

Where, $Z1=1+\left(\frac{W_m}{W_s*L_P}\right)$, $Z2=\frac{(pct\ P)}{L_P}+\frac{[pct\ P]*W_m}{W_S*L_P}$, $Z3=A*rhom*k_O*W_m$,and $Z4=Z1*[\%P]_O-Z2$.

where L_P denotes the equilibrium phosphorous partition ratio and is discussed in later stage. The area term (A) in Z3 is the average interfacial area of all the droplets during their motion through the slag after ejection. Average area of a single droplet is multiplied by the total number of droplets in emulsion to get the total interfacial area in emulsion. The total number of droplet is calculated by dividing the total mass of droplets generated (R_B x Residence time) by mass of a single droplet. The overall mass transfer coefficient for mixed control dephosphorization, k_o, is calculated by equation 10. The values for k_s and $k_m$were taken from Wei et al[14].

$$k_o=\frac{1}{\frac{\rho_m}{k_S\rho_S L_P}+\frac{1}{k_m}} \quad (10)$$

As discussed by Monaghan, the equilibrium concentration is difficult to predict and is usually taken equal to the minimum phosphorus content or equal to the concentration at thermodynamic equilibrium. Basu et al [17] has calculated the phosphorous distribution with time in running converter and concluded that after 10 minutes of blow the phosphorous partition is 40-60% of the equilibrium partition ratio. Similar observation were made by Choudhary and co-workers [18] where they have suggested that Turkdogan's [19] model fits best for end phosphorous calculation for industrial furnaces. However, Manning and Fruehan [16] used the equation of Ide and Fruehan [20] to calculate the L_P. This calculated L_P is used in equation 9. In equation 11, the T is the temperature in Kelvin and [%O] is the oxygen concentration calculated from equation 13 where (%FeO) is the FeO wt % in slag. Similarly, in equation 12 and 14, the quantities in parentheses are wt. % of components in slag. Apart from this, in present study, the mixed control of mass transfer was compared with slag controlled mass transfer for the dephosphorization.

$$log(L_P)_{Turkdogan}=\frac{21740}{T}-9.87+0.071BO+2.5log([\%O]) \quad (11)$$

$$BO=(\%CaO)+(\%CaF_2)+0.3(\%MgO) \quad (12)$$

$$\frac{[\%O]}{(\%FeO)}=(0.1T-155.3)\times10^{-4} \quad (13)$$

$$log(L_P)_{Ide} = \frac{10730}{T} + 4.11\,log(\%CaO + 0.015\%MgO + \%CaF_2 - 0.05\%Fe_tO) + 2.5\,log(\%FeO) + 0.5\,log(\%P_2O_5) - 13.87 \quad (14)$$

3. Results and Discussion

The residence time and average interfacial area were calculated from the bloated droplet model and the slag model fed the physical properties of the slag in the bloated droplet model to account for changing slag characteristics with time.Fig. 3 shows the variation of density and viscosity with time. The residence time and average interfacial area was then fed in to the dephosphorization model. The dephosphorization model used mixed transport control, along with two methodologies to calculate the equilibrium phosphorous concentration, namely, the Turkdogan's equation and Ide and Fruehan's equation. The mechanism and equilibrium equations were clubbed together one by one and compared. Fig. 4shows the final result obtained from the developed model. The input data for the model was from Millman et al [11].While comparing the different equations of L_P, Ide and Fruehan's as well as Turkdogan's equation gave similar results. However, in initial part, none of the models can be said to be completely successful in predicting the phosphorous in emulsion.

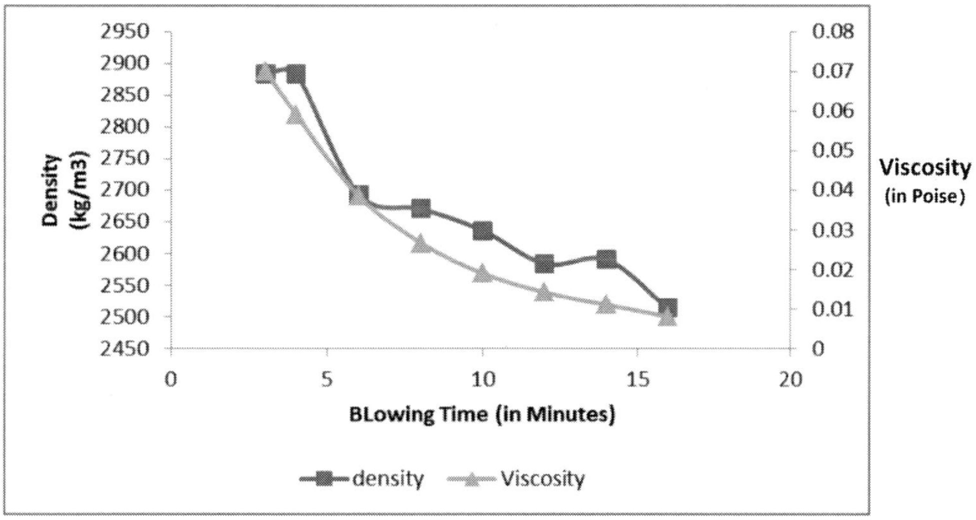

Figure 3: Variation of Density and Viscosity with blowing time

Fig. 5 is the temporal evolution of phosphorous content in metal droplet in the emulsion zone, which is similar to the results of Gaye [21]. One possible explanation for variation between the results of the model and experiments is the difference in conditions. The models used above had been developed at laboratory scale which were not much close to the industrial process. One more possibility that was not considered was the breakage or bursting of droplets. If evolution of carbon monoxide is sufficiently vigorous, the droplet may break into smaller droplets which should increase the effective surface area of the droplet. The droplets may later coalesce or remain unaffected while continuing their downward motion. This sudden increase of surface area (and diameter) may increase dephosphorization rate and the effective residence time of the droplets which in turn would change the mass of metal in the emulsion. It is difficult to model the breakage and coalescence of metal droplets in emulsion because of high temperature conditions which makes model validation an issue. Efforts are being made in author's laboratory to model the breakage of droplets using stochastic approaches.

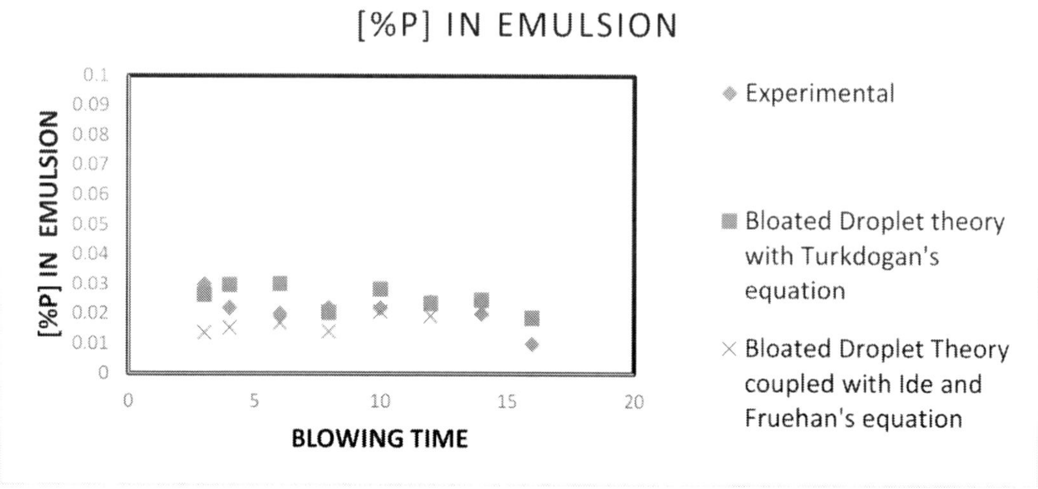

Figure. 4: Variation of final Phosphorous content in metal droplet calculated by model in emulsion with blowing time. The bloated droplet theory was coupled with the dephosphorization models mentioned in legend. The experimental data is from Millman et al [11]

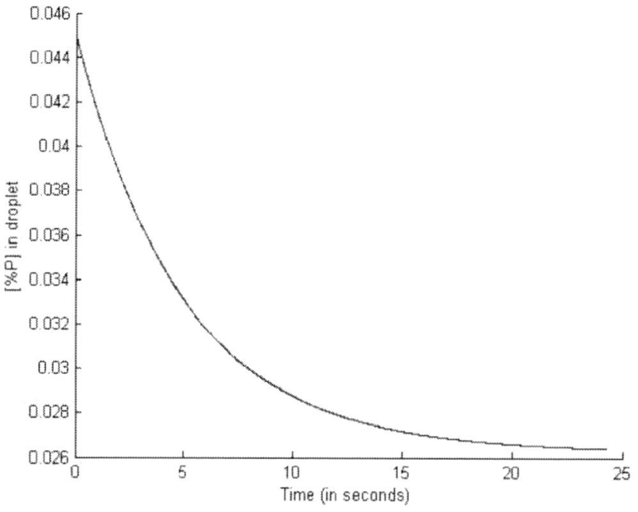

Figure 5: Evolution of Phosphorous content of metal with residence time at t=6 minutes, Turkdogan's Model with mixed transport control.

4. Conclusion

A formal approach of incorporation of interfacial area changes due to bloating of droplet interaction kinetics of dephosphorization has been developed in the study. The result also quantifies the extent of dephosphorization occurring in emulsion of an oxygen steelmaking reactor. Both Turkdogan's and Ide and Fruehan's equation for L_P along with the mixed mass transport controlled mechanism are found to fit satisfactorily with the experimental conditions.

5. References

[1] Molloseau C L and Fruehan R J 2002 *Metall. Mater. Trans. B* **33B** 335
[2] Brooks G A, Pan Y, Subagyo, and Coley K S 2005 *Metall. Mater. Trans. B* **36B**, 525
[3] Dogan N, Brooks G A, and Rhamdhani A 2011 *ISIJ Int.* **51** 1086
[4] Dogan N, Brooks G A, and Rhamdhani A 2011 *ISIJ Int.* **51** 1093
[5] Dogan N, Brooks G A, and Rhamdhani A 2011 *ISIJ Int.* **51** 1102
[6] Subagyo, Brooks, GA, Coley, and K.S., Irons, G.A. 2003 *ISIJ Int.* **43** 983
[7] Subagyo, Brooks GA, and Coley, K.S. 2005 *Can. Met. Quart.* **44** 119
[8] Lin, Z and Hollapa J 2003 *ISIJ Int.***43** 1698
[9] Keene B J and Mills K C 1995 *Slag Atlas* (VerlagStahleisen GmbH, Dusseldorf) 313.
[10] Mills K C 1995 *Slag Atlas* (VerlagStahleisen GmbH, Dusseldorf) 353.
[11] Millman M S et al 2013 *T. Indian. I. Metals* **55** 525.
[12] Chatterjee A and Ghosh A 2013 *Ironmaking and Steelmaking* (PHI, India)
[13] Mori, K, Doi, S, Kameko, T, and Kawai. Y 1988 *T. Iron. Steel. I. JPN.* **28** 261.
[14] Wei P, Sano M, Hirasawa M, and Mori K 1993 *ISIJ Int.* **33** 479.
[15] Monaghan B J, Pomfret R J, and Coley K S 1998 *Metall. Mater. Trans. B* **29B** 111
[16] Manning, C P and Fruehan, R 2013 *Metall. Mater. Trans. B* **44B** 37
[17] Basu S, Lahiri A K , Seetharaman S, and Halder J 2007 *ISIJ Int.* **47** 766
[18] Choudhary S K, Lenka S N, and Ghosh A 2007 *Ironmak. Steelmak.* **34** 343.
[19] Turkdogan E.T. 2000 *ISIJ Int.* 40 964.
[20] Ide, K and Fruehan R J 2000 *Trans. ISS AIME* December 65.
[21] Gaye H and Riboud P V 1977 *Metall. Mater. Trans B* 8B 409.

IOP Conf. Series: Materials Science and Engineering **75** (2015) 012023 doi:10.1088/1757-899X/75/1/012023

Morphological and Photoluminescence analysis of Zinc Oxide thin films deposited by RF sputtering at different substrate temperatures

S P Ghosh[1], K C Das[1], N Tripathy[1], G Bose[2], T Lee[3], J M Myoung[4] and J P Kar[1]*

[1]Department of Physics and Astronomy, National Institute of Technology, Rourkela, India 769008

[2]ITER, SOA University, Bhubaneswar, India 751030

[3]Department of Bio-Nanotechnology, Gachon University, Gyeonggi-do, Republic of Korea, 461701

[4]Department of Materials Science and Engineering, Yonsei University, Seoul, Republic of Korea, 120749

E-mail:suryaprakash.ghosh@gmail.com

Abstract. Zinc oxide (ZnO) thin films were prepared using reactive RF magnetron sputtering of a pure metallic zinc target onto n-type (100) silicon substrates. The evolution of the surface morphology and the optical properties of the films were studied as a function of the substrate temperature, which was varied from ambient to 300°C. X-ray diffraction pattern of ZnO thin film shows the appearance of c-axis oriented (002) peak for all samples shows varying degrees of crystallinity of films. Photoluminescence studies were also carried out (350-700 nm) to study the crystallinity and optically active defects in the films. PL spectra of the film shows UV emission peak depicts good crystallinity of ZnO film where as the intensity of deep level emission band decreases with increase in substrate temperature due to the formation of stoichiometric ZnO film which causes decrease in defect.

1. Introduction

Zinc Oxide (ZnO) films have engrossed considerable interest now a days, due to its excellent material properties like piezoelectricity, large bandgap of 3.37eV at room temperature and higher exciton binding energy (60MeV) which is nearly two and half times greater than other semiconductors like Gallium Nitride (GaN) [1,2]. Therefore, ZnO is currently used as multifunctional material having application in numerous areas such as light emitting diodes, laser diodes, varistors, piezoelectric transducers, field-effect transistors, touch display panels, flat panel displays, photovoltaic solar cells, and gas sensors due to its high electrochemical stability, absence of toxicity, and ease of fabrication [3]. The synthesis of ZnO thin films of better quality is highly required in order to increase the performance of the electronic devices.

The properties of the thin films not only depend on the deposition techniques, but also depend on the deposition parameters. In current days, several physical and chemical deposition techniques such as spray pyrolysis, chemical vapor deposition (CVD), sol–gel process, sputtering techniques, chemical spray, plasma enhanced CVD, and pulsed laser deposition (PLD) are being adopted for the synthesis of ZnO thin film [4]. Among the growth techniques RF magnetron sputtering has several advantages such as its simplicity, low thermal budget, low-cost, non-toxic and its ability to produce high quality films with desired properties [5]. Although an attempt have been made for improving the quality of the ZnO

Content from this work may be used under the terms of the Creative Commons Attribution 3.0 licence. Any further distribution of this work must maintain attribution to the author(s) and the title of the work, journal citation and DOI.
Published under licence by IOP Publishing Ltd

thin film for various applications, there is still a need for a systematic study of morphological evolution of the sputtered film under various deposition conditions. In this paper we focus on the structural, morphological and optical properties of ZnO films deposited at various substrate temperatures by RF sputtering method.

2. Experimental work

ZnO thin films were deposited by RF magnetron sputtering technique on n-type silicon (100) substrates 1-10 Ωcm. Sputtering was carried out using a pure (99.99%) Zn target in a mixture gas of Ar and O_2 (Ar: O_2 = 2:3) and RF power of 150W at different substrate temperatures ranging from ambient to 300°C. The substrate holder is rotated during sputtering in order to obtain uniform films. After deposition of ZnO film, crystalline properties was investigated by X-ray diffraction (XRD, Rigaku/UltimaIV) with a CuKα (λ=0.154nm) radiation. FESEM (JEOL JSM-7001F) was used to study the surface morphology of the film. In order to investigate the optical properties of ZnO film PL measurement was carried out on a luminescence spectrometer (Ramboss 500i) using He-Cd laser of wavelength 325 nm as the excitation source.

3. Results and discussion

3.1. Structural properties

Figure 1. represents the X-ray diffraction pattern of ZnO thin film at different growth temperatures. C-axis oriented (002) peak is observed at 34° for all samples shows better crystallinity of film. The sample grown at room temperature (RT) shows one additional peak (103) at 62°. This peak disappeared for the higher substrate temperature. The c-axis orientation in sputtered ZnO films is due to the lowest surface free energy of the (002) plane [6]. The (002) peak becomes prominent at 200°C, which depicts an improvement in crystallinity of thin film. The decrease in crystallinity at higher temperature may be due to non-stoichiometric oxygen, which increases the defect level in the ZnO layer. The desorption of oxygen on the surface of thin film governs the change in stoichiometry ratio between Zn and O, and as a result of which c-axis orientation was deteriorated at higher temperature. Figure 2. shows the FESEM image of ZnO thin film deposited at a substrate temperature of 200°C. A crack free and uniform film is observed having almost homogeneous microstructure with average grain size of about 35-45 nm.

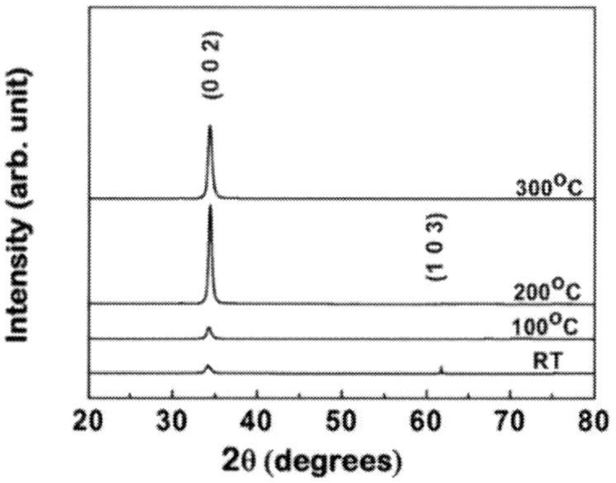

Figure. 1. X-ray diffraction patterns of ZnO thin films deposited at different substrate

Figure. 2. FE-SEM image (100nm scale) of ZnO thin film deposited on n-type silicon (100) substrate at 200°C

3.2. Optical properties

Figure. 3. Shows the PL spectra of ZnO thin film deposited at various substrate temperature ranges from RT to 300°C. One strong emission peak is observed in UV range and another broad luminescence band is observed in the visible range. The UV emission can be referred to as the ''near-band-edge'' emission (NBE), generated by the recombination of the free excitons [7]. The broad band in visible range arises due to the deep level emissions which increases the oxygen vacancies, zinc interstitials or zinc vacancies [8]. The increase in the intensity of UV peak with increase in deposition temperature indicates enhancement of crystallinity of the film. At low substrate temperature the adatoms may not get sufficient energy to form stoichiometric ZnO film, so there is a possibility of formation of interistitial Zn or Oxygen vacancy. Therefore, the intensity of deep level emission decreases with increase in substrate temperature.

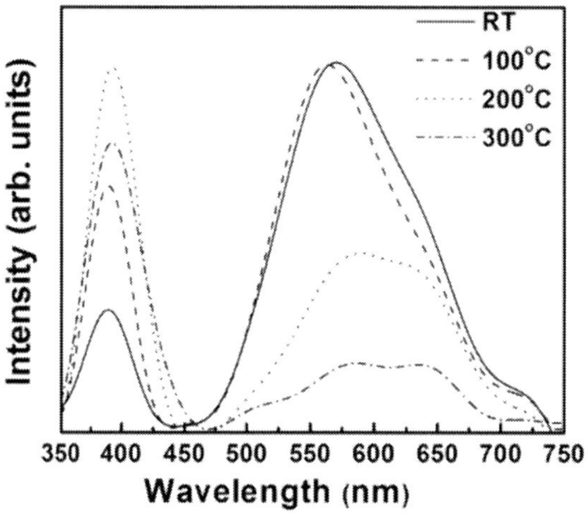

Figure. 3. PL spectra of ZnO thin film deposited at different substrate temperature

4. Conclusion

ZnO thin films were grown on n-type silicon (100) substrates 1-10 Ωcm by RF magnetron sputtering. From XRD measurements the films were found to be c- axis oriented along the (002) direction and the best crystallinity occurs at 200°C. FESEM results depicts a crack free and uniform ZnO thin film deposited at a substrate temperature of 200°C having almost homogeneous microstructure with average grain size of about 35-45 nm. PL spectra of the film shows UV emission peak depicts good crystallinity of ZnO film where as the intensity of deep level emission band decreases with increase in substrate temperature due to the formation of stoichiometric ZnO film which causes decrease in oxygen vacancy.

Acknowledgments

This work was supported by Indo-Korea joint research program of Department of Science and Technology, India.

References

[1] Mangalaraj P, Paik P, Raj A, and Ramakrishna A 2012 *Appl. Surf. Sci.* **258** 6765–71
[2] Kim H, Jung E, Lee W, Kim J, Ryu S and Choi S 2008 *Ceram. Int.* **34** 1097–1101
[3] Senay V, Pat S, Korkmaz S, and Balbag M 2014 *Appl. Surf. Sci.* **318** 2-5
[4] Hsu C and Cheng T 2010 *Thin Solid Films* **518** 1953–57
[5] Kar J, Kim S, Shin B, and Myong J 2010 *Solid-State Electronics* **54** 1447–50
[6] Fujimura N, Nishihara T, Goto S, Xu J and Ito T 1993 *J. Cryst. Growth* **7** 130- 269
[7] Kumar A, Jeedigunta S, Tarasov I and Ostapenko S 2010 *AZo J. of Mater.Online* **6** 1833
[8] L Zhao, J Lian, Y Liu and Q Jiang 2006 *Appl. Surf. Sci.* **252** 8451–55

Nickel coating on high strength low alloy steel by pulse current deposition

S Nigam[1], S K Patel[1], S S Mahapatra[1], N Sharma[3] and K S Ghosh[2]
[1]Mechanical Engineering Department, National Institute of Technology, Rourkela-769008, Odisha, India
[2]Metallurgical and Materials Engineering Department, National Institute of Technology, Durgapur-713209, West Bengal, India
[3]Metallurgical and Materials Engineering Department, National Institute of Technology, Rourkela-769008, Odisha, India

E-mail:shrutee.nigam@gmail.com,ksghosh2001@yahoo.co.uk

Abstract. Nickel is a silvery-white metal mostly used to enhance the value, utility, and lifespan of industrial equipment and components by protecting them from corrosion. Nickel is commonly used in the chemical and food processing industries to prevent iron from contamination. Since the properties of nickel can be controlled and varied over broad ranges, nickel plating finds numerous applications in industries. In the present investigation, pulse current electro-deposition technique has been used to deposit nickel on a high strength low alloy (HSLA) steel substrate.Coating of nickel is confirmed by X-ray diffraction (XRD) and EDAX analysis. Optical microscopy and SEM is used to assess the coating characteristics. Electrochemical polarization study has been carried out to study the corrosion behaviour of nickel coating and the polarisation curves have revealed that current density used during pulse electro-deposition plays a vital role on characteristics of nickel coating.

1. Introduction

Refinement of grains is an important aspect that improves the mechanical properties such as strength, wear and impact resistance [1-3]. However, the effect on corrosion resistance due to variation in grain size is still not much explored. Also in many cases, contradictory conclusions have been drawn that decrease in grain size can either increase or decrease corrosion resistance [4-7]. The nanostructured materials with significant amount of surfaces and interfaces have attracted much interest because of their demonstrated unique properties compared to conventional materials [8].The coating of nanocrystalline deposit on steel leads to significant improvement of corrosion resistance [9-11]. As such, there are various nano-coating processing techniques. In this work, pulse-current electro-deposition has been used with an attempt to deposit finer grains as compared to which is obtained by conventional electroplating technique. An optimum amount of saccharin has been used in the nickel bath as grain refinement agent. Till date, this area is not much explored and there is lots of scope to work further.Corrosion, a deteriorating phenomenon of materials, particularly in metals, often dictates the life span of a product. By careful observations and proposing newer methods to inhibit corrosion, material life of any product could be improved and thus preventing loss to the society. Usually corrosion resistance of metals and alloys can be achieved by applying coatings over them, just as nickel films which have been proved suitable in this regard [12-17]. In the present research the corrosion behavior of the coating is evaluated with the help of potentiodynamic polarization tests. The resistance of the coatings towards corrosion is evaluated on the basis of the corrosion parameters obtained from the polarization curves viz. corrosion potential, corrosion current density, charge transfer resistance, double layer capacitance, corrosion rate, etc. Experiments have been done to analyze the effect of pulse current parameter such as current density on surface texture of the Nickel coating[18-20].

Content from this work may be used under the terms of the Creative Commons Attribution 3.0 licence. Any further distribution of this work must maintain attribution to the author(s) and the title of the work, journal citation and DOI.
Published under licence by IOP Publishing Ltd

2. Experimental Procedure

2.1. Material.
A high strength low alloy (HSLA)steel used in this investigation was obtained from TATA Steel, Jamshedpur in plate form of thickness 6.0 mm. The composition of this alloy is given in Table 1. The alloy plate was hot rolled and the hot rolling parameters of the HSLA were: soaking temperature 1185 ^{0}C, finish rolling temperature 910 ^{0}C and coiling temperature 580 ^{0}C.

Table 1. Chemical composition [wt %] of the HSLA Steel.

C	Mn	Si	P	S	Al	Ti	V	Nb	Cr	Fe
0.08	0.95	0.024	0.013	0.006	0.04	0.025	0.01	0.04	0.021	balance

2.2. Specimen Preparation.
The coupons of HSLA steel of approximate dimensions of 15 mm x 15 mm were prepared. The surface of the specimens was polished by SiC emery paper ranging from grit size of 220-1500. The samples were then further cloth polished using alumina abrasives to get scratch free mirror-like surface. Washing with ethanol or acetone helps to remove any dust or grease on the surface of the specimen. The specimen was wrapped in the Teflon tape exposing only the polished area. Over the Teflon, an organic polymer (nail polish) coat was put on in order to completely insulate every part of the specimen except the polished portion. Then electroplating on the sample was carried out in glass beaker containing the electrolyte/bath using a Potentiostat.

2.3.Pulse-current electro-deposition technique.
Pulse electro-deposition (PED) is a novel technique for producing nano-structured coatings on metallic substrates. It is associated with many advantages such as reduction in porosity, low level of inclusions, and higher rate of deposition as compared to conventional direct current (DC) electro-deposition process. There is much more flexibility involved in varying three basic parameters such as pulse peakcurrent density, pulse frequency, on time and off time in pulse electro-deposition resulting into unique composition and microstructure of the coating being deposited.
Pulse current electro-deposition (PED) is done using a Potentiostat/Galvanostat/FRA unit of model PARSTAT 2273 manufactured by Advanced Electrochemical Systems. The electro-deposition is carried out at room temperature in a nickel bath. The composition of the bath is given in Table 2.The set-up for pulse current electro-deposition showing all the three electrodes inserted in a glass beaker is connected to Potentiostat as shown in Figure 1.Terminals of the Potentiostat/galvanostat/FRA e.g. working electrode was connected to sample, counter electrode with 99% pure nickel strip and a saturated calomel electrode (SCE) as a reference electrode. The pulse frequency is 5 Hz and the on and off time is 1 s each.

Table 2. Nickel Bath Composition.

Substrate	Nickel Bath (gm l^{-1})				Bath Temp(^{0}C)	pH
HSLA Steel	Nickel sulphate	Sodium chloride	Boric acid	Saccharin		
	200	20	30	1.5	25	4

The pulse current electro-deposition was performed at different current densities of 7A/dm^2 - 50 A/dm^2.The solution/electrolyte used for carrying out nickel deposition was prepared as given in Table 2.

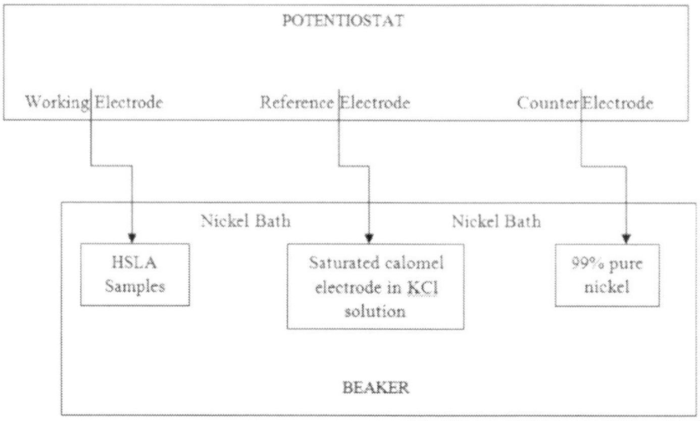

Figure1. Set up for nickel deposition.

2.4Observation of microstructures of the substrate and nickel deposits.
A Leica DM2500M optical microscope with an in-built QWIN3 image analyser and a scanning electron microscope (SEM) model Hitachi S-3000Nwere used to observe the microstrotures of the HSLA steel and the nickel deposits.

2.5X-Ray Diffraction (XRD).
X-ray diffraction (XRD) was carried out to acquaint with the phases present in the alloy of different tempers. XRD study was carried out using a PANlyticalXPert Pro Basic X-ray diffractometer. A scanning electron microscope (SEM) model HITACHI S-3000 N with an EDAX attachment unit is used for microstructural observation of nickel coating and for elemental analysis as well.

2.6. Electrochemical study.
Electrochemical polarisation study was carried out on the nickel coated HSLA steel at room temperature using a computercontrolled potentiostat/galvanostat/FRA model of PARSTAT 2273 Advanced electrochemical systems model with an inbuilt Power suite software. The experiments were performed at room temperature in 3.5 wt. % NaClsolution which is a standard saline solution used to study the corrosion behaviour of steel and Al-alloys.The standard three electrodes configuration: saturated calomel electrode (SCE) as a reference, platinum electrode as counter and the sample as the working electrode have been used for potentiodynamic polarisation study of the nickel coated steel in 3.5 wt. % NaCl solution. Polarisation scan was carried out towards more noble value at a scan rate of 0.5mV/s, after 10 minutes of immersion of sample into the electrolyte or allowing a steady state potential to develop. Polarisation tests for each sample were repeated at least twice to confirm the reproducibility of the test results.

3. Results and Discussions

3.1. Optical Micrographs.
Figure 2 shows the microstructure of the HSLA steel (as-received HSLA alloy plate was solutionised at 1050°C for 3 hours followed by air cooling). Microstructure reveals mostly fine ferrite grains and a small amount of very fine pearlite at the grain junctures. Figures 3 and 4 show the micrographs of the nickel deposit carried out at pulse current density 10 A/dm^{2}and the coating thickness, respectively. The micrograph (Figure 3) displays fine grains structure containing a few numbers of pits as well.

3.2. SEM and Energy Dispersive X-Ray Analysis.

Figure 4, the SEM micrograph displays the coating thickness approximately100μm produced by the pulse current metal deposition. The figure 5 and figure 6 exhibit the SEM microstructure of the nickel coating at current densities of 30A/dm^2 and 50A/dm^2. The microstructures shown in the figures 4-5 reveal that the coating consists of polyhedral grain of nickel. Further, there is presence of prominent cracks (Figure 6) in the coating done at current density 50 A/dm^2, but the tendency of cracking of less when pulse coated at current density of 30 A/dm^2, although a very few fine cracks are visible (Figure 4). This suggests that pulse current electro-deposition using a proper and appropriate current density will results in producing submicron and or nano-structure nickel deposit on steel substrate.

Figure 7 shows the EDAX analysis done at different positions on the nickel coating. EDAX analysis confirms the presence of nickel, carbon, oxygen, manganese and iron. The presence of nickel is evident in all the points as shown in the figures 7(b-e).

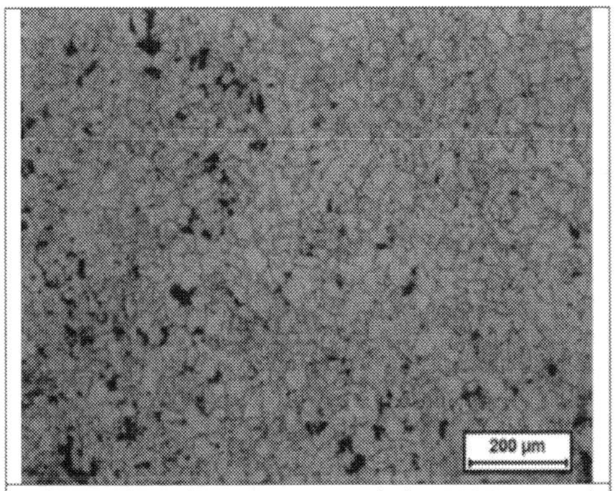

Figure 2. Microstructure of the HSLA steel substrate before coating.

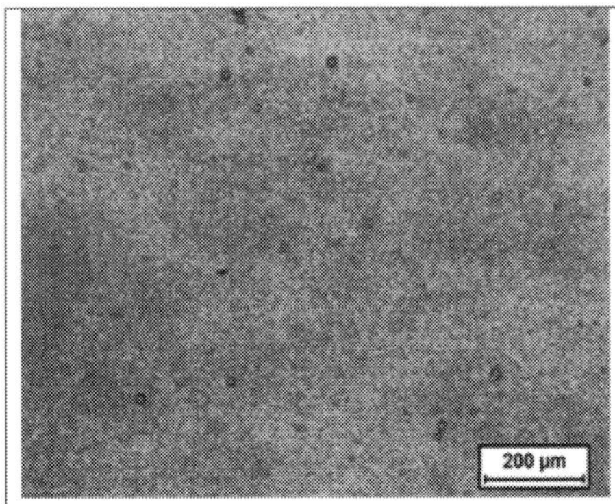

Figure 3. Nickel coating onto HSLA steel, coated at 10A/dm^2.

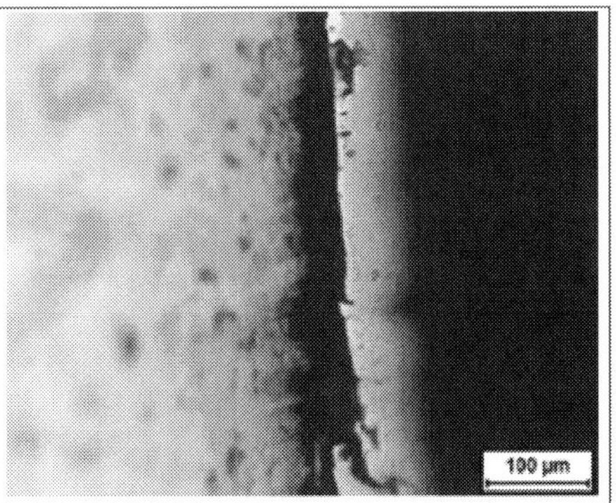

Figure 4. Micrograph of samples done at 10A/dm^2 at higher magnification.

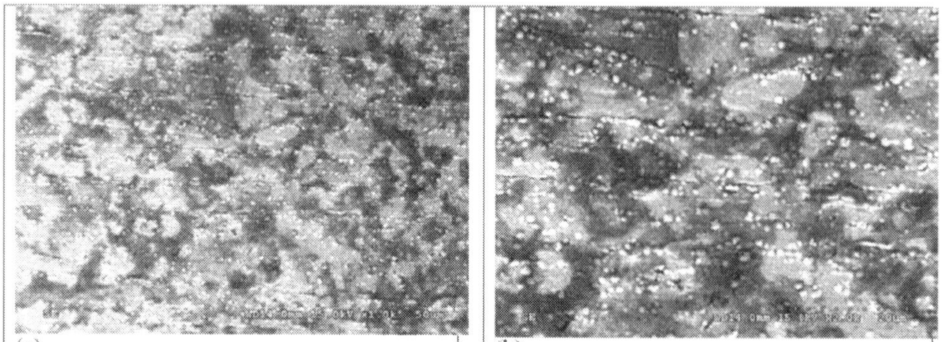

(a) (b)

Figure 5(a-b). SEM microstructure of Ni deposit at 30 A/dm^2.

(a) (b)Crack

Figure 6(a-b). SEM microstructure of Ni deposit at 50A/dm^2.

The percentage of the nickel is quite high as compared to other elements present, which is as obvious. Line scan EDAX analysis was made as shown in figure 8(a-b). The figure 8b also shows the presence of all these elements, and the nickel count is the maximum amongst all the other elements. Small quantity of Mn and Fe present in the coating may be considered as impurities. Further, the nickel coating is not uniform enough and contains numerous cracks as well and these account the peaks of Fe and Mnarisenfrom the substrate.

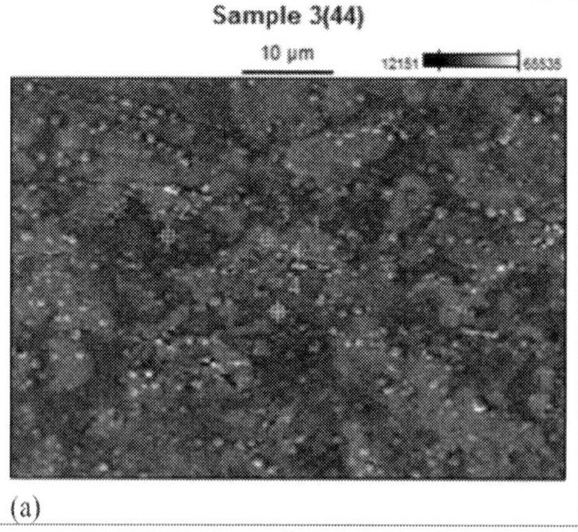

(a)

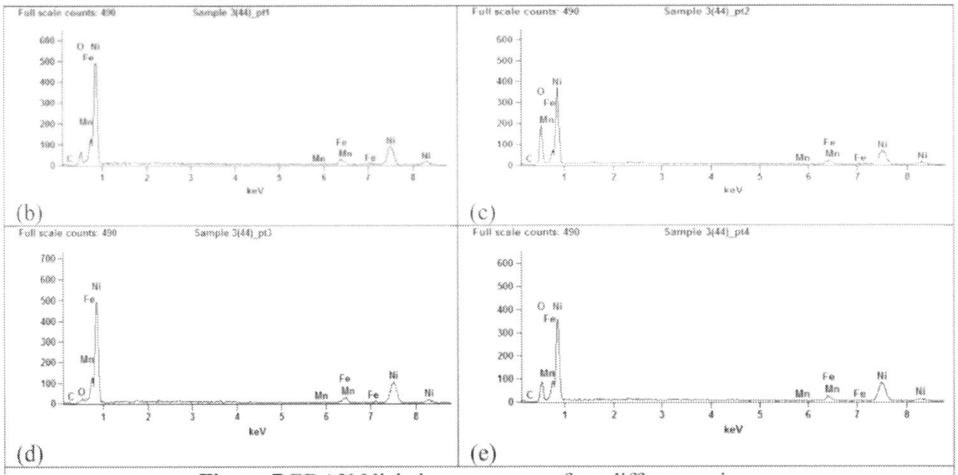

Figure 7.EDAX:Nickel percentage at four different points.

Table 3. Weight percentageof elements obtained by EDAX analysis

Element	C	O	Mn	Fe	Ni
Point 1	0.47	5.69	0.00	10.45	83.39
Point 2	0.39	18.04	0.47	8.00	73.09
Point 3	0.40	2.3	0.08	9.00	88.21
Point 4	0.00	8.92	0.09	8.00	82.99

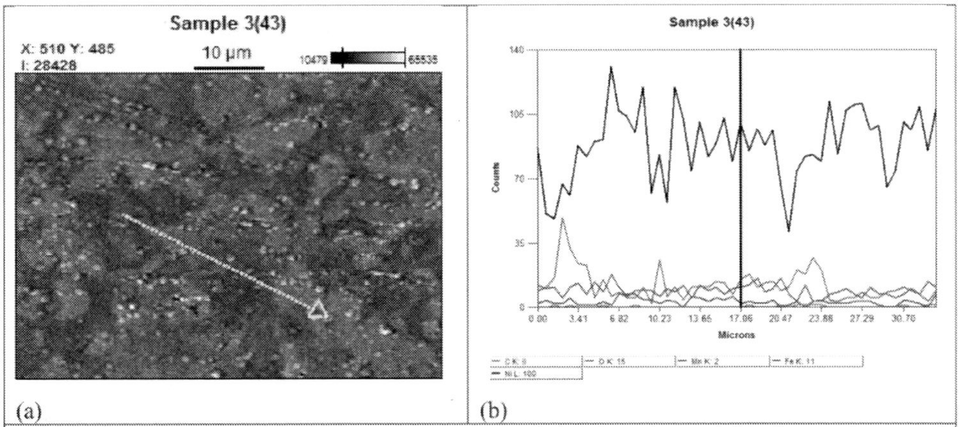

Figure 8.EDAX line scan on the (a) microstructure of nickel coating and (b) elementalcounts of the coating.

IOP Conf. Series: Materials Science and Engineering **75** (2015) 012024 doi:10.1088/1757-899X/75/1/012024

3.3.X-Ray Diffraction (XRD).

Figure 9 shows the X-ray diffractograms using CuK_{α} radiation of the coating done at different current densities. The diffractograms showing the peaks of $Ni_{(111)}$ and $Ni_{(002)}$ confirms the presence of nickel in the coating.

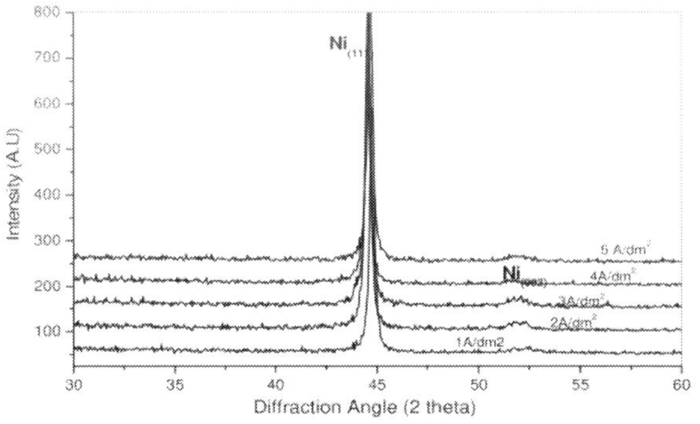

Figure 9. X-ray diffractograms of nickel deposits produced by pulse electro-deposition

3.4. Potentiodynamic Polarization.

Figure10 shows the potentiodynamic polarization curvesof the nickel deposit coated by different pulse current electro-deposition technique on HSLA steelin 3.5 wt. % NaClsolution. The shape of the polarization curves (Figure10)is similar for all the nickel deposits at different densities. In all the cases, the cathodic and anodic branches of the curves show typical *E-i* characteristics of a metal and or alloy in 3.5 wt.% NaCl solution at near neutral pH. The cathodic branches of the polarization curves showed the typical Tafel behaviour enabling to evaluate cathodic slope or Tafel constant (β_c) as well as corrosion current density (i_{corr}) by Tafel extrapolation method. The anodic branch of the polarization curves in 3.5 wt. % NaCl and 3.5 wt.% NaCl solutions exhibited typical active metal dissolution behaviour with an initial sharp increase of corrosion current density with applied anodic overvoltage.

These polarisation curves clearly display that the corrosion current density (i_{corr}) is very low when the deposition was carried out at 20 A/dm^2 compared to pulse deposition carried out at other current densities. This could be attributed to the fact that the nickel deposition at the 20 A/dm^{2}of current density is free from cracking. However, the higher corrosion current (i_{corr}) for the nickel deposits at other pulse current deposition is due to the fact that the deposits are not free from cracking, as observed in figure4 & 5. Thus, it can be inferred from the polarisation curves and SEM micrographs that an optimum pulse current density will result is producing defect free micro and/or sub-micron nickel deposit with an improved corrosion resistance.

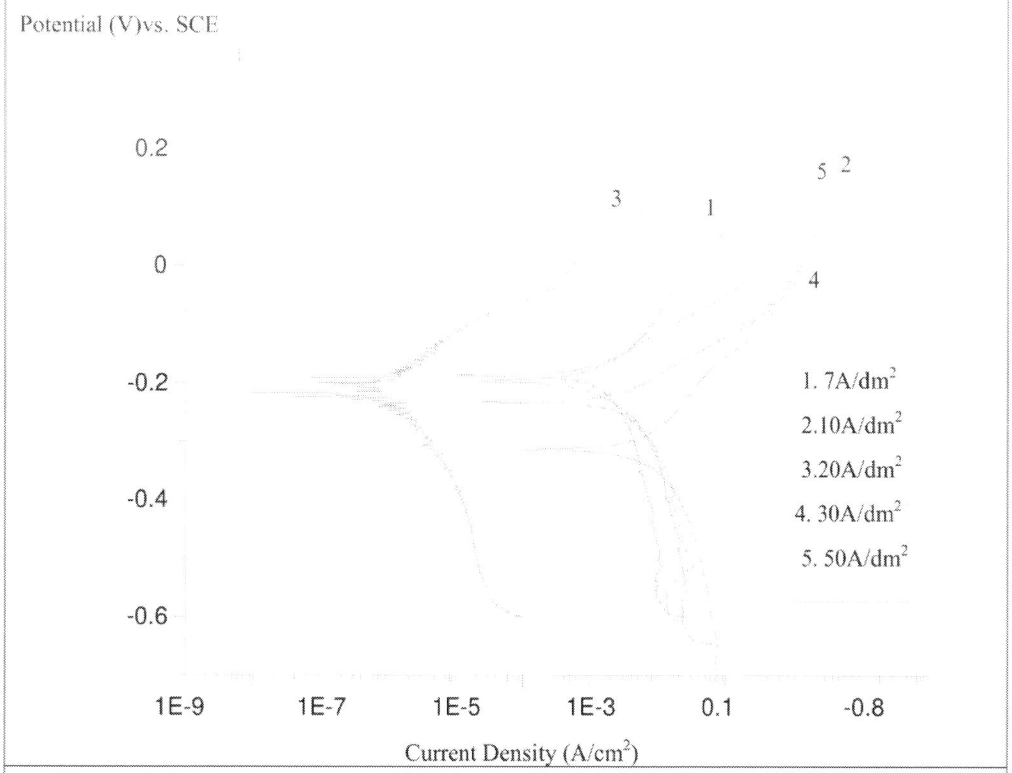

Figure 10. Potentiodynamic polarization curves of nickel coating in 3.5 wt. % NaCl solution.

Table 4. Electrochemical data of the Ni-coating on HSLA steel in 3.5 wt. % NaClsolution.

Coating Current Density	Environment	E_{corr}vs SCE (mV)	i_{corr} (μAcm^{-2})	β_c (mVdec^{-1})	β_a (mVdec^{-1})
7	3.5 wt. % NaCl solution	–444	34.53	–453	260
10		–388	20.67	–920	60
20		–417	0.02	–394	20
30		–515	89.5	–648	138
50		–462	31.7	–490	210

E_{corr} is the corrosion potential and i_{corr} is the corrosion current density at E_{corr} determined by Tafel extrapolation techniqueand β_c and β_a Tafel constants.

4. Conclusions

Pulse current electro-deposition method at different current densities in a suitable nickel bath exhibited micro-or sub-micron nickel deposit over a HSLA substrate. Electro-deposition at current densities of 20-30 A/dm^2 was found out to be crack free compared to coating deposited at higher current densities.The coating done at lower current density also does not seem to be very prominent enoughas the deposit is becoming very thin.XRD and EDAX analyses confirmed the presence of nickel deposit. Potentiodynamic polarisation study of the nickel coating in 3.5 wt. % NaCl solution exhibited characteristics *E-i* curves. Further, it has been observed that the corrosion rate is least of the nickel coating when pulse electro-deposition was performed at current density of 20 A/dm^2.

Acknowledgement

The authors would like to acknowledge M/s Tata Steel, Jamshedpur, India for providing the steel and CMERI, Durgapur for providing facility to use SEM.

References

1] K S, Swygenhoven H V and Suresh S 2003 Mechanical behavior of nanocrystalline metals and alloys *ActaMaterialia*, **51**5743

2] Northwood D O, and Alpas A T 1998. Mechanical and tribological properties of nanocrystalline and nanolaminated surface coatings *Nanostructured materials* **10** 777

3] Wang Z B, Tao N R, Li S, Wang W, Liu G, Lu J, and Lu K 2003 Effect of surface nanocrystallization on friction and wear properties in low carbon steel *Materials Science and Engineering: A***352**144

4] Kim H S, and Bush M B 1999 The effects of grain size and porosity on the elastic modulus of nanocrystalline materials *Nanostructured materials***11**361

5] Balyanov A, *et.al* 2004 Corrosion resistance of ultra fine-grained Ti *ScriptaMaterialia* **51**225

6] Rofagha R, Langer R, E Sherik, A MErb, Palumbo G and Aust K. T 1991 The corrosion behaviour of nanocrystalline nickel *Scriptametallurgica et materialia*, **25**2867

7] Wang L, Gao Y, Xu T, and Xue Q 2006 A comparative study on the tribological behavior of nanocrystalline nickel and cobalt coatings correlated with grain size and phase structure *Materials Chemistry and Physics* **99**96

8] Natter H, Schmelzer M, and Hempelmann R 1998 Nanocrystalline nickel and nickel-copper alloys: Synthesis, characterization, and thermal stability *Journal of Materials research* **13** 1186

9] Badawy W A, Al-Kharafi F M, & Al-Ajmi J R 2000 Electrochemical behaviour of cobalt in aqueous solutions of different pH *Journal of applied electrochemistry* **30**693

10] Wang L, Gao Y, Xue Q, Liu H, and Xu T 2005 Graded composition and structure in nanocrystalline Ni–Co alloys for decreasing internal stress and improving tribological properties *Journal of Physics D: Applied Physics***38** 4339

11] Yang C C, and Cheh H Y 1995 Pulsed Electrodeposition of Copper/Nickel Multilayers on a Rotating Disk Electrode I. Galvanostatic Deposition *J. of the Electroc. Soc.***142** 3034

12] Badawy W A, Ismail K M, and Fathi A M 2005 Effect of Ni content on the corrosion behavior of Cu–Ni alloys in neutral chloride solutions *Electrochimicaacta* **50** 3603

13] Huang C A, Chen C Y, Hsu C C, and Lin C S 2007 Characterization of Cr–Ni multilayers electroplated from a chromium (III)–nickel (II) bath using pulse current *Scriptamaterialia***57** 61

14] McIntyre N S, and Cook M G 1975 X-ray photoelectron studies on some oxides and hydroxides of cobalt, nickel, and copper *Analytical Chemistry***47** 2208

15] Heidari G, Tavakoli H, and Khoie S M 2010 NanoSiC-Nickel composite coatings from a sulfamat bath using direct current and pulsed direct current *J. of materials engineering and performance* **19**1183

16] Brenner A, and Riddell G E 1947 Deposition of nickel and cobalt by chemical reduction *J. of Research of the National Bureau of Standards***39**385

17] Chandrasekar M S, and Pushpavanam M, 2008 Pulse and pulse reverse plating—Conceptual, advantages and applications *ElectrochimicaActa*, **53**3313

18] Bradley P E, and Landolt D 1999 Pulse-plating of copper–cobalt alloys *Electrochimicaacta* **45** 1077

19] Landolt, D., &Marlot, A. 2003 Microstructure and composition of pulse-plated metals and alloys. *Surface and Coatings Technology* **169** 8

20] Gohar S, and Pathak R K 2014 Potentiostatic Electrochemical Preparation and Characterisation of Aluminium Containing Nickel Selenide *Oriental Journal of Chemistry* **29**1469

On the microstructural behavior of titanium alloy during non-optimal regime of superplastic deformation

R.Sahoo, B.B.Jha and T.K.Sahoo

Department of Surface Engineering, CSIR - Institute of Minerals and Materials Technology, Bhubaneswar, India-751013.

E-mail: raja.igit@gmail.com

Abstract. In industrial superplastic forming, maintenance of the narrow ranges of condition required for optimal superplastic flow from beginning to end is a very difficult and challenging task. When these conditions required for optimal superplastic flow lie beyond a certain limit during deformation process, non-optimal process of superplastic deformation is observed during which the microstructure of the material changes very significantly. As the microstructure changes, the stress required to deform the material changes accordingly and the response of the material becomes dependent on the history of loading in this regime of superplastic deformation. A near alpha titanium alloy has been used in this study to find out those parameters of microstructure, which are varying significantly during non-optimal regime of superplastic deformation. Superplasticity tests have been carried out on tensile specimens at 930^0C with a constant strain rate of $1*10^{-4}$ s^{-1} and successfully stopped at two different percentages of elongation. Results indicated that percentage of alpha phase, number of alpha grains per unit area, size of alpha grain, parameter of non-uniaxiality of alpha grain and grain boundary area of alpha grain, etc. varied significantly during non-optimal regime of superplastic deformation.

1. Introduction

Modern day technology demands materials which must have low density, high specific strength, excellent corrosion resistance and good biocompatibility. In this respect, titanium alloys are most favorable candidate materials and are being widely used for many industrial applications such as aerospace, automotive, biomedical etc [1-2]. Unfortunately, these titanium alloys are very hard and brittle in nature. Most of the above mentioned industries have already taken advantages to produce complex shaped components out of these materials in a limited number of manufacturing steps using superplastic deformation [3]. It is now well established that superplasticity is generally achieved in almost all polycrystalline materials with a fine grained microstructure when deformation of these materials is carried out under narrow ranges of strain rate and temperature [4]. Even, the much known brittle materials i.e. titanium alloys exhibit large elongations without fracture or formation of necks at these special regimes of superplastic deformation. These optimal regimes of superplastic condition are found experimentally by many researchers which are almost empirical in nature. To achieve superplastic flow in titanium alloys, three basic requirements must be fulfilled [5]. First, the average grain size should be smaller than 10 micron and these grains should be equiaxed. Second, superplastic deformation requires a relatively high temperature typically at or above $0.5T_m$, where T_m is the

Content from this work may be used under the terms of the Creative Commons Attribution 3.0 licence. Any further distribution of this work must maintain attribution to the author(s) and the title of the work, journal citation and DOI.
Published under licence by IOP Publishing Ltd

absolute melting temperature of the material in degree Kelvin. Finally, these alloys must be deformed within a controlled strain rate mainly in the range of 10^{-4} to 10^{-2} s^{-1}. Of late, most of the research efforts on study of superplastic behavior of titanium alloys are mainly focused to find out the optimal regime of deformation for each alloy system [3, 6].

If all the above mentioned conditions are satisfied adequately during the superplastic deformation, optimal superplastic flow will be established. This behavior is mainly described by strain rate sensitivity parameter (m), the value of which must be greater than ~ 0.3 [7-9]. Main feature of optimal superplastic deformation is the occurrence of intensive grain boundary sliding associated with different accommodation mechanisms. The accommodation mechanisms observed in optimal superplastic deformation are mainly grain boundary migration, recrystallization, grain rotation, diffusional mass transport and slip in grains, etc [10]. As a result, the microstructure of the materials doesn't change actively during optimal superplastic flow and the flow stress is a function of strain, strain rate and the grain size of the material [11-12]. Based on various deformation mechanisms operating during superplastic forming, the constitutive equation as described by many researchers is as follow:

$$\sigma = f\,(\varepsilon, \frac{d\varepsilon}{dt}, T, d)$$

Where σ is the flow stress, ε is the strain, dε/dt is the strain rate, T is the temperature and d is the average grain size of the material [12-13]. In real technological applications, maintenance of these narrow ranges of condition required for optimal superplastic flow from beginning to end is a very difficult and challenging task [14-15]. In superplastic forming industry, the deformation process is carried out not at optimal, but close to the boundary of optimal superplastic condition that is called near super plastic regimes for which microstructure of the material changes significantly during the deformation process [14-17]. As the microstructure changes, the stress required to deform the material changes accordingly and the response of the material becomes dependent on the history of loading in this regime of superplastic deformation [15-16].For simulation and modeling of these industrial superplastic forming processes, it is very much essential to develop a constitutive equation which can also take into account the major changes in microstructure of materials that takes place under non-optimal process of superplastic deformation.

In 1986, Ghosh et al. reported that the distribution of grain size of materials can change due to either grain growth or grain refinement occurring during superplastic deformation [18]. In superplastic deformation, strain is known to produce grain growth when a polycrystalline material deforms predominantly by diffusion creep and grain refinement when the same deforms by dislocation creep. In 1996, Zhou et al. developed constitutive equations taking grain size in to account based on deformation mechanisms i.e. diffusional creep, grain boundary sliding, dislocation creep and grain growth to describe the superplastic behavior of Ti-6Al-4V Alloy [19]. In 2009, Bkhattacharya et al. carried out superplasticity tests on Ti-6Al-4V alloy in non-optimal regime and reported that the average grain size of Ti-6Al-4V alloy is changed from its initial values during this regime of superplastic deformation [14, 20]. In 2012, Bhaskaran et al. developed a constitutive equation taking grain size distribution into account to describe the behavior of various titanium alloys (i.e. near alpha titanium alloy and VT-9 titanium alloy) during near superplastic / non-optimal regime of deformation [15]. They concluded that the constitutive equation should also include complex microstructural transformation other than changes in grain size occurring in the non-optimal regime of superplastic deformation in order to improve its robustness and accuracy. As an input to the development of the constitutive equation for this regime of superplastic deformation, it is very much important to find out the microstructural parameters other than average grain size which are varying during the deformation processes. To the best of author's knowledge, no serious research efforts have been undertaken till date on this matter. In this present investigation, therefore, an attempt has been made to find out those parameters of microstructure which are varying significantly during non-optimal regime of superplastic deformation of titanium alloy.

2. Research methodology

Development of an appropriate constitutive equation is the core subject of mechanics of solids. The investigations in this area are becoming much more complex when mechanical properties are strongly dependent on the microstructure of materials which changes significantly during the process of deformation. Such type of dependence is observed in the non-optimal regime of superplastic deformation. The present study aims to investigate the microstructural behavior of titanium alloy when the conditions of experiments are close to the boundary of optimal superplastic condition i.e. non-optimal regime of superplastic deformation. To satisfy the above condition, therefore, the limitations of the conditions required for superplastic flow need to be violated in one or more than one parameters.

Out of the three controllable parameters (i.e. grain size, temperature and strain rate), it is very much convenient to control the strain rate externally. We also have control on the initial microstructure of the materials. As a result, the average grain size of titanium alloy must be less than 10 micron in order to achieve superplastic deformation in these alloy system. Temperature is another parameter which is most difficult to be actively controlled in the experiments due to large thermal inertia of the system. Hence, the temperature has fixed value from the range of temperatures required for superplastic deformation. The strain rate is the only parameter which can be controlled externally during the course of experiment by controlling the velocity of cross head of the universal testing machine [11]. Therefore, the experiment need to be carried out at a constant strain rate of $1*10^{-4}$ s^{-1} (which is close to the boundary conditions required for optimal superplastic flow) and strictly adhering to a constant temperature of 930^{0}C (fixed value from the range of temperatures required for superplastic deformation).

3. Material and methods

3.1. ***Superplasticity test in non-optimal regime***

The superplasticity tests have been carried out on tensile specimens made from as received rods of near alpha titanium alloy. These rods have undergone the mill process of cold rolling and annealing and its chemical composition is given in Table 1.

Tab.1 Chemical composition of as received titanium alloy

Element	**Al**	**V**	**Mo**	**Fe**	**Si**	**C**	**Ti**
Weight %	8.57	1.83	0.14	0.2	0.06	0.08	Balance

The beta transus temperature of this alloy was found experimentally to be 1040^{0}C. Cylindrical specimens with gauge length 15 mm and diameter 5 mm have been made out of 12.7 mm diameter rod and tested on a universal testing machine (Instron 8801) having an electric heating furnace. The measurement of temperature of the specimen was made by thermocouples placed in immediate proximity to the gauge part of the specimen. It has been observed that the difference in temperature indicated by thermocouples attached with the furnace wall and the same placed at the surface of the specimen is about ± 2^{0}C. The specimens were tested with a constant strain rate of $1*10^{-4}$ s^{-1} at 930^{0}C and successfully stopped at two different percentages of elongation i.e. 160 and 466 percentages respectively. The constant strain rate was maintained by programming the exponential law for the velocity of crosshead of the universal testing machine [11]. At the end of experiment, the specimen was water quenched immediately in order to preserve the microstructure. Metallographic samples were prepared from the central portions of the tensile specimen after quenching. Kroll's chemical reagent is used in order to reveal the microstructure of deformed specimens of titanium alloy [21].

3.2 ***Material characterization***

The microstructure of both as received and deformed titanium alloy was characterized through optical microscope. To know the various phases present in the deformed tensile specimens after quenching,

XRD analysis was carried out on representative metallographic samples in a XRD Instrument (X'pert PRO PANalytical make) using copper as target. Hardness values for both initial and deformed titanium alloy have been measured using a micro-Vickers hardness tester (Leitz make) with a load of 300 gram on metallographycally prepared samples. Minimum five numbers of readings have been taken on each sample in order to get reasonable statistics of the measured hardness values.

3.3 *Calculation of microstructural parameters*

Various types of microstructure are found in titanium alloys such as Lamellar, Equiaxed and Bimodal depending upon the different combination of the alpha and beta phases [22-23]. Size, shape and arrangement of these two phases primarily control the microstructural features of this alloy system. The geometrical parameter of these two phases such as size of alpha grain, parameter of non-uniaxiality of alpha grain, relative length of alpha grain boundary, orientation of alpha grain, volume percentage of alpha phase, and size of transformed beta grain, etc. are used to describe the microstructure of titanium alloys [24-25]. In this present study, we calculated percentage of alpha phase, number of alpha grains per unit area, size of alpha grain, parameter of non-uniaxiality of alpha grain and grain boundary area of alpha grain in a optical microscope (Leica DMI 3000M) by taking each individual alpha grain into account.

Three metallographic photographs were taken at about 1000x magnification along axial as well as transverse direction of material before and after superplasticity test. The shape of the alpha grain is assumed to be elliptical in nature and the length of both the major and minor axis was measured [26]. Length of the major axis was reported to be the size of the alpha grain. Parameter of non-uniaxiality of alpha grain was calculated as the ratio of length of major axis to that of minor axis. In case of two dimensional figures, the perimeter of alpha grain is reported as its grain boundary area. The average value with respect to minimum of three metallograph is reported.

4. Results and discussion

4.1 *Microstructural features and XRD analysis*

The microstructures of the as received rod of titanium alloy in longitudinal and transverse direction are shown in the Fig.1. and Fig 2. These micrographs show that the original material contains small elongated alpha grains which are uniformly distributed throughout the beta matrix. The grain size distribution of alpha phase of as received titanium alloy in longitudinal direction is shown in Fig.3.

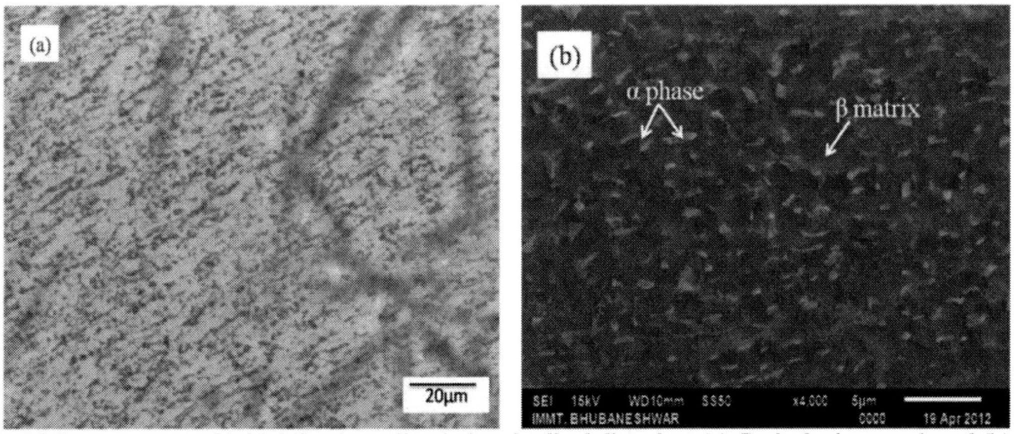

Fig.1 Microstructure of initial titanium alloy in longitudinal direction: (a) Optical micrograph and (b) SEM micrograph

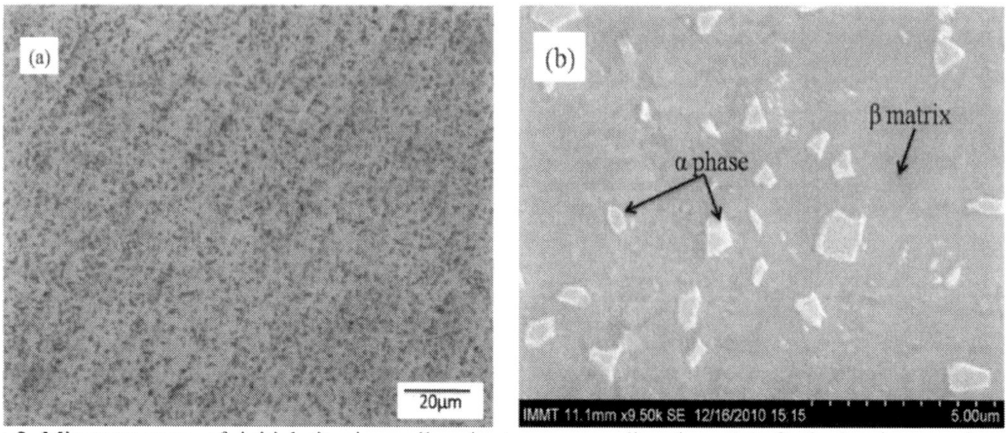

Fig.2 Microstructure of initial titanium alloy in transverse direction: (a) Optical micrograph and (b) SEM micrograph

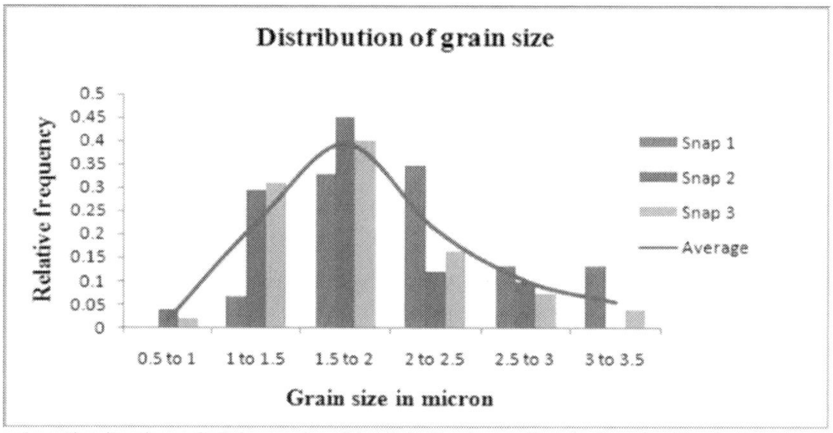

Fig.3 Grain size distribution of alpha grain of initial titanium alloy in longitudinal direction

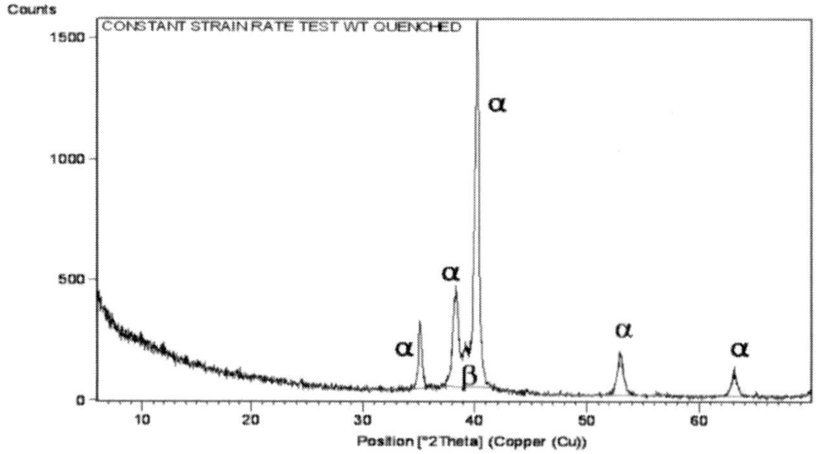

Fig.4 XRD plot of deformed titanium alloy after 160% elongation

XRD analysis has been carried out on representative samples to determine the phases present in the deformed material and are shown in Fig.4. From the XRD results, it has been confirmed that the deformed materials contain alpha phases in a beta matrix. All the peaks are showing the presence of alpha phases; but only the peak corresponding to the 2θ value equal to 39^0 shows the presence of beta phases. These results are in conformity with the previously reported results of commercially pure titanium, Ti-6Al-4V and Ti-6Al-7Nb alloys [21].Detailed microstructural parameters (with their standard deviation values) of the initial titanium alloy in both longitudinal and transverse direction are shown in Table 2 and Table 3 respectively.

Tab. 2 Detailed microstructural parameters of initial and deformed titanium alloy in longitudinal direction

Microstructural parameters	**Initial material**	**After 160% elongation**	**After 466% elongation**
Percentage of alpha phase	22.5 ± 0.5	48 ± 0.5	55 ± 0.5
Number of alpha grains per unit area(100 μm X100μm)	3962 ± 12	306 ± 9	130 ± 6
Average size of alpha grain (μm)	1.9 ± 0.3	12.43 ± 0.41	8.13 ± 0.38
Parameter of Non-uniaxiality of alpha phase	1.55 ± 0.22	1.39 ± 0.01	1.35 ± 0.04
Average grain boundary area of alpha phase (μm)	7.86 ± 0.68	38.31 ± 0.73	28.6 ± 0.49
Micro Vickers hardness (VHN)	305 ± 2	318 ± 2	350 ± 3

Tab. 3 Detailed microstructural parameters of initial and deformed titanium alloy in transverse direction

Microstructural parameters	**Initial material**	**After 160% elongation**	**After 466% elongation**
Percentage of alpha phase	20 ± 0.5	45 ± 0.5	47.5 ± 0.5
Number of alpha grains per unit area (100 μm X100μm)	2423 ± 9	270 ± 7	174 ± 7
Average size of alpha grain (μm)	1.72 ± 0.34	9.67 ± 0.39	6.63 ± 0.34
Parameter of Non-uniaxiality of alpha phase	1.41 ± 0.02	1.29 ± 0.02	1.26 ± 0.02
Average grain boundary area of alpha phase (μm)	6.29 ± 0.75	31.30 ± 0.71	24.16 ± 0..72
Micro Vickers hardness (VHN)	301±2	315 ± 1	342±2

After constant strain rate ($1*10^{-4}$ s^{-1}) test, the elongation has been found to be 160 and 466%. The microstructure in longitudinal direction as well as in the transverse direction at the central portion of the tensile specimen tested at constant strain rate of $1*10^{-4}$ s^{-1} is shown in Fig.5 and Fig.6. After comparing the microstructure shown in Fig.1 and Fig.2 with that of Fig.5 and Fig.6, it is clearly observed that there is a significant change in microstructure of titanium alloy during deformation in non-optimal regime. Detailed microstructural parameters of the deformed materials for different percentages of elongation in both longitudinal and transverse direction after the constant strain rate ($1*10^{-4}$ s^{-1}) test are shown in Table 2 and Table 3. Our results indicated that the microstructural parameters, those varying significantly during superplastic deformation of titanium alloy in non-optimal regimes are percentage of alpha phase, number of alpha grains per unit area, average size of the alpha grain, parameter of non-uniaxiality of alpha grain, and the grain boundary area of alpha grain. It was clearly observed that all these microstructural parameters get changed from initial material to the deformed material in both longitudinal as well as transverse directions. It has been revealed that the alpha grains grow at the expense of the beta phases in both longitudinal and transverse direction during the process of deformation. The size of the alpha grain is found to be higher after 160% elongation as compared to the same after 466% elongation in both longitudinal and transverse directions.

Fig.5. Microstructure of deformed titanium alloy after 160% elongation: (a) longitudinal direction and (b) transverse direction

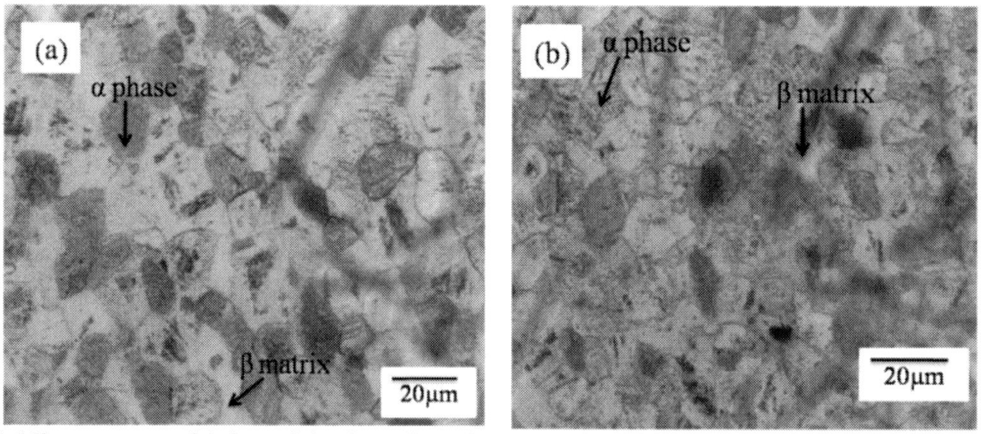

Fig.6. Microstructure of deformed titanium alloy after 466% elongation: (a) longitudinal direction and (b) transverse direction

4.2 *True stress vs. True strain behavior*

Typical True stress vs. True strain plot for constant strain rate ($1*10^{-4}$ s^{-1}) test, carried out at 930^0C for both 160% and 466% elongation is shown in the Fig.7. For 160% elongation, this plot clearly indicates the occurrence of strain hardening during superplastic deformation in which the requirement of true stress for further deformation is gradually increasing with increase in the extent of deformation. The requirement of continuously increasing true stress to carry out further deformation could possibly be due to grain growth occurring in this material during deformation [27-30]. This has been clearly observed by comparing the initial microstructure of titanium alloy (Fig.1 and Fig.2) with that of deformed one (Fig.5). For 466% elongation, this plot clearly indicates the occurrence of strain hardening and strain softening during superplastic deformation in which the requirement of true stress for further deformation is gradually increasing with increase in the extent of deformation up to a certain percentage of elongation (i.e. 352%) and after that the requirement of true stress for further deformation is gradually decreasing. The requirement of continuously decreasing true stress to carry out further deformation after reaching a certain percentage of elongation (i.e. 352%) could possibly be due to recrystallization occurring in this material during deformation [15].

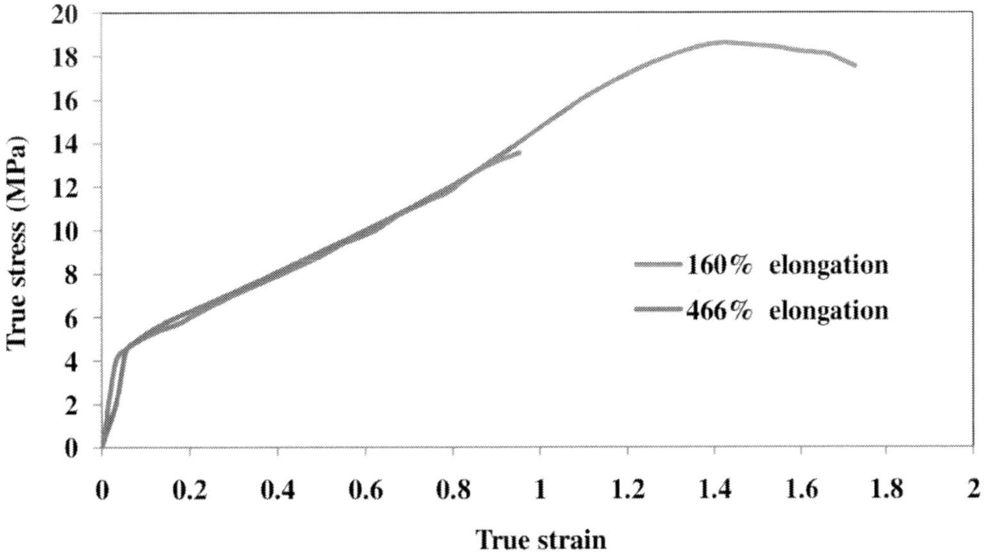

Fig.7. True stress vs. True strain curve for the superplasticity test of titanium alloy at 930^0C with a strain rate of $1*10^{-4}s^{-1}$

4.3 ***Strain induced grain growth and recrystallization***

To achieve large elongation in superplastic materials, it is very much essential to maintain a small and equiaxed grain structure during the whole range of superplastic deformation [4, 29-31]. The mechanisms of superplastic deformation of materials are now well documented [5, 10]. Among all the mechanisms, most important is the intensive grain boundary sliding which is mainly contributing for such large elongations without the formation of neck [4, 10, 31]. Though, the majority of experimental data suggests that grain boundary sliding plays the dominant role in the superplastic deformation process, but it is reported that some additional accommodation processes (either diffusion controlled or dislocation motion) must accompany the grain boundary sliding to prevent the formation of neck [32]. Since diffusional flow is strongly dependent on grain size and dislocation creep is independent of the same, small grains are generally deformed by diffusion creep and large grains are generally deformed by dislocation creep during superplastic deformation [18]. Wilkinson et al. reported that grains can increase in size by several microns for every unit of applied strain when they are deformed at very slow strain rate [18]. There are already many literatures which report that strain induced grain growth taking place at a very slow strain rate during superplastic deformation [31, 33]. Since the contribution of diffusional creep to superplastic deformation in small grain materials is more significant and grain growth is inevitable at slower strain rate, it is attributed that diffusional creep leads to strain induced grain growth as observed in our experiment up to 352% of elongation [18, 31, 33]. Bhaskaran et al. reported that all the grains grow until they reach a critical size and as soon as the grain achieves that critical size, it undergoes refinement [15]. With the formation of dislocation walls, large grains after reaching the critical size tend to break into small ones or new grain nucleate and grow on the surface of an existing large grain [16]. As the contribution of dislocation creep to superplastic deformation in large grain materials is more significant, it is attributed that dislocation creep leads to strain induced recrystallization as observed in our experiment after 352% of elongation.

4.4 *Deformation induced phase transformation*

After superplastic deformation of titanium alloy in non-optimal regime, the percentage of alpha phase in both longitudinal and transverse directions of the deformed material has been found to be increased in comparison to the same present in the initial material. This change in microstructural parameter, especially transition from beta to alpha phase could be facilitated in the presence of strain. Xue et al. observed deformation induced phase transformation in polycrystalline zinc sulfide at a temperature range of 900^0C to 1150^0C. They attributed that the conversion of hexagonal closed pack structure to body centered cubic structure below the allotropic transformation temperature i.e. 1020^0C as strain dependent [34]. The percentage of alpha phase in the initial material was 22.5% and 20% in longitudinal and transverse direction respectively. However, after deformation it was found that in both longitudinal and transverse directions, the percentage of alpha phase has been increased significantly in comparison to the same present in the initial material as shown in Table 2 and Table 3. As the beta transus temperature of this titanium alloy is 1040^0C and the experiments have been carried out at much lower temperature i.e. 930^0C, this type of phase transformation may possibly be attributed to deformation induced phase transformation, which has been reported in other alloy system as well [35-39]. It is also observed that with increase in percentage of elongation, the percentage of alpha phases in the deformed material increases.

4.5 *Microstructural parameters*

A detail of microstructural parameters of both as received and deformed titanium alloy is given in Table 2 and Table 3. Microstructure plays a very important role in determining the mechanical properties of the titanium alloy, because practically all the mechanical properties are structure sensitive [22]. After superplasticity test, the micro-Vickers hardness values have increased from as received material to deformed material. As hardness is a structure sensitive mechanical property, it clearly indicates that this increase in hardness is the effect of change in the microstructure during superplastic deformation of titanium alloy in non-optimal regime. From this preliminary investigation, it is clearly observed that the microstructural parameters varying significantly are the percentage of alpha phase, number of alpha grains per unit area, size of the alpha grain, parameter of non-uniaxiality of alpha grain and grain boundary area of alpha grain. Hence, all these microstructural parameters need to be incorporated in the constitutive equation enabling the same to describe the materials behavior during superplastic deformation in non-optimal regime.

5. Conclusions

- Significant changes in microstructure have been observed in both longitudinal and transverse direction during superplastic deformation of titanium alloy in non-optimal regime.
- After superplastic deformation of titanium alloy in non-optimal regime, the percentage of alpha phase in both longitudinal and transverse directions of the deformed material has been found to increase in comparison to the percentage of alpha phase in the initial material. This type of phase transformation below the beta transus temperature has been attributed to deformation induced phase transformation. Strain induced grain growth and recrystallization is also observed during superplastic deformation of titanium alloy in non-optimal regime.
- The microstructural parameters those varying significantly are percentage of alpha phase, number of alpha grains per unit area, size of the alpha grain, parameter of non-uniaxiality of alpha grain, and grain boundary area of alpha grain.

- All these microstructural parameters other than average grain size need to be incorporated in the constitutive equation describing the superplastic behavior of materials in order to increase its robustness and accuracy.

References

[1] Tan H et al 2001*Materials and Manufacturing Processes* **16** 331
[2] Xiaomei L and Soo S 2010 *Materials and Manufacturing Processes* **25** 1470
[3] Zhanglong Z et al 2009 *Rare Metals* **28** 523
[4] LangdonT G 2009 *Journal of Material Sciences* **44** 5998
[5] Giuliano G 2008 *Journal of Materials and Design* **29** 1330
[6] Lin D and Sun F 2004 *Intermetallics* **12** 875
[7] Wang G and Fu M W 2007 *Journal of Materials Processing Technology* **192-193** 555
[8] Carrino L et al 2003 *Finite Elements in Analysis and Design* **39** 1083
[9] Khaleel M A et al 2001 *International Journal of Plasticity* **17** 277
[10] Sieniawski J and Motyka M 2007 *Journal of Achievements in Materials and Manufacturing Engineering* **24** 123
[11] Pan W et al 2005 *Proceedings of the Institution of Mechanical Engineers, Part L: Journal of Materials: Design and Applications* **219** 149
[12] Chandra N 2002 *International Journal of Non-Linear Mechanics* **37** 461
[13] Khaleel M A et al 1998 *International Journal of Plasticity* **14** 1133
[14] Bylya O I et al 2008 *AIP Conference Proceedings* **1063** 349
[15] Bhaskaran K et al 2012 *Applied Mechanics and Materials* **110-116** 163
[16] Bylya O I et al 2012 *AIP Conference Proceedings* **1461** 132
[17] Bylya O I et al 2013 *Materials Science Forum* **735** 232
[18] Ghosh A K and Raj R 1986 *Acta metallurgica* **34** 447
[19] Zhou M and Dunne F P E 1996 *The Journal of Strain Analysis for Engineering Design* **31** 187
[20] Bkhattacharya S S et al 2009 *Mechanics of Solids* **44** 951
[21] Ajeel S A et al 2007 *Eng. & Technology* **25** 431
[22] Lutjering G 1998 *Materials Science and Engineering A* **243** 32
[23] Rack H J and Qazi J I 2006 *Materials Science and Engineering C* **26** 1269
[24] Tiley J et al 2004 *Materials Science and Engineering A* **372** 191
[25] Collins P C et al 2009 *Materials Science and Engineering A* **508** 174
[26] Wu G Q et al 2011 *Materials science and Technology* **27** 1777
[27] Ghosh A K and Hamilton C H 1979 *Metallurgical and materials transactions A* **10 (a)** 699
[28] Seidensticker J R et al 1998 *Acta materialia* **46** 4883
[29] Cheong B H et al 1999 *The Journal of Strain Analysis for Engineering Design* **35** 149
[30] Cheong B H et al 2001 *Journal of Materials processing technology* **119** 361
[31] Wilkinson D S and Caceres C H 1984 *Acta Metallurgica* **32** 1335
[32] Geckinil A E 1983 *Metal science* **17** 12
[33] Caceres C H and Wilkinson D S 1984 *Acta metallurgica* **32** 415
[34] Xue L A and Raj R 1990 *Journal of Materials Science Letters* **9** 818
[35] Zhu Y H et al 2003 *Materials Research Bulletin* **38** 1851
[36] Mukhopadhyay C K et al 1993 *Journal of Materials Science* **28** 145
[37] Das A et al 2011 *Materials Science and Technology* **27** 366
[38] Han H N et al 2004 *Acta Materialia* **52** 5203
[39] Ziętek G and Mroz Z 2011 *International Journal of Structural Changes in Solids – Mechanics and Applications* **3** 21

Production of semi-solid feedstock of A356 alloy and A356-5TiB$_2$ *in-situ* composite by cooling slope casting

S Deepak Kumar[1,*], **Mihira Acharya**[2], **A Mandal**[2], **M Chakraborty**[3]
[1]School of Mechanical Sciences, Indian Institute of Technology Bhubaneswar, Bhubaneswar-751013, Odisha, India
[2]School of Minerals, Metallurgical and Materials Engineering, Indian Institute of Technology Bhubaneswar, Bhubaneswar-751007, Odisha, India
[3]Department of Metallurgical and Materials Engineering, Indian Institute of Technology Kharagpur, Kharagpur-721032, India

*E-mail: dks10@iitbbs.ac.in

Abstract. Cooling Slope casting process has gained significant importance for manufacturing of semi-solid feedstock Al alloys and composites which find applications in automotive and aerospace industries. The primary objective of this research is to generate the semi-solid feedstock suitable for thixoforming process. The current work discusses the phenomena involved in the evolution of microstructure of the semi-solid feed stock by Cooling Slope casting process and the effect of various parameters. The process parameters like the angle and length of inclined plate affect the size and morphology of α-Al phase. The Cooling slope process resulted in the formation of fine grain size of about 38 μm of α-Al phase, compared to that of 98 μm processed conventionally. Further, the pouring temperature played a crucial role in the generation of semi-solid microstructures in case of both the alloy and composites. Moreover, fine TiB$_2$ particles of size 0.2-0.5 μm are uniformly distributed in the almost spherical α-Al grains and at the grain boundaries in the composite feed stock. Cooling slope casting route is a suitable and economical route for semi-solid slurry generation by effectively varying the process parameters.

Keywords: Semi-solid processing; Rheocasting; Cooling Slope; A356 alloy; TiB$_2$ *in-situ* composite;

1. Introduction

Semi-solid Metal Processing (SSMP) has been posed to be a potential route to form near-net-shape products in the automotive, aerospace and other engineering industries.The key element in the SSMP is the globular microstructure rather than dendritic microstructure and it is the thixotropic behaviour of the semi-solid slurry which reduces the segregation and porosity in the castings [1-2].Compared with conventional techniques, SSMP provides several advantages like higher product quality, lower forming temperature, higher production rate and improvement of the mechanical properties through microstructural refinement [3].SSMP methods require thixotropic feedstock materials as starting material for further thixoforming process [4].Over the past four decades, much research is underway

* E-mail: dks10@iitbbs.ac.in

Content from this work may be used under the terms of the Creative Commons Attribution 3.0 licence. Any further distribution of this work must maintain attribution to the author(s) and the title of the work, journal citation and DOI.
Published under licence by IOP Publishing Ltd

around the globe to develop cost effective feedstock production methods and recently in the last decade, Cooling slope (CS) casting has emerged as viable route for generation of feedstock [5]. Extensive research on CS casting process was initially carried out by T. Haga and research group and they reported that CS casting process is a simple, has minimal equipment requirements and which is able to produce feedstock material for semisolid processing [6]. Later, Birol [7] reported that when the feedstock is reheated to the semisolid temperature range, nondendritic, spherical solid particles in a liquid matrix suitable for thixoforming are obtained [8]. Robert et al. [9] produced semi-solid A356 alloy by CS route and stated that thixotropic property is greatly influenced by CS processing route. In addition, Legoretta et al. [10] developed a thorough understanding of the microstructural development of A356 alloy during the CS process and demonstrated the effect of process parameters on the CS cast feed stock. Taghavi and Ghassemi [11] investigated on the effects of the length (20, 30, 40, 50, 60 cm) and angle of inclined plate (20°, 30°, 40°, 50°) on the thixotropic microstructure of A356 aluminum alloy. Their results indicated that 40° and 40 cm condition is optimum inclined plate angle and length presenting the minimum grain size of 28μm and maximum shape factor of 0.70 with highest uniformity. Das et al. [12] has conducted experiments and studied the effects of slope angle and addition of grain refiner on microstructure evolution of the solidifying melt through rheocasting process employing cooling slope (CS) casting route. They reported that there is a significant increase in the tensile properties of the rheocast feed stock processed by CS casting route.
Recently, Deepak et al. [13] conducted a review on Cooling Slope casting process of semi-solid Aluminium alloys and reported that CS casting process is gaining popularity for the generation of metal matrix composite feedstock with globular microstructure for further thixoforming. From the literature survey, it is clear that very few works exists on the semi-solid feedstock production of Al based metal matrix composites. Thus, the present work is an attempt to investigate the feasibility of this Cooling Slope casting process to generate semi-solid A356-5TiB_2 *in-situ* composite feedstock suitable for thixoforming process.

2. Materials and Methods

2.1. Materials

A commercial A356 alloy and A356-5wt%TiB_2 *in-situ* composite synthesized by Flux assisted synthesis (FAS) technique were used in the present study. The chemical composition of A356 alloy is shown in table 1.

Table 1. Chemical composition of A356 alloy (wt %)

Alloy	Si	Mg	Fe	Cu	Mn	Al
A356	6.87	0.297	0.12	0.01	0.003	92.70

2.2. Processing of A356-5TiB_2 in-situ composite

The A356-5wt%TiB_2 *in-situ* composite was fabricated by FAS technique. Initially, an appropriate amount of Al-50Si master alloy and commercial pure aluminium (CPAL) was melted in a graphite crucible in a resistance furnace to prepare the Al-7Si alloy. The K_2TiF_6 and KBF_4 halide salts were added to the melt which undergo an exothermic, *in-situ* reaction with molten Al-7Si alloy to create titanium-diboride (TiB_2) dispersoids in the melt. The melt was kept at a temperature of 800 °C and stirred for every 10 minutes for homogenous formation of TiB_2 particles. After a reaction time of one hour, the dross which floats on the top of the melt was decanted and appropriate amount of Al-20Mg master alloy was added so as to make the net composition of the composite as Al-7Si-0.3Mg- 5TiB_2. A schematic diagram of synthesis of *in-situ* composites is shown in figure 1.Further details of FAS technique can be found elsewhere [14-15].

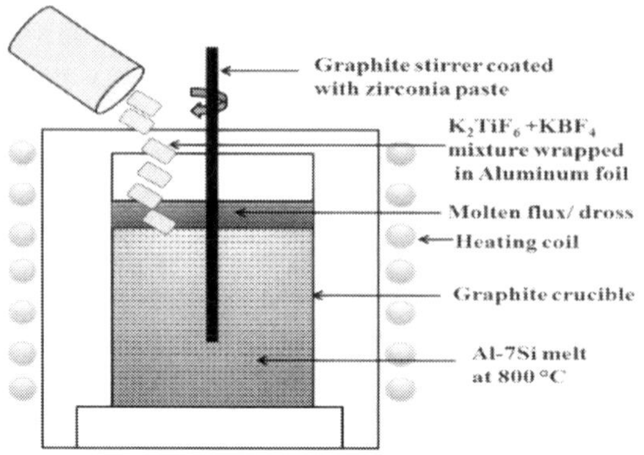

Figure 1. Schematic diagram of synthesis of A356-TiB_2 composites

2.3. *Cooling Slope casting*

The Cooling slope (CS) casting process was employed as a feedstock production method for both the alloy and composites. The A356 alloy and A356-5TiB_2 *in-situ* composites were melted at 720 °C and 800 °C respectively in a resistance furnace. Previous studies in our group [14-16] suggests that the complete reaction of K_2TiF_6 and KBF_4 halide salts with the molten aluminium alloy takes place at an optimum reaction temperature of 800°C for reaction time of 60 minutes resulting in formation of only TiB_2 particles in the molten aluminium alloy. Hence the composites were melted at a higher temperature of 800°C while A356 alloy was melted at 720°C.The melt is then degassed with 1 wt% hexa-chloroethane and was then allowed to cool to desired pouring temperatures. The CS casting experiments involved pouring the molten A356 alloy and A356-5TiB_2 *in-situ* composite over a 50 mm wide and 400 mm long, inclined mild steel plate into a permanent mold with a diameter of 30 mm and a depth of 135 mm.The cooling plate was adjusted at different angles varying from 30° to 60° with respect to the horizontal plane and was cooled by water circulation underneath. The experimental setup of the CS casting process is shown in figure 2 and the experimental parameters are given in Table 2. The flow rate of water circulation was measured using a single jet water meter (Model: LXC-15DS) and is kept constant of 2 L/min throughout the experiments. The surface of the plate was coated with a thin layer of zirconia paste in order to avoid sticking of the melt on the plate. A pre-heated graphite funnel was used as a pouring cup and a graphite cone was used to allow the melt to fill in the mold more easily. Temperatures were monitored with a K-type thermocouple fixed at the start of the slope and one inserted in the mold to record the start and exit temperatures of the melt along the slope. The semi-solid feedstock of the alloy and composites thus generated, is illustrated in figure 3.Thereafter, the composites were characterized by X-ray diffractometer (Model: PW 1710, PHILIPS, Netherlands) using Cu-K_α radiation (wavelength =1.54 Å) to identify the phases, as shown in figure 4.

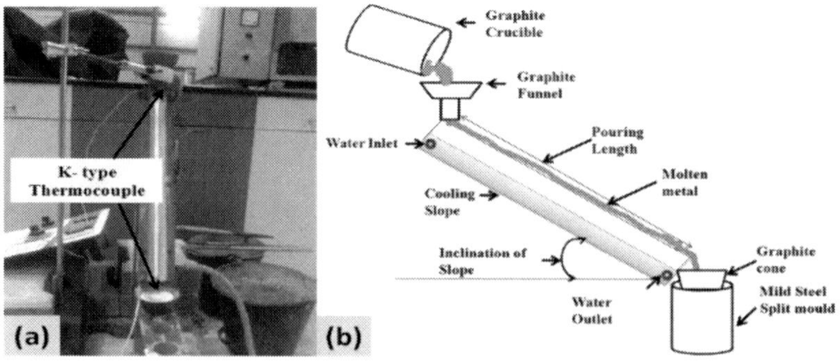

Figure 2. Cooling slope casting (a) Experimental setup and (b) schematic illustration

Table 2. Experimental conditions of CS casting process

CS Process parameters	Range
Cooling slope plate	Length = 400 mm , 300 mm and 200 mm
Material of plate	Mild steel
CS Plate Dimensions	Length =400 mm; width = 50 mm; Radius = 12.5mm (U-Type)
Coating on the inclined plate	Zirconia paste
Pouring Temperatures	630 °C, 640 °C, 650 °C
Cooling slope plate angle	30°,45°,60°
Material of Mould	Mild steel split type
Mould Dimensions	Φ30 mm x 135 mm

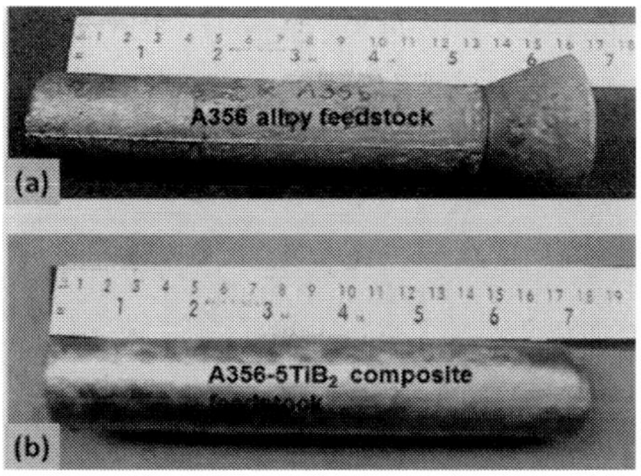

Figure 3. Semi-solid feedstock of (a) A356 alloy (b) A356-5TiB$_2$ composite

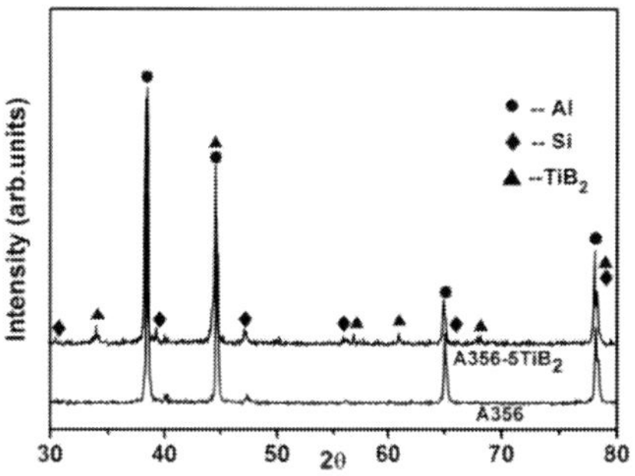

Figure 4. XRD patterns of A356 alloy and A356-5TiB$_2$ composites

3. Results and discussion

3.1. Microstructural Investigation of A356 alloy

The morphologies of A356 alloy processed by conventional casting have dendritic structure as shown in figure 5a, while that processed via CS process typically consists of mixture of rosettes and nearly spherical grains as shown in figure 5b.This is attributed due to the inclined plate which acts as a source of nucleation sites, generates crystal nuclei in the melt during its flow along the slope [9-10]. The α-Al crystals nucleate and grow on the slope wall and further, they are detached from the surface of the plate, because of shearing action generated during the melt flow. The solidified strip and the morphology of A356 alloy on CS strip at exit end, poured at 60° angle is shown in figure 6.

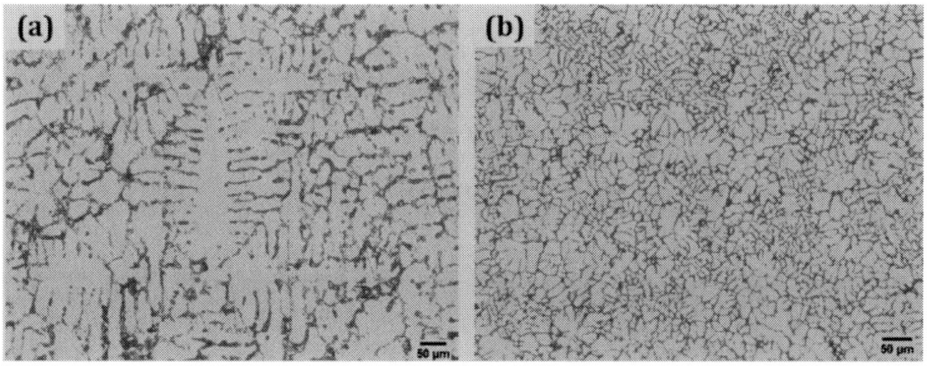

Figure 5. Microstructures of A356 alloy (a) Conventionally cast (b) Cooling Slope casting

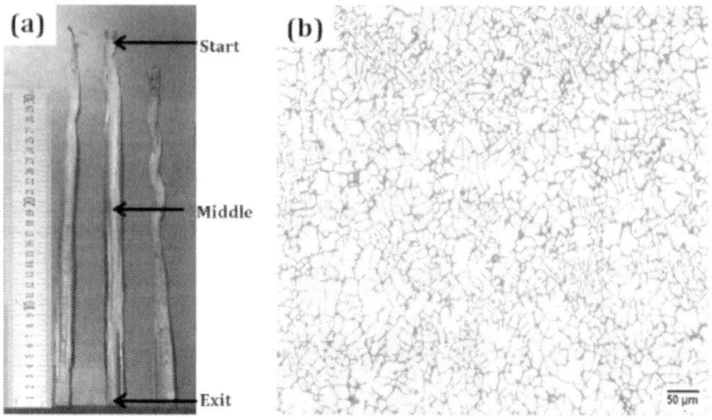

Figure 6. (a) Solidified strip (b) Morphology of A356 alloy on CS strip at exit end, poured at 60° angle

In order to investigate the effectiveness of the cooling slope, a study on the effect of angles of the cooling slope viz. (30°,45°,60°) on the microstructure of the A356 alloy was investigated and the corresponding microstructures were shown in figure 7(a-c).
Further, inorder to quantify the morphology and size of primary solid phase (α-Al), the grain size, i.e circular diameter (C.D) and the degree of spheriocity, i.e Shape factor (S.F) are calculated according to the following equations Eq. (1) & Eq. (2), reported elsewhere [11,12]. The variation of Grain size and S.F of α-Al phase of A356 alloy at different processing conditions is shown in figure 8.

$$CD = 2\sqrt{\frac{A_\alpha}{\pi}} \quad (1)$$

$$SF = \frac{4\pi A_\alpha}{P_\alpha^2} \quad (2)$$

Where, A_α and P_α represent the area and perimeter of α-Al phase, respectively.
It is evident from figure 7(a-c) that, with the increment in the angle of cooling slope from 30° to 45° leads to a decrement of the size of α-Al phase from 74µm down to 52µm and further, increment in the angle of cooling slope from 45° to 60° resulted in the size of α-Al phase to be 38µm, which is comparatively less than that of 98 µm of conventionally cast A356 alloy. Moreover, it has also been observed that there is gradual increase in the S.F. from 0.55 in the as cast condition to 0.82 in the C.S. cast at 60°.

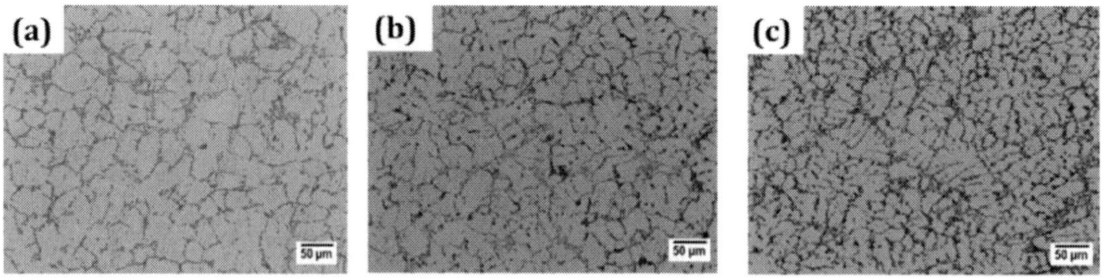

Figure 7. Microstructures of Semi-solid slurry A356 alloy poured at (a) 30° (b) 45° (c) 60° angles

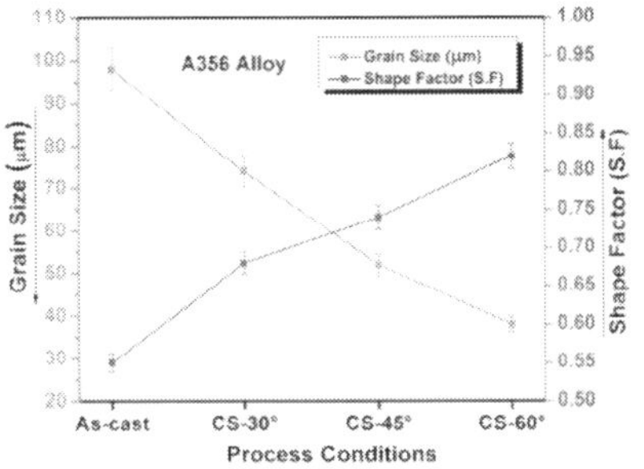

Figure 8. Variations of Grain size and S.F of α-Al phase of A356 alloy

The pouring temperature has great impact on the microstructure of the alloy and composites. The microstructures of the CS cast A356 alloy poured at different temperatures is shown in figure 9. It has been observed that in the current study, for a slant angle of 60°, the temperature suitable for desired semi-solid microstructure was 640°C, while it was found to be 650°C for 30° cooling slope angle.
It has been observed from the experiments that higher the inclined plate angle lower the pouring temperature is required.

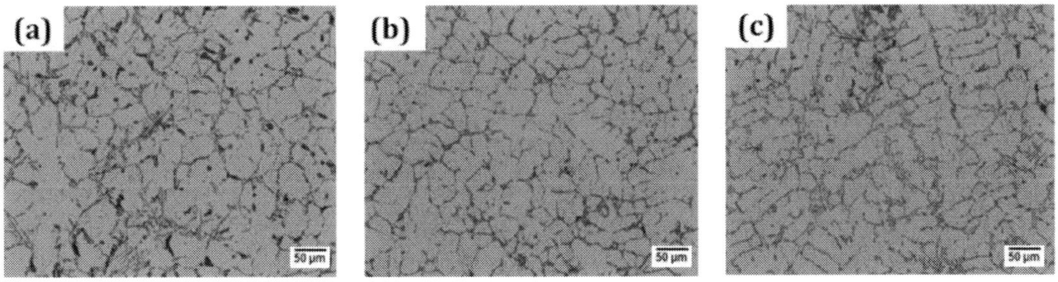

Figure 9. Microstructures of the CS cast A356 alloy poured at (a) 630° C (b) 640° C(c) 650° C

3.2. Microstructural evolution of rheocast A356-5TiB$_2$ in-situ composite

The feasibility of generation of feed stock of A356-5TiB$_2$ composite was also conducted, and the microstructure of A356-5TiB$_2$ composite obtained through CS casting process is shown in figure 10.It can be observed that TiB$_2$ particles having a range of 0.2-0.5μm along with eutectic silicon are found in the grain boundary region of α-Al matrix.

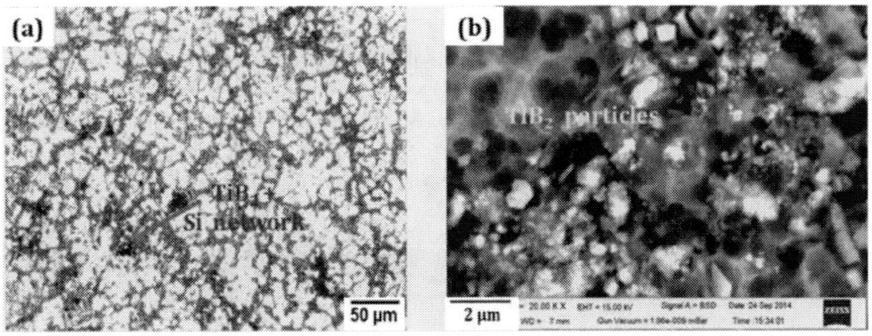

Figure 10. (a) Microstructures of CS cast A356-5TiB$_2$ composite, poured at 60° angle and 640° C
(b) SEM micrograph showing minute TiB$_2$ particles

4. Conclusions

The study in this research has discussed the application of Cooling Slope casting process for the generation of semi-solid feedstock of A356 alloy and its feasibility as a viable route for generation of non-dendritic feedstock of A356-5TiB_2 *in-situ* composite, which subsequently acts as a starting material for further thixoforming.

The following can be concluded from the present study:

- The dendritic primary α-Al phase of 98 μm of conventionally cast alloy transformed to fine and non-dendritic structure of 38 μm by the application of cooling slope casting process.
- The α-Al grains readily transformed to mixture of spherical and rosette type structure after being flowed through cooling slope casting route.
- A significant reduction in grain size from 74 μm to 38 μm was observed as the cooling slope plate angle increased from 30° to 60°.
- A maximum sphericity of 0.82 of α-Al grains is obtained for a 60° cooling slope angle.
- The optimal combination of cooling slope casting process parameters were found to be, a pouring temperature of 640 °C, slope angle of 60° and cooling length of 300 mm.
- Moreover, minute TiB_2 particles of size 0.2-0.5 μm are uniformly distributed in the spherical α-Al structure and at the grain boundaries in the composite feedstock.
- Cooling slope casting route is thus offered as an attractive and alternative route in the production of semi-solid feedstock of both alloy and composites by effectively varying the process parameters.

Acknowledgment

The authors gratefully acknowledge the financial support received from Indian Institute of Technology Bhubaneswar under the SEED project grant for fabricating the "Cooling Slope casting" experimental setup.

References

[1] Flemings M C 1991 *Metall. Trans. A* **22** 957
[2] Kirkwood D H 1994 *Int. Mater. Rev.* **39** 173
[3] Fan Z 2002 *Int. Mater. Rev.* **47** 37
[4] Haga T and Suzuki S 2001 *J. Mater. Process. Tech.* **118** 169
[5] Haga T and Kapranos P 2002 *J. Mater. Process. Tech.* **130-131** 594
[6] Haga T and Kapranos P 2002 *J. Mater. Process. Tech.* **130-131** 581
[7] Birol Y 2007 *J. Mater. Process. Tech.* **186** 94
[8] Birol Y 2008 *J. Mater. Process. Tech.* **207** 200
[9] Robert M H, Zoqui E J, Tanabe F and Motegi T 2007 *J. Ach. Mater. Manuf. Eng.* **20** 19
[10] Legoretta E C, Atkinson H V and Jones H 2008 *J. Mater. Sci.* **43** 5456
[11] Taghavi F and Ghassemi A 2009 *Mater. Des.* **30** 1762
[12] Das P, Samanta S K, Ray T and Venkatpathi B R K 2012 *Adv. Mater. Res.* **585** 354
[13] Deepak Kumar S, Mandal A and Chakraborty M 2014 *Int. J. Engg. Res. Tech.* **3** 269
[14] Mandal A, Chakraborty M and Murty B S 2008 *Mater. Sci. & Eng. A* **489** 220
[15] Kumar S, Chakraborty M, Subramanya Sarma V, Murty B S 2008 *Wear* **265** 134
[16] Deepak Kumar S, Pandu Ranga V, Mantry S, Mandal A and Chakraborty M 2014 *Proc. Mater. Sci.* **5** 232
[17] Das P, Samanta S K, Chattopadhyay H and Dutta A 2012 *Acta Metall. Sin.* **25** 329

IOP Conf. Series: Materials Science and Engineering **75** (2015) 012027 doi:10.1088/1757-899X/75/1/012027

Ratcheting induced cyclic softening behaviour of 42CrMo4 steel

R Kreethi, A K Mondal and K Dutta[1]

Department of Metallurgical and Materials Engineering, National Institute of Technology, Rourkela – 769008, India

Email- krisiitk@gmail.com

Abstract. Ratcheting is an important field of fatigue deformation which happens under stress controlled cyclic loading of materials. The aim of this investigation is to study the uniaxial ratcheting behavior of 42CrMo4 steel in annealed condition, under various applied stresses. In view of this, stress controlled fatigue tests were carried out at room temperature up to 200 cycles using a servo-hydraulic universal testing machine. The results indicate that accumulation of ratcheting strain increases monotonically with increasing maximum applied stress however; the rate of strain accumulation attains a saturation plateau after few cycles. The investigated steel shows cyclic softening behaviour under the applied stress conditions. The nature of strain accumulation and cyclic softening has been discussed in terms of dislocation distribution and plastic damage incurred in the material.

1. Introduction

The continually increasing safety and quality levels for engineering and automotive industry, foreseen by global regulations, make compulsory the design and implementation of quality administration scheme to the entire supply series for failure prevention and continuous improvement. The performance and reliability of an engineering structure or assembly is a function of the quality of the various components [1]. It is understood that critical engineering structures and machine components often subject to various kinds of load transients, which may be static as well as fluctuating in nature; thus prone to fatigue failure [2, 3]. Situation becomes more critical when these load transients become asymmetric in nature. It is well established that the conventional low cycle fatigue life of engineering components are predicted by Coffin – Manson relationship [4], but application of load transients of asymmetric nature leads to increased damage accumulation in components which leads to reduced fatigue life of the component [5]. The phenomenon of strain accumulation in a material under the influence of positive and/or negative mean stress is known as ratcheting. Investigations on ratcheting behaviour of materials were started during 1911 when Bairstow did some experiments on carbon steel [6]; it was extensively studied in the last few decades as reviewed by Ohno [7] and Yoshida [8]. The ratcheting–fatigue interaction was experimentally investigated by Rider et al. [9], Xia et al. [5], and Kang et al. [10]. Investigators have studied the ratcheting behaviour of various materials like stainless steel [11, 12], aluminum [13], and aluminum alloys [14], copper alloys [13], composites as well as polymers [15, 16]. Kang et al. studied uniaxial ratcheting and fatigue behaviour of tempered 42CrMo

[1] To whom any correspondence should be addressed.

Content from this work may be used under the terms of the Creative Commons Attribution 3.0 licence. Any further distribution of this work must maintain attribution to the author(s) and the title of the work, journal citation and DOI.
Published under licence by IOP Publishing Ltd

steel on constitutive modeling [17]. But limited works have been directed to understand the ratcheting behaviour of 42CrMo4 steel as per the knowledge of current investigation.

In the present study, ratcheting behaviour of 42CrMo4 steel is investigated under different applied stresses. These tests were carried out up to 200 cycles with different maximum applied stress. Experimental observation, which shows influence of material behaviour on cyclic stress strain response as well as effect of maximum applied stress of investigated steel also, has been discussed.

2. Experimental Procedures

2.1. Material and Heat Treatment

The selected 42CrMo4 steel was available in the form of solid cylindrical bars of diameter 16 mm. The chemical composition of the steel was assessed using optical emission spectrometer (Model: ARL 3460 Metals Analyzer, Thermo Electron Corporation Limited, Switzerland) and it is given in Table 1. To remove any pre-existing anomalies of material properties, all samples were subjected to annealing. This is done by heating the steel up to a temperature of 900°C followed by soaking for 60 mins and ultimately cooled inside the furnace.

Table 1: Chemical composition of the investigated 42CrMo4 steel

Element	C	Cr	Mo	Mn	Si	Al	P	Fe
Wt%	0.384	0.926	0.222	0.678	0.210	0.035	0.020	Balance

2.2. Tensile and Fatigue Test

Heat treated samples were mechanically tested for tensile and fatigue properties. Round bar specimens, for tensile and fatigue tests having a gauge lengths of 25 mm and 13 mm with diameters of 6 mm and 7 mm respectively were fabricated as per ASTM standards E 8M and E 606 respectively [18].These tests were done on servo-hydraulic universal testing machine. All tensile tests were carried out at a crosshead speed of 1mm/min. Samples for ratcheting tests were with very well surface finish. All these tests were done at a constant stress rate of 50 MPa/s. Whole tests carried on varying maximum applied stress (σ_{max}). A summary of the test conditions is shown in Table 2. All fatigue tests were done up to 200 cycles.

Table 2: Selected σ_{max} values for ratcheting tests

Sl. No.	σ_{max} /MPa
1	400
2	430
3	460
4	490
5	520

3. Results and Discussion

As described above, samples were subjected to annealing heat treatment. Nature of microstructure, tensile and ratcheting behaviors of 42CrMo4 steel after this heat treatment is presented below in detail. All mechanical tests were performed at room temperature.

3.1. Microstructural and Tensile Properties

Figure 1 reveals the microstructure of annealed steel. It consists of pearlite colonies (dark contrast) and ferrite grains (light contrast). It is evident that the pearlite lamellar colonies are clearly existed. The average grain size was evaluated as 20.66 μm, using linear intercept method [19]. Typical engineering stress strain plot of the investigated steel was shown in Figure 2. It shows continuous yielding behaviour from elastic to plastic region and therefore, the yield strength has been calculated by offset method (offset = 0.2%) as specified in ASTM standard E8M [20]. The tensile properties of the investigated steel are summarized in Table 3.

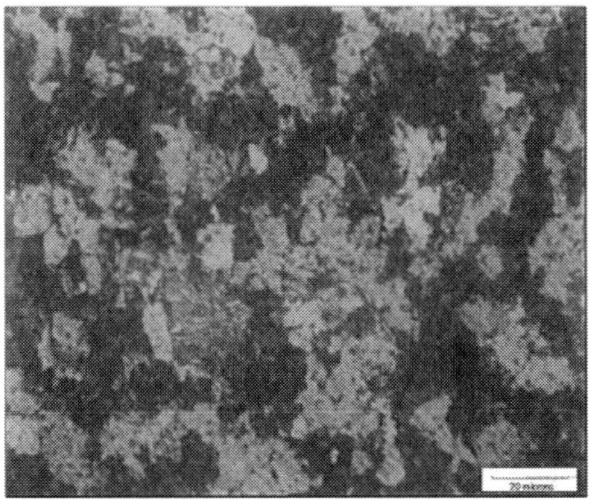

Figure 1: Typical optical microstructure of the annealed 42CrMo4 steel.

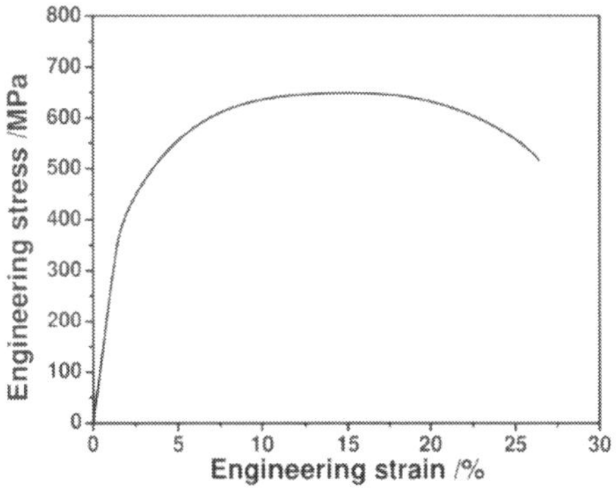

Figure 2: Engineering stress-strain plots of the investigated42CrMo4steel

Table 3: Tensile properties of the selected 42CrMo4 steel

Yield Strength (MPa)	Tensile Strength (MPa)	%Uniform Elongation	%Total Elongation
387	650	15.56	26.4

3.2. Ratcheting Behaviour

Uniaxial ratcheting experiments were carried out under asymmetrical cyclic loading conditions up to 200 cycles. Figure 3 shows the characteristic evolution of ratcheting strain as a function of number of cycles for different maximum applied stress (σ_{max}) values. It can be seen from these graphs that strain accumulation takes place in distinct stages with varying rates. It is reported in earlier investigations that there exists three different stages of strain accumulation during ratcheting; analogous to creep curves [20]. In the primary stage, the ratcheting strain rate decreases until it reaches a steady-state; attainment of steady state represents secondary stage. The strain accumulation curves presented in Figure 3 indicate presence of primary and secondary stages, as cyclic loading was stopped after 200 cycles. It is worthy to mention here that the tertiary stage of strain accumulation starts when the material under investigation accumulates large strain at a high rate and accelerates until the final failure. However, this stage is absent in the current investigation. Moreover, it can be noted from Figure 3 that strain accumulation during ratcheting depends on the maximum applied stress (σ_{max}) values, and the evolution of ratcheting is accelerated remarkably by the increase of maximum applied stress. The strain accumulation varies from 0.85% to 7.26% for applied stress of 400 MPa and 520 MPa respectively.

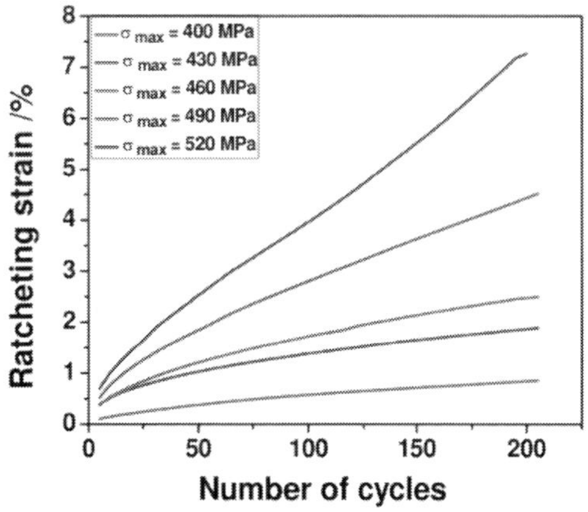

Figure 3: Variation of ratcheting strain with number of cycles

3.3. Analysis of Hysteresis Loops

The nature of hysteresis loops is one of the better ways to explain the ratcheting behavior of materials. The hysteresis loops are the representation of stress strain response during each cycle. Figure 4 (a) shows the representative stress-strain hysteresis loops (N = 1) obtained in ratcheting tests of investigated steel for different maximum applied stresses. This figure indicates that, the total deformation zone gets increased with increasing maximum applied stress. Moreover, the plastic

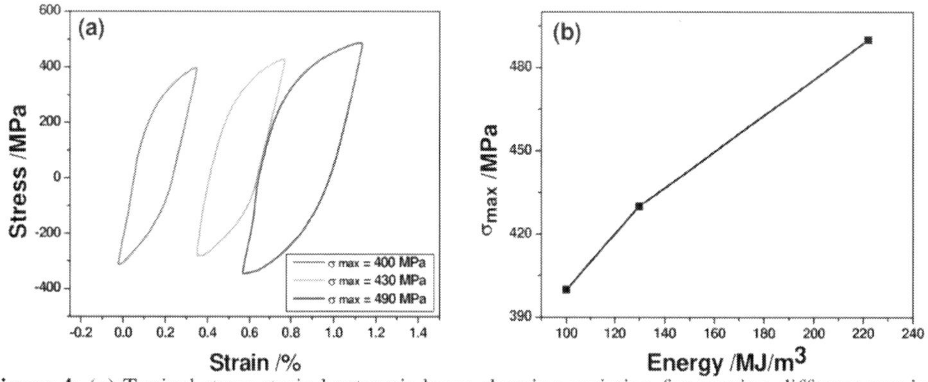

Figure 4: (a) Typical stress strain hysteresis loops showing variation for varying different maximum stress and **(b)** Variation of σ_{max} with corresponding variations in their hysteresis loop energy.

damage increases with increasing σ_{max}. Figure 4 (b) show that hysteresis loop energy increases with increase in σ_{max}, indicating increased accumulation of cyclic damage in the specimens. It can be stated that any increase in ratcheting strain would induce higher extent of plastic deformation, consequently strain accumulation increases with increasing maximum stress. However, all the discussion made on strain accumulation is based upon mechanistic aspects of cyclic deformation. The phenomenon can be discussed in terms of materialistic point of view. Dutta et.al and Kang et.al have discussed in their recent report that strain accumulation due to ratcheting in material depends on dislocation formation and their redistribution. In the forward cycle, dislocation gets generated; a few of this dislocations gets annihilated in opposite direction due to asymmetric of loading. Hence, considerable numbers of dislocations remain in the substructure of material in each cycle [21]. Thus, strain accumulates during cyclic loading. The amount of remnant dislocations increases with increasing maximum applied stress of loading. Therefore accumulation of ratcheting strain increases with increasing maximum stress.

3.4. Cyclic Softening Behaviour

Figure 5 shows the stress-strain hysteresis loops generated during asymmetric cyclic loading of the investigated 42CrMo4 steel at constant σ_{max}. The 1st, 50th, 100th, 160th and 200th loops are shown here. One can note that widths of the hysteresis loops increase with increasing number of cycles. This is attributed to cyclic softening behaviour of the material. For better understanding of damage accumulated during cyclic loading, a quantitative assessment of hysteresis loop area was done and this shows that the loop area varies from 24.49 MJ/m^3 to 90.740 MJ/m^3, as summarized in Table 4. The phenomenon of cyclic softening in the investigated material can be explained following Hussain et al. [22]. According to Hussain et al., in a high strength material, initially the dislocation density is high and cyclic loading cause's re-arrangement of the dislocations into a new configuration that offers less resistance to deformation and therefore material shows cyclic softening feature.

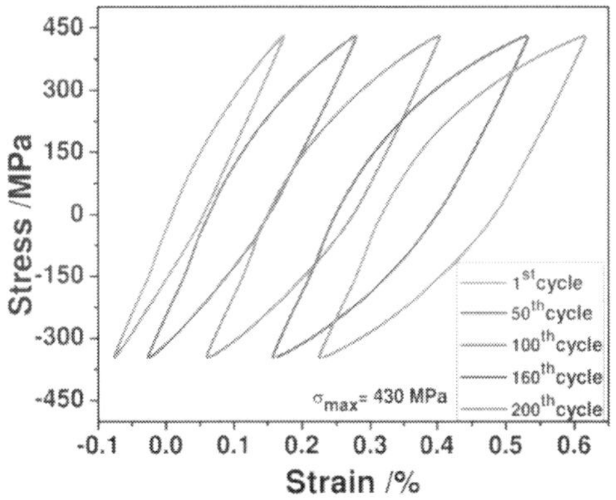

Figure 5: Representative stress strain hysteresis loop obtained from uniaxial ratcheting fatigue tests.

Table 4: Energy values of the cyclic softening hysteresis loops

Sl. No	No. of Cycles	Energy /MJ-m^{-3}
1	1	24.49
2	50	48.013
3	100	66.352
4	160	81.576
5	200	90.740

4. Conclusions

The following conclusions can be drawn from the experimental results and pertinent discussion. The obtained results and their relevant analyses lead to infer that:

(1) Accumulation of ratcheting strain increases with increasing maximum applied stress (σ_{max}). The strain accumulation varies from 0.85% to 7.26% for applied stress of 400 MPa and 520 MPa respectively. The increase in ratcheting strain is explained in terms of cyclic damage measured as the area under hysteresis loops, which increases with increase in maximum applied stress, indicating increased accumulation of cyclic damage in the specimens. Further, it may be inferred that increase in strain accumulation occurs due to increased remnant dislocation density with increased maximum stress, followed by Dutta et. al [21].

(2) The investigated annealed 42CrMo4 steel has shown cyclic softening feature. The phenomenon is explained in terms of dislocation activities during asymmetric cyclic loading; this kind of cyclic loading reduces the resistance to the motion of dislocations in the material and thus material shows cyclic softening.

References

[1] Vazdirvanidis A, Pantazopoulos G and Louvaris A 2009 *Eng. Failure. Analysis* **16** 1033–1038.
[2] Ohno N and Abdel-Karim M 2000 *J. Eng. Mater. Technol* **122** 35–41.
[3] Ringsberg J W 2001 *Int. J. Fatigue* **23** 575–586.
[4] George E. Dieter, Mechanical Metallurgy, McGraw Hill Book Company lmt.
[5] Xia Z, Kujawski D and Ellyin F 1996 *Int. J. Fatigue* **18** 335–341.
[6] Bairstow L 1911 *Phil. Trans. Royal. Soc* **210** 359–386.
[7] Ohno N 1990 *Appl. Mech. Rev* **436** 283–295.
[8] Yoshida F 1990 *Int. J. Press. Vessels. Pip* **44** 207–223.
[9] Rider R J, Harvey S J and Chandler H D 1995 *Int. J. Fatigue* **17** 507–511.
[10] Kang G Z and Liu Y J 2008 *Mater. Sci. Eng* A **472** 258–268.
[11] Ruggles M D and Krempl E 1990 *J. Mech. Phy. Solids* **38** 575–585.
[12] Kang G Z, Gao Q and Yang X 2002 *J Int. J. Mech. Sci* **44(8)** 1645–1661.
[13] Kang G, Liu Y, Dong Y and Gao Q 2011 *J. Mater. Sci. Technol* **27(5)** 453–459.
[14] Zhao T and Jiang Y 2008 *Int. J. Fatigue* **30(5)** 834–849.
[15] Kang G 2006 *Comp. Sci. Technol* **66** 1418–1430.
[16] Chen X and Hui S 2005 *Polym. Test* **24** 829-833.
[17] Kang G, Liu Y, Ding J and Gao Q 2009 *Int. J. Plasticity* **25** 838–860.
[18] E 8M-03, ASTM Annual Book of Standards, Standard test method for tension testing of metallic materials (Metric), 03.01; (2003), West Conshohocken, PA.
[19] George F Vandervoort, Metallography: Principles and Practice 1984 ASM internationals.
[20] Zhao P and Xuan F Z 2011 *Mech. Mater* **43** 299–312.
[21] Dutta K and Ray K K 2012 *Mater. Sci. Eng* A **540** 30–37.
[22] Hussain K 1993 *Metall. Sci. Technol* **11(1)** 2–30.

Recovery of Cu and Zn from Complex Sulphide Ore

Trinath Talapaneni[1], S.Sarkar[1], N. Yedla[1], Dr. P.L.N Reddy[2]

[1]Department of Metallurgical and Materials Engineering, National Institute of Technology, Rourkela, Orissa, India.
[2]Department of Metallurgical and Materials Engineering, National Institute of Technology, Surathkal-575075, Karnataka, India.

E-mail: trinath.talapaneni@gmail.com

Abstract

Complex Sulphide Ores are often found to be a close mutual association with each other and with the nonmetallic gangue. The beneficiation experiments showed that it would be very difficult to recover Cu and Zn from the lean complex Sulphide ores using traditional ore beneficiation methods. In the present work, leaching of complex sulfide ores in sulfuric acid was investigated by the Electro hydrometallurgy process. The lab-scale experiments were conducted to investigate the influences of pulp-density, electrolyte concentration, particle size, current density and time on recovery of Cu and Zn. The leach liquor obtained after electrolysis was subjected to Atomic Absorption Spectroscopy analysis for the recovery of minerals.

1. Introduction

The processing of complex Sulphide ores depends on the nature of the gangue material and percentage composition of various nonferrous metals present in the ores or in the bulk concentrates. As Zinc and copper often occur together, the separation of one metal from the other has presented problems to the nonferrous metallurgists [1].

Sulphide ores are extremely complicated mineralogical associations of the inter grown of their constituent minerals [2] comprising mainly chalcopyrite (CuFeS2), sphalerite (ZnS), and galena (PbS), these ores do not allow the recovery of metal by direct chemical leaching due to their low solubility in most leaching reagents[3]. In such cases, the ore must be pre-treated in order to release the desired metals physically and to make the ore amenable to leaching reagents. Processes available for the treatment of complex Sulphide ores involve Sulphide oxidation to form water soluble sulphates or oxides. The sulphur oxidation processes include roasting, pressure oxidation, chemical oxidation and bioleaching. The processes that are traditionally used are complex and expensive. Hence attention has been drawn to the use of easily available and cheap reagents like sulphuric acid for base metal recovery and means of optimizing the dissolution kinetics. Although sulphuric acid may not be as effective as other stronger oxidizing acids like nitric acid, it is, however, preferred to all other reagents in terms of cost, corrosion wear and the ease of regeneration during electro winning [4].

The ore minerals in poly-metallic and complex deposits are often found to be in close mutual association with each other and with the nonmetallic gangue. In many cases, attempts to concentrate these ores have run in to problems because it is either difficult to obtain a suitable grade of the

Content from this work may be used under the terms of the Creative Commons Attribution 3.0 licence. Any further distribution of this work must maintain attribution to the author(s) and the title of the work, journal citation and DOI.
Published under licence by IOP Publishing Ltd

concentrate by conventional flotation methods or the recovery of metals in the respective concentrate is poor. Zinc that finds its way to a copper concentrate is always discarded in the slag as a waste, while copper in a lead concentrate leads to serious smelting problems.

The major step in hydrometallurgical methods involves the absorption of the nonferrous metallic values in a solution and leaving behind a major part of the iron in the leach residue. The next steps are solution purification, separation of the nonferrous metals by electrolysis. During the metal extraction, the bulk of the metal is separated from impurities present in the compound which invariably comprises more than one phase. The driving force for the reactions involved in metal extraction depends on the differences in the chemical potentials of the elements.

The scanning electron microscope (SEM) is one of the most versatile and widely used tools of modern science as it allows the study of both the morphology and composition of materials [5]. It has been extensively applied in mineralogical studies of sulphide ores and their residues [6, 7, 8, 9,10]. The main purpose of this work is to study the various parameters (time, particle size, pulp density, current density, concentration) on recovery of Cu and Zn form complex Sulphide ores.

2. Experimental procedure

A standard 500 ml beaker is taken in which graphite (anode) and copper (cathode) electrodes are placed. The positive terminal of the D.C. source is connected to the ammeter the then connected to graphite anode. The negative terminal of D.C. source is connected to cathode copper sheet by a rheostat. The inter-electrode distance was maintained at 6cm.

For each experiment a known weight of sample of specific size fraction was taken. The thermometer and electrodes were set in the position. As soon as the sample was added a constant amount of current was passed from D.C. power supply. The stirring of the slurry was provided with Teflon covered magnetic stirring bar. The slurry was stirred in just fast enough to maintain suspension of the solids. The leaching is continued for a specific amount of time, after leaching it was filtered and the leach liquor was analyzed for copper, lead and zinc. The effect of current density, the effect of time, the effect of concentration, effect of pulp density and the particle size were studied.

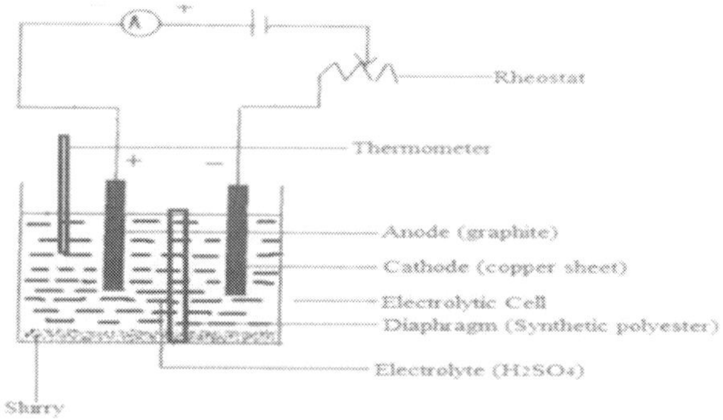

Figure 1. Experimental setup used for electro leaching

$$MS + H_2SO_4 + 0.5O_2 \longrightarrow MSO_4 + S^0 + H_2O$$ [11] Where M can be Zn, Cu

2.1 Atomic absorption spectroscopy (AAS)

The technique of atomic absorption developed by Sir Alan Walsh of C.S.I.R.O in the mid 1950 has become the preferred method of elemental analysis. Walsh discovered that the majority of free atoms in the commonly used flame were in the ground state, but that the flames did not also have enough energy to excite these atoms. A light source emitting a narrow spectral line of the characteristic energy is used to excite the free atoms formed in the flame. The decrease in energy (absorption) is then measured. The absorption is proportional to the concentration of free atoms in the flame, given by the Lamberts Beer Law.

$$\text{Absorbance} = \log_{10}\frac{I_0}{I_t} = \text{K.C.L} \tag{1}$$

Where,

I_0= intensity of incident radiation emitted from the light source
I_t= intensity of transmitting radiations
C= concentration of sample
K= constant
L= path length.

Atomic Absorption Spectroscopy uses the absorption of light to measure the concentration of gas phase atoms. Since samples are usually liquids or solids, the analyte atoms or ions must be vaporized in a flame or graphite furnace. The atoms absorb UV or Visible light and make the transition to higher electronic energy levels. The analyte concentration is determined from a working curve after calibrating the instrument with standards of known concentration. Atomic absorption is similar to Ultraviolet Spectroscopy in that similar wavelength and the same relationship are used, but differs in the fact that a line source is used, and the sample compartment (a flame or electro thermal atomizer) is positioned before the monochromatic and not after it. In practical atomic absorption analysis the Lamberts Beers Law can be reduced to

$$\text{Absorbance} = \log_{10}\frac{I_0}{I_t} \propto c \tag{2}$$

2.2 Analytical procedure

2.2.1 Calibration Curve:
This is the most common method where interference effects are known to be absent. Usually at least three standards and a blank are used to cover the range 0.1 to 0.8 ABS. The calibration is performed by using the blank solution to zero the instrument. The standards are then analyzed with the lowest concentration first, and the blank run between standards, to ensure the base line (zero point) has not changed. Samples are then analyzed and their absorbance recorded. A graph of absorbance VS concentration is plotted.

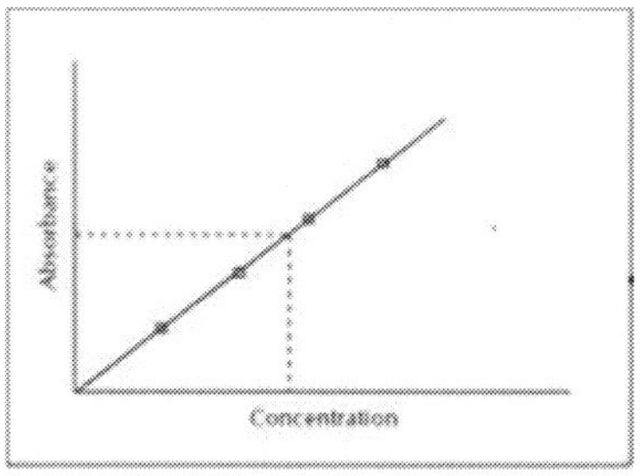

Figure 2. Calibration curve

IOP Conf. Series: Materials Science and Engineering **75** (2015) 012028 doi:10.1088/1757-899X/75/1/012028

3. Results and Discussions:

The SEM image and the chemical composition of complex sulphide ore is mentioned below.

Figure 3. SEM image of complex sulphide ore

Table 1: Chemical composition of sulphide ore

Element	(Kev)	Mass%	Error%	At%	K
C K	0.277	30.17	0.08	52.34	9.2782
O K	0.525	26.55	0.11	34.58	26.4518
Mg K	1.253	1.31	0.06	1.12	1.0728
Si K	1.739	2.97	0.05	2.21	3.7332
S K	2.307	6.50	0.04	4.23	11.2963
Fe K	6.398	2.62	0.14	0.98	4.0928
Cu K	8.040	0.83	0.29	0.27	1.2601
Zn K	8.630	6.18	0.37	1.97	9.4144
Pb M	2.342	22.87	0.18	2.30	33.4003
Total		100.00		100.00	

3.1 Atomic Absorption Spectroscopy Readings in PPM for all Parameters

After electrolysis, the samples are subjected to Atomic Absorption Spectroscopy to recover Cu and Zn from leach liquor.

Table 4: AAS readings for time

Time (HR.)	Cu	Zn
1	1.61	0.876
2	1.754	1.915
3	2.18	2.361
4	2.364	3.69

Table 5: AAS readings for particle size

Particle size (μm)	Cu	Zn
150	0.88	0.708
105	1.16	1.199
75	1.46	2.454
53	1.56	3.092

Table 6: AAS readings for pulp density

Pulp density (g/l)	Cu	Zn
10.10	1.46	4.32
20.40	1.22	4.14
30.92	1.20	3.08
41.65	1.09	2.33

Table 7: AAS readings for concentration

Concentration (M)	Cu	Zn
1	2.69	5.40
2	1.557	2.13
3	1.247	1.82

Table 8: AAS readings for Current density

Current density (amps/dm²)	Cu	Zn
3.75	1.085	2.45
5	1.116	2.615
6.25	1.51	2.940

3.2. Percentage *recovery of Cu and Zn for various parameters*

3.2.1. Effect of Time on Electro Oxidation

Operating conditions:

- Pulp Density: 30.92 g/l
- Electrolyte concentration: 1 M H_2SO_4
- Temperature: 30°C
- Particle size: 150 µm
- Current Density: 2.5 amps/dm²

Table 9: % recovery of Cu and Zn verses Time

Time (hrs)	% Copper Recovery	% Zinc Recovery
1	32.38	4.730
2	35.24	10.33
3	43.89	12.74
4	47.48	19.95

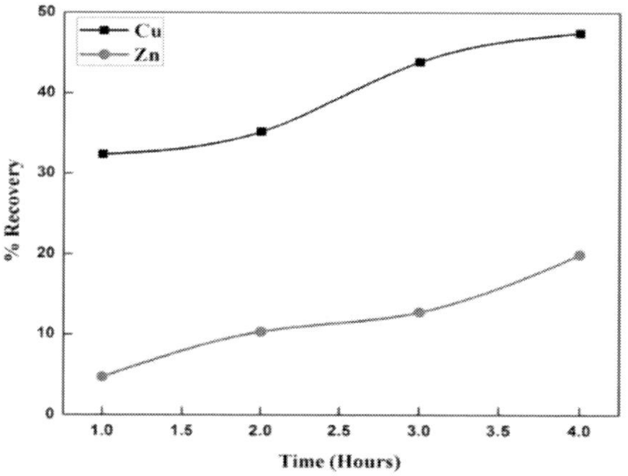

Figure 4. Percentage recovery of copper and zinc Versus Time (hrs)

3.2.1.1. Effect of Time:

The effect of time on the rate of copper and Zinc recovery was investigated at different Time intervals of 1, 2, 3, and 4 hrs respectively. The experiments were conducted at 30 (room temperature), 1 M H_2SO_4 concentration at a pulp density of 30.92 g/l and particle size of 150 µm. Increase in time of leaching increased recovery of both copper and Zinc from 32.38% to 47.48% and 4.730 % to 19.95%. The increase in dissolution with the increase in time at 30 is a direct consequence of the prolonged exposure of the mineral surface to the acid medium.

IOP Conf. Series: Materials Science and Engineering **75** (2015) 012028 doi:10.1088/1757-899X/75/1/012028

3.2.2. Effect of particle size on Electro Oxidation

Operating conditions:

- Pulp Density : 30.92 g/l
- Time: 1hours
- Temperature: 30°C
- Electrolyte Conc: 1M H_2SO_4
- Current Density: 2.5amps/dm^2

Table 10: % recovery of Cu and Zn verses particle size

Particle size (μm)	%Copper Recovery	%Zinc Recovery
100	17.69	3.824
140	23.37	6.472
200	29.47	13.24
270	31.40	16.68

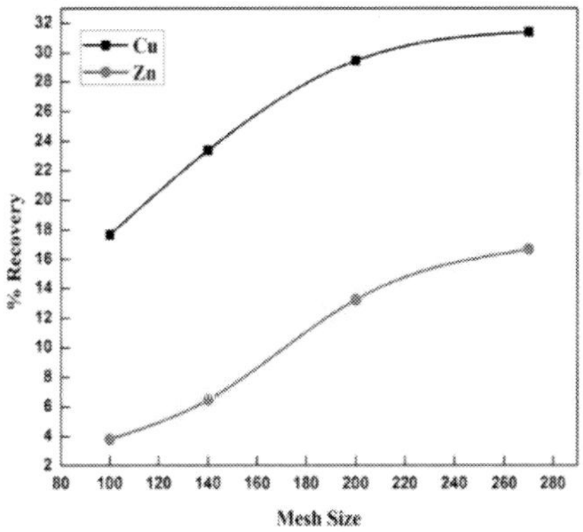

Figure 5. Percentage recovery of copper and zinc versus particle size in μm

3.2.2.1. Effect of particle size:

The effect of time on the rate of copper and Zinc recovery was investigated at different mesh size 100, 140, 200, and 270 μm respectively. The experiments were conducted at 30⍰ (room temperature), 1 M H_2SO_4 concentration at a pulp density of 30.92 g/l for 1hr. The finer the particle size, recovery of both copper and Zinc increased from 17.69 to 31.40% and 3.824 % to 16.68%. As particle size decreases more surface area is available for reaction, so recovery will be more.

3.2.3. Effect of Pulp Density on Electro Oxidation

Operating conditions:

- Current Density: 2.5 amps/dm^2
- Time: 1 hours
- Temperature: 30°C
- Electrolyte concentration: 1 MH_2SO_4
- Particle Size: 100 μm

Table 11: % recovery of Cu and Zn verses pulp density

Pulp Density (g/l)	%Copper Recovery	% Zinc Recovery
10.10	17.59	13.98
20.40	7.32	6.69
30.92	4.81	3.32
41.65	3.28	1.88

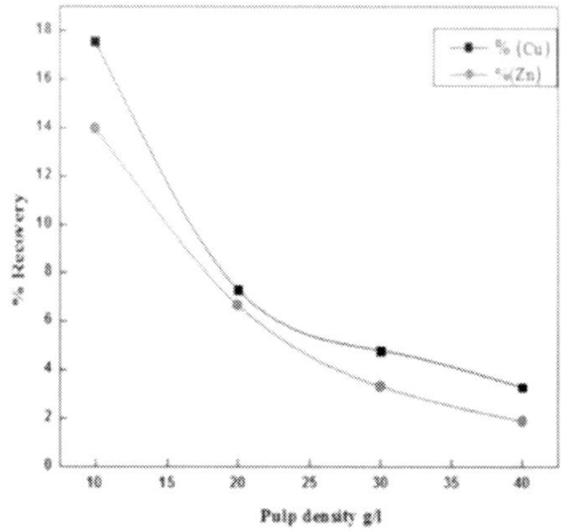

Figure 6. Percentage recovery of copper and zinc versus Pulp Density

3.2.3.1. Effect of pulp Density:

The effect of pulp density on the rate of copper and Zinc recovery was investigated at different pulp density of 10.10, 20.40, 30.92 and 41.65 g/l. The experiments were conducted at temperature 30°C 1 M H_2SO_4 concentration at a current density of 2.5amps/dm^2 and particle size of 100 μm. Pulp density had the minimum effect on the Cu and Zn extraction, an increase in pulp density will cause difficulties for agitation, slurry filtration operation etc.

3.2.4. Effect of Concentration on Electro Oxidation

Operating conditions:

- Pulp Density: 30.92 g/l
- Time: 1 hours
- Temperature: 30° C
- Particle size: (100 mesh) 150 μm
- Current Density: 2.5 amps/dm^2

Table 12: % recovery of Cu and Zn verses concentration

Concentration (M)	%Copper Recovery	% Zinc Recovery
1	54.05	29.14
2	3.28	11.51
2.5	25.05	9.84

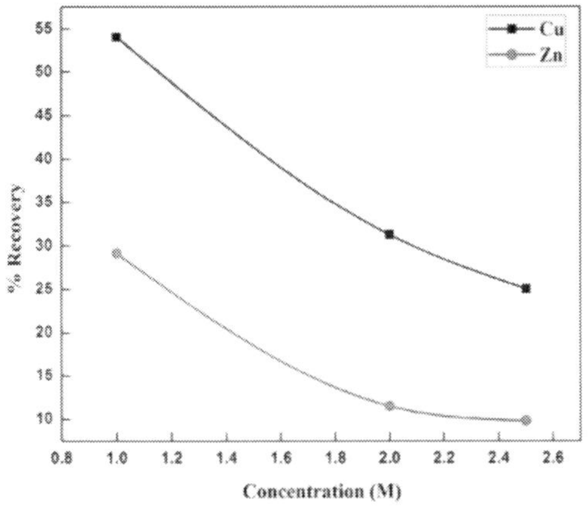

Figure 7. Percentage recovery of copper and zinc versus electrolyte concentration

3.2.4.1. Effect of Concentration:

The effect of concentration on the rate of copper and Zinc recovery was investigated at different concentration intervals of 1, 2, and 2.5 molars respectively. The experiments were conducted at 30? (room temperature), 1 M H_2SO_4 concentration at a pulp density of 30.92g/l and particle size of 150μm. The dissolution of both Cu and Zn decreases, as the concentration increases. The rate was independent of the concentration of the acid.

3.2.5. Effect of Current Density on Electro Oxidation

Operating conditions:

- Pulp Density: 30.92 g/l
- Time: 1 hours
- Temperature: 30°C
- Electrolyte concentration: 1 M H_2SO_4
- Particle Size: (100 mesh) 150 μm

Table 13: % recovery of Cu and Zn verses current density

Current density amps/ dm^2	%Copper Recovery	% Zinc Recovery
3.75	21.80	13.25
5	22.42	14.11
6.25	30.46	15.86

IOP Conf. Series: Materials Science and Engineering **75** (2015) 012028 doi:10.1088/1757-899X/75/1/012028

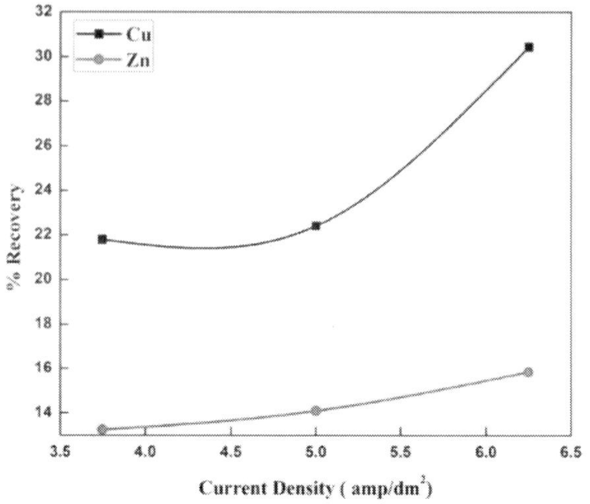

Figure 8. Percentage recovery of copper and zinc versus current density

3.2.5.1. Effect of Current Density:

The effect of the rate of current density on the rate of copper and Zinc recovery was investigated at different current density of 3.75, 5, and 6.25 amps/ dm^2. The experiments were conducted at 30°C and 1 M H_2SO_4 concentration at a pulp density of 30.92g/l and particle size of 150μm. As current density increases ionic movement will be more which favors the reaction to take place faster and recovery will be more.

4. Conclusions:

- The increase in dissolution with the increase in time is a direct consequence of the prolonged exposure of the mineral surface to the acid medium; intern increases the recovery of Cu and Zn.
- As particle size decreases more surface area is available for reaction, so recovery will be more.
- Pulp density had the minimum effect on the Cu and Zn extraction, an increase in pulp density will cause difficulties for agitation, slurry filtration operation etc.
- The dissolution of both Cu and Zn decreases, as the concentration increases. The rate was independent of the concentration of the acid.
- As current density increases ionic movement will be more which favors the reaction to take place faster and recovery will be more.

References:

[1] Ray H S, Sridhar R and Abraham K P 2001 Extractions of Non-ferrous metals *Affiliated East-West press private limited* New Delhi 397-399

[2] Gomez C, Limpo J L, Luis D, A, Blazquez M, L, Gonzalez F and Ballester A 1997 Hydrometallurgy of bulk concentrates of Spanish complex sulfides: Chemical and Bacterial leaching *Canadian Metallurgical Quarterly* **26** 15-23

[3] Hiskey J B and Wadsworth M E 1975 Galvanic conversion of chalcopyrite *Metallurgical Transactions.* **B** 6 183–190

[4] Biswas A K and Davenport, W G 1980 *Extractive Metallurgy of Copper Pergamon Press Oxford* 263–269

[5] Xiao Z and Laplante A R 2004 Characterizing and recovering the platinum group minerals–a review. *Minerals Engineering* **17** 961–979

[6] Hey P V 1999 The effects of weathering on the UG2 chromitite reef of the Bushveld Complex with special reference to the platinum-group minerals *South African Journal of Geology* **3** 251-260

[7] Gornostayev S and Mutanen T 2003 The platinum-group minerals in chromitites of the Akanvaara deposit, Northern Finland and in their processing products. *Minerals Engineering* **16** 1307–1312

[8] Kayanuma Y, Okabe T, H, Mitsuda Y and Maeda M 2004 New recovery process for rhodium using metal vapor. *Journal of Alloys and Compounds* 211–220.

[9] Cabri L J, Beattie, M, Rudashevsky N S and Rudashevsky V N 2005 Process mineralogy of Au, Pd and PT ores from the Skaergaard intrusion, Greenland, using new technology *Minerals Engineering.* **18** 887–897

[10] Grieco, G Diella, V Chaplygina, and Savelieva G N 2006 Platinum group elements zoning and mineralogy of chromitites from the cumulate sequence of the Nurali massif (Southern Urals, Russia), Ore Geology Reviews, in-press

[11] Peng P 2005. Leaching of a sphalerite concentrates with H_2SO_4-HNO_3 Solutions in the presence of C_2Cl_4. Hydrometallurgy **4** 265-271

Sea Water Ageing of GFRP Composites and the Dissolved salts

A P Chakraverty*[1], U K Mohanty[1], S C Mishra[1] and A Satapathy[2]
[1]Department of Metallurgical and Materials Engineering,
National Institute of Technology, Rourkela-769008, Odisha, India
[2]Department of Mechanical Engineering
National Institute of Technology, Rourkela-769008, Odisha, India

*E-mail: anantac8@gmail.com

Abstract. This paper houses the effect of sea water immersion on glass fibre reinforced polymer (GFRP) composites. The major sources of interest are study of sea water absorption, penetration of the dissolved salts, details of chemical and physical bonds at the interface, variations of mechanical properties and study of failure mechanisms as revealed through SEM fractographs. Eighteen ply GFRP composites are immersed in sea water for a period of one year in steps of two months durations. It is revealed that the moisture absorption transforms from a Fickian to non-Fickian behavior with lapse of time. The dissolved salt 'K' shows highest depth of penetration after one year of immersion while 'Na' shows a least depth of penetration, as revealed from the EDS spectra. It is also revealed that 'Ca' seems to have a sudden burst in the rate of penetration even surpassing that of 'K'. This trend can be attributed to the combined effect of ionic mobility of the various dissolved salts and the probable interaction between 'K' and the –OH group of epoxy resin. This interaction between dissolved 'K' and the –OH group in the polymer could have arrested the further advancement of 'K' salts in the polymer, resulting in comparatively high rates of 'Ca' penetration. The mechanical properties such as inter laminar shear stress (ILSS), stress and strain at rupture, glass transition temperature (T_g) and elastic modulus show a decreasing trend with the increased duration of immersion. As revealed from the SEM fractographs pot-holing, fiber pull-out, matrix crack etc. are seen to be the major reason for failure of the immersed samples under load.

Keywords: GFRP composites, depth of penetration, ILSS, glass transition temperature (T_g), SEM fractographs

1. Introduction

Glass fibre reinforced polymer (GFRP) composites are recommended for use in marine infrastructures. These materials, due to their excellent corrosion resistance and light weight property have gathered tremendous attention in marine sectors for applications in ship building industry, water storage vessels, desalination plants, manufacture of leisure boats etc [1]. The cost of installation and maintenance of these composite structures are reportedly lower than that of some conventional materials like steel and aluminium [2]. However, the marine ship structures composed of GFRP

* To whom any correspondence should be addressed.

Content from this work may be used under the terms of the Creative Commons Attribution 3.0 licence. Any further distribution of this work must maintain attribution to the author(s) and the title of the work, journal citation and DOI.
Published under licence by IOP Publishing Ltd

composite laminates are subjected to immersion in sea water for years together necessitating accumulation of experimental data in this respect. Polymeric composites during immersion (fully or partly) in sea water may lead to mechanical property deterioration by plasticization and swelling of polymer [3]. Also, various salt components of the sea water may annihilate the moisture absorption rate compromising some properties of FRP systems subjected to sea water immersion [4].

The effect of moisture coupled with temperature on Glass/Epoxy composite in the presence of the dissolved salts in sea water has been reported by several workers [5-10]. Sea water absorption causes changes in the thermo-physical, mechanical and chemical characteristics of the matrix by the processes of both plasticization and hydrolysis. It is reported [11] that the effect of sea water on GFRP composites differs with matrix material and conditioning environment. In Glass/Epoxy composite, in particular, the mode of failure is altered from a brittle matrix and ductile fibre to ductile matrix and brittle fibre.

The ingressed moisture tend to result in plasticization and swelling of the matrix where as the dissolved salts penetrating to different depths in the composite body influence the moisture ingression, thereby affecting the processes of deterioration of mechanical properties. On the other hand, these adverse effects are minimized by the dual mechanisms of stress relaxation and swelling [12]. Moisture absorption may also cause cracking, blistering, chemical degradation, debonding, hydrolysis, oxidation and leaching, the processes being irreversible [13]. Thus, some durability data with sea water aging for prolonged periods of immersion have to be obtained in order to precisely decide the utility of such composite material when immersed in sea water.

Keeping the above in mind, GFRP composite samples are immersed in sea water for a period of one year maximum in batches of two months each. The effect of moisture ingression as influenced by the penetration of different dissolved salts is studied at length. Specifically, the effect of sea water immersion on the mechanical properties of GFRP composites such as ILSS, Stress and Strain at rupture, elastic modulus and Glass-transition temperature (T_g) are reported on the basis of the experimental data. Lastly, the mode of failure on load is determined and reported as obtained from the SEM fractographs.

2. EXPERIMENTAL

GFRP composites composed of woven E-Glass fibre and epoxy resin (Lapox L-12) are fabricated using conventional hand lay-up method. Diamnine (Lapox K-6) is used as the hardener/curing agent. After 48 hr of room temperature curing, short beam shear specimens (50 mm length X 12 mm breadth) are prepared as per the ASTM D 2344-84 (2013) standards [14]. Complete removal of moisture is ensured by alternate heating (55-60°C) and weighing of as-cured composite specimens till a fixed weight is obtained. Some typical properties of epoxy resin as provided by ATUL limited (polymer division) are listed in Table 1.

Table 1: Typical Properties of Epoxy Resin (Lapox L-12)

Properties	
Tensile strength	55-70 MPa
Flexural Strength	120-140 MPa
Elastic Modulus	3.2 -3.6 GPa
Glass Transition Temperature (T_g)	110-130°C (120°C as mid value)
Co-efficient of Thermal Expansion	64-68 K^{-1}

The as-cured composite specimens are immersed in sea water at room temperature for 2, 4, 6, 8, 10 and 12 months. The percentage of moisture gain $M(t)$ after each period of sea water immersion is determined from the weight gain of the specimen as a function of time [9] using the following formula.

$$M_t = \frac{w_t - w_0}{w_0} \times 100 \qquad (1)$$

where 'w_0' and 'w_t' are the weights of the dry and wet specimens at a time 't'.

Data pertaining to Table 2 reveal the major salt components dissolved in the sea water as obtained from literature [11].

The immersed specimens are subjected to EDS (Energy Dispersive Spectroscopy) analysis to determine the depth of penetration of salt components of sea water into the composite body. Line EDS technique is adopted to determine the depth of penetration in micrometer range. This technique is performed across the specimen (which is cut after sea water treatment) from exposed end, till the reference salt components are absent in the EDS spectra. The length of this depth is measured according to the reference scale in the SEM micrograph.

Table 2: Concentrations of salt components in sea water

Element	Parts per million
Cl (Chloride)	21,700
Na (Sodium)	19,497
SO_4 (Sulphate)	2880
Mg (Magnesium)	1838
K (Potassium)	743
Ca (Calcium)	602
HCO_3 (Bicarbonate)	200

All the Samples (as-cured and sea water immersed) are then subjected to 3-point bend test by the help of INSTRON-1195 to record ILSS, stress at rupture, strain at rupture and the elastic modulus. Inter laminar shear strength values are calculated by using following formula [15].

$$\text{ILSS} = 0.75 \times 1000 \times \frac{p_b}{bt} \text{ MPa} \qquad (2)$$

Where p_b = breaking load (load at rupture in KN)
b = width of the specimen
t = thickness of the specimen.

Low temperature DSC technique is adopted to record the glass transition temperature (T_g) of immersed samples. The T_g measurements are performed on a Mettler-Toledo 821 with intra cooler using STAR software in DSC module. The experiments are performed in the temperature range of 30-150°C with scan rate 10° C/min. Scanning electron microscope (SEM; JEOL, JSM-6480 LV) fractographs are investigated to study the mode of failure in extreme cases of sea water conditioning at room temperature.

3. RESULTS AND DISCUSSION

3.1 Moisture gain

Figure 1 illustrates the variation of moisture gain with square root of sea water immersion periods. The percentage of moisture gain is increased with increase in immersion period. The rate of moisture absorption assumes a high value initially followed by a comparatively slower rate of variation up to a period of 8 months and then increases again. The initial increase in moisture gain is due to concentration driven moisture absorption. In this case, the absorption behaviour is fickian. But, sea water ageing of FRP composites still remains an uncertain factor [1]. Ingression of pure water in to the epoxy polymer is quite more feasible than that of saline water due to presence of dissolved salts. A concentration driven osmotic pressure causing moisture ingression in to the polymeric composite body is less of a problem in saline solutions than in pure water. The bulky nature of various salt components of sea water, with their osmotic effect, influences the rate of moisture ingression with time. The flattened trend of moisture gain may be due to deposition of trace elements on the surface of composite body which intervene with the diffusion processes.

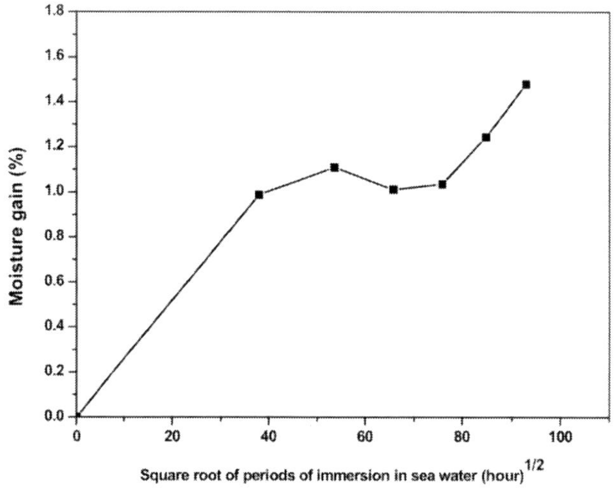

Figure 1. Variation of moisture gain with square root of immersion period

Gallert and Turley [1] opine that sea water immersion may result in fibre related mechanisms for transport of moisture in to the composite body. These mechanisms include transport of moisture along the interface and/or continuing diffusion through the matrix with accumulation of fibre. Asbee and Wyatt [12] report the etching of glass fibre surface with water. Prolonged immersion in water can degrade E-glass fibre due to leaching of alkali oxides (sodium and potassium oxides) by forming water skin. This may lead to formation of surface micro-cracks which can act as stress concentrators. It is reported [16] that the aggressiveness of leaching process is more prominent in case of de-ionized water than that of plane or sea water. Hence, after prolonged immersion in sea water (8 months), the leaching of glass fibre would have enhanced the rate of moisture absorption adopting a non-Fickian mechanism.

3.2 Effect of sea water ageing on mechanical properties

3.2.1 ILSS

The variation of ILSS with respect to period of immersion in sea water is illustrated in Figure 2. ILSS decreases with increased periods of sea water immersion. The continuous decrease of ILSS may be attributed to the exchange of hydroxyl groups among the neighboring epoxy chains as a consequence of moisture ingression [17]. This may lead to breaking of organic bonds in the polymer network resulting in segmentation of epoxy chain structure. In addition to this, prolonged sea water immersion is likely to cause mechanical property deterioration concomitant to plasticization and swelling leading to interfacial debonding assisting further moisture uptake. The present findings are well in agreement with the work of Wei [18]. As reported by Wei, the bending strength as well as the tensile strength of glass fibre/epoxy composite continuously decreases with increased immersion period in sea water.

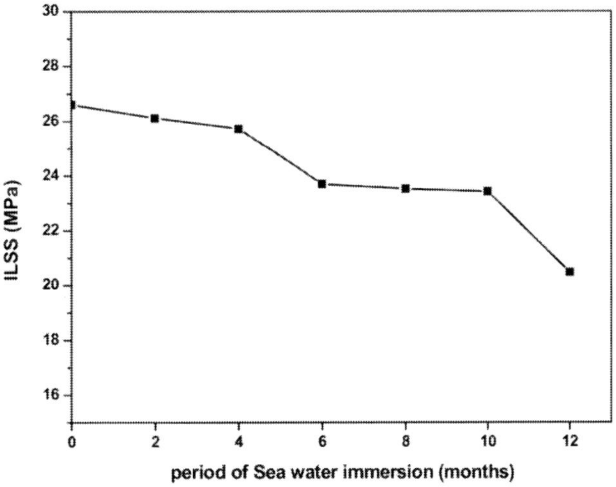

Figure 2. Variation of ILSS with periods of immersion

Figure 3 illustrates the variation of ILSS with percentage of moisture gain. The variation of ILSS with percentage of moisture gain exhibits a non-equilibrium zig-zag trend up to 8 months of immersion. This may be due to a complex deformation behavior as a consequence of local swelling in the matrix and/or matrix/fibre interface when the absorbed moisture forces apart the polymer micro-molecules causing an expansion of the glass fibre reinforced composites. Probably, the initial time periods of immersion are not sufficient to bring about an average response to the external stimulation.

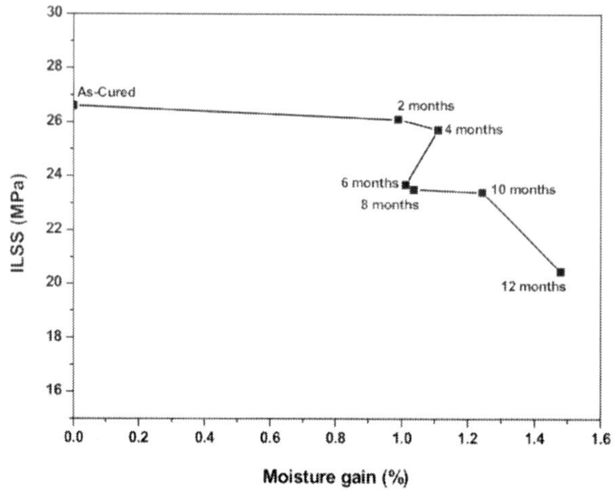

Figure 3. Variation of ILSS with percentage of moisture gain

In general, composite samples show a decreasing trend in ILSS with increased moisture content. There is a 10% reduction in ILSS after 6 months of immersion followed by 23% reduction after 1 year of immersion. This increasing trend can be attributed to the leaching of the glass fibres as a result of a prolonged exposure to sea water ageing [16].

3.2.2 Stress and Strain at Rupture

The combined variations of stress and strain at rupture with different periods of immersion in sea water are illustrated in Figure 4. Both stress and strain at rupture decrease with increased periods of immersion in sea water. Up to 4 months of immersion, the variations of stress and strain at rupture values show a zig-zag trend. The initial decrease after 2 months of immersion may be attributed to the non-homogeneous distribution of water in the composites. During this period, the water distribution in the glass/epoxy composite would not have reached equilibrium. This is likely to create localized stresses which act as regions of defect leading to lower strength. The inconsistent deviation in the strength property is likely to be due to certain localized defects.

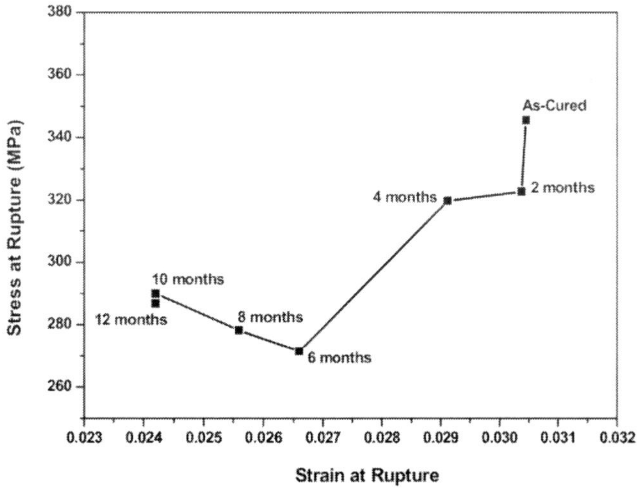

Figure 4. Combined variations of Stress and Strain at rupture with periods of immersion

Stress at rupture reduces to 21% of that for as-cured sample after 6 months of sea water immersion and 16% after 1 year showing somewhat recovery in the stress at rupture values. Similarly, the strain at rupture values decrease 12% after 6 months and 20% after 1 year revealing a continued decreasing trend within the limits of experimentation. It has to be clearly understood that any recovery in stress at rupture values adds to the strength properties where as the decrease in strain at rupture values may indicate the stiffening of the matrix. This varied trend can be attributed to several competing phenomena occurring simultaneously. The prolonged sea water immersion causes damages in the composite material due to plasticization and swelling [9] which affect the stress and strain values at failure and the tolerance capacity of the composite is severely affected.

3.2.3 Modulus

Figure 5 shows the variation of elastic modulus with respect to the period of sea water immersion. The modulus decreases for the first 2 months of immersion then increases slightly and again decreases reaching the minimum value after 6/8 months of immersion, being decreased by 25%. The extent of reduction in the elastic modulus is then compensated, the final reduction being made up from an initial value of 25% to 5% after the lapse of 1 year.

The initial reduction in modulus is due to plasticization of the matrix and the ability of the matrix molecules to move against each other [11]. However, the increase of the modulus in the later stages may be attributed mostly to swelling which improves mechanical adhesion between fibre and matrix. These two reversing trends result in decrease of the modulus by 5% only, making it up from 25% to 5% after 1 year.

To sum up the increasing/decreasing trend of modulus may be attributed to a combined effect to the following competing phenomena:

(i) Initial non-homogeneous water distribution creating localized defect
(ii) Relaxation of residual stress
(iii) Plasticization resulting in lower modulus and high strain to failure
(iv) Swelling leading to increase in modulus and lowering of strain to failure

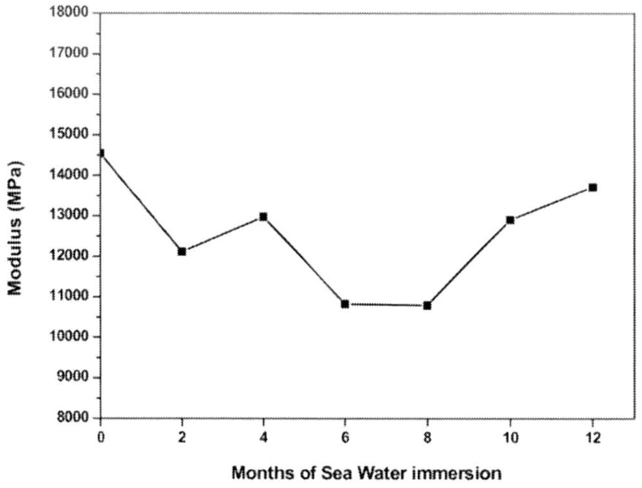

Figure 5. Variation of modulus with periods of sea water immersion

3.2.4 Glass Transition Temperature (T_g)

The DSC curves of sea water immersed samples are illustrated in Figure 6. The onset values of T_g are indicated in the graph. Figure 7 represents the variation of T_g with the duration of immersion.

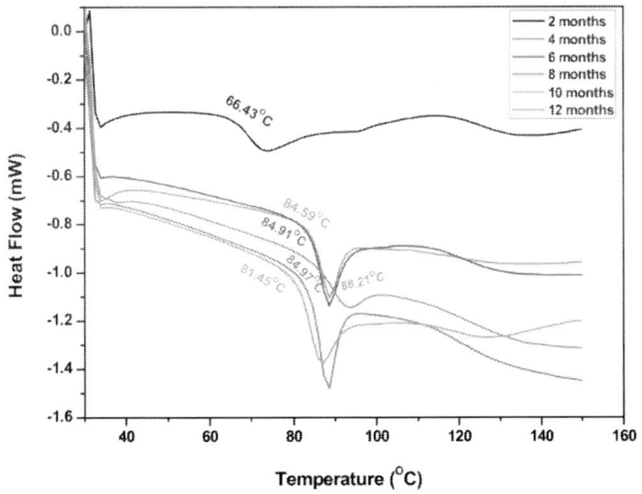

Figure 6. DSC curves of sea water immersed composite samples

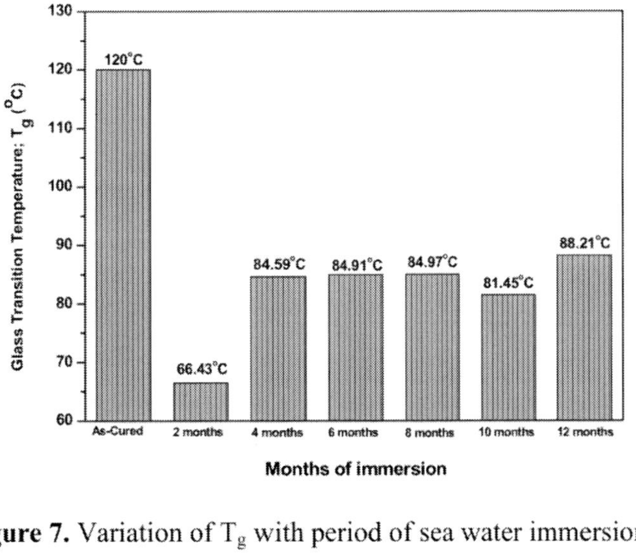

Figure 7. Variation of T_g with period of sea water immersion

T_g values of sea water immersed samples are lower compared to that for as-cured samples, lowest T_g being recorded for the samples after 2 months of sea water immersion. T_g of polymeric materials depends on the extent of cross-linking density [19]. Due to imperfect curing/cross-linking of the polymer, some voids could have been created in the polymer. These voids, referred to as free volume, occupy 1/40th of total polymer volume at/below glass transition temperature [20]. Thus, the reason for early lowering of glass transition temperature is due to the increase of internal voids/free volumes in the epoxy polymer. This could have happened due to less confinement of epoxy network in a particular volume.

Probably, the bulky nature of various salts present in sea water with their osmotic effect [1] promotes the epoxy chain extension by breaking the polymeric network. This leads to the creation of free volumes in epoxy resin. This phenomenon is prominent for lower period of immersion (2 months). Hence, lowest T_g is recorded for this sample. For higher periods of immersion, however, increased salt penetration could have restricted the above phenomenon by accommodating the free volumes of epoxy polymer giving rise to increase in T_g.

3.2.5 Depth of penetration of dissolved salts

Figure 8 and Figure 9 illustrate some EDS spectra indicating depth of penetration of various salt components of sea water. Figure 10 represents the variation of depth of penetration of various salt components with period of sea water immersion.

Following observations are evident from the results.

(i) Various salt components penetrate in to the composite body to different depths.
(ii) 'K' shows highest depth of penetration after 1 year of sea water immersion.
(iii) Depth of penetration is found to increase continuously for 'K'.
(iv) 'Ca' shows instant increase in penetration after 6 months of sea water immersion and then decreases slightly.
(v) 'Na' shows least depth of penetration.

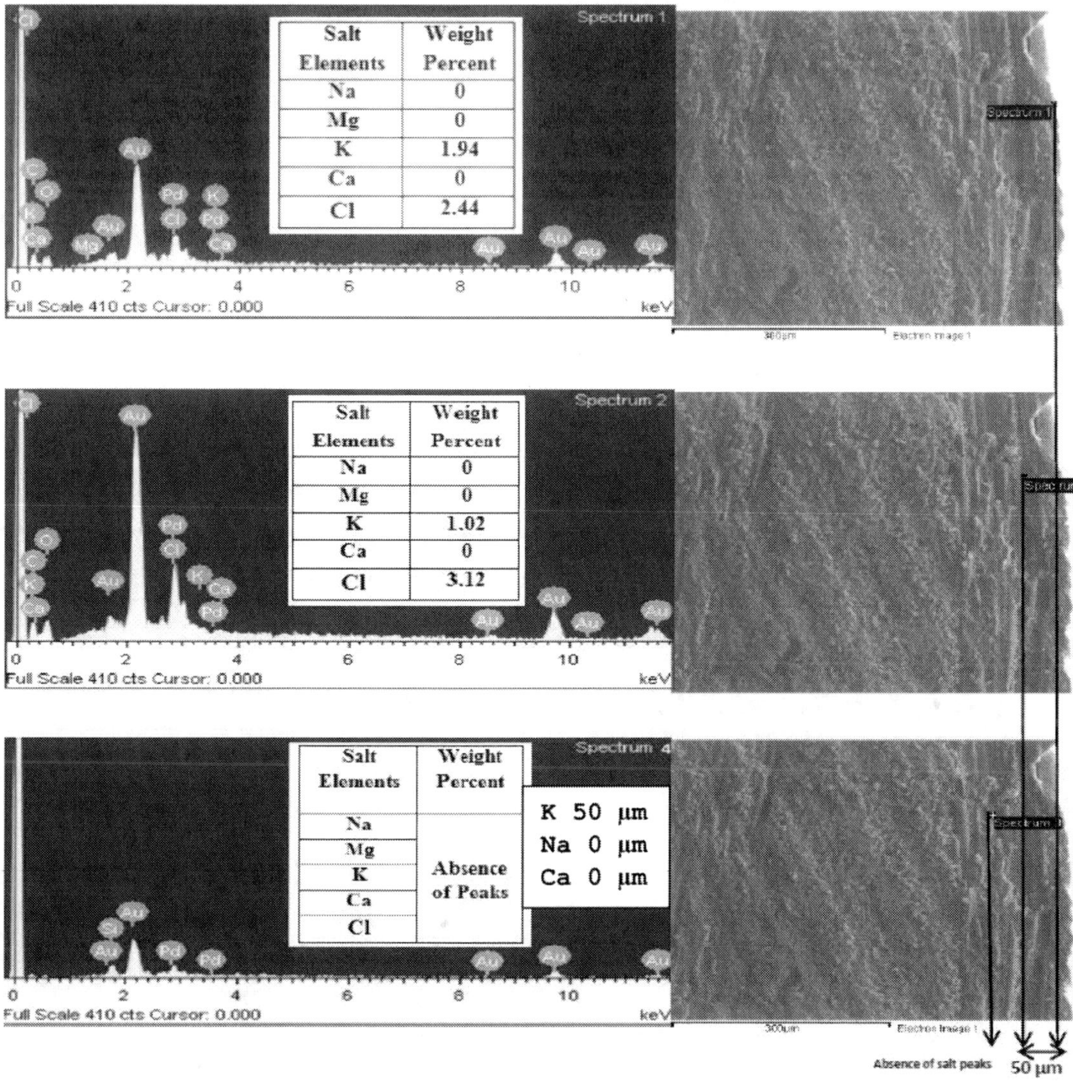

Figure 8. EDS Spectra indicating depth of penetration of salt components after 2 months of immersion period

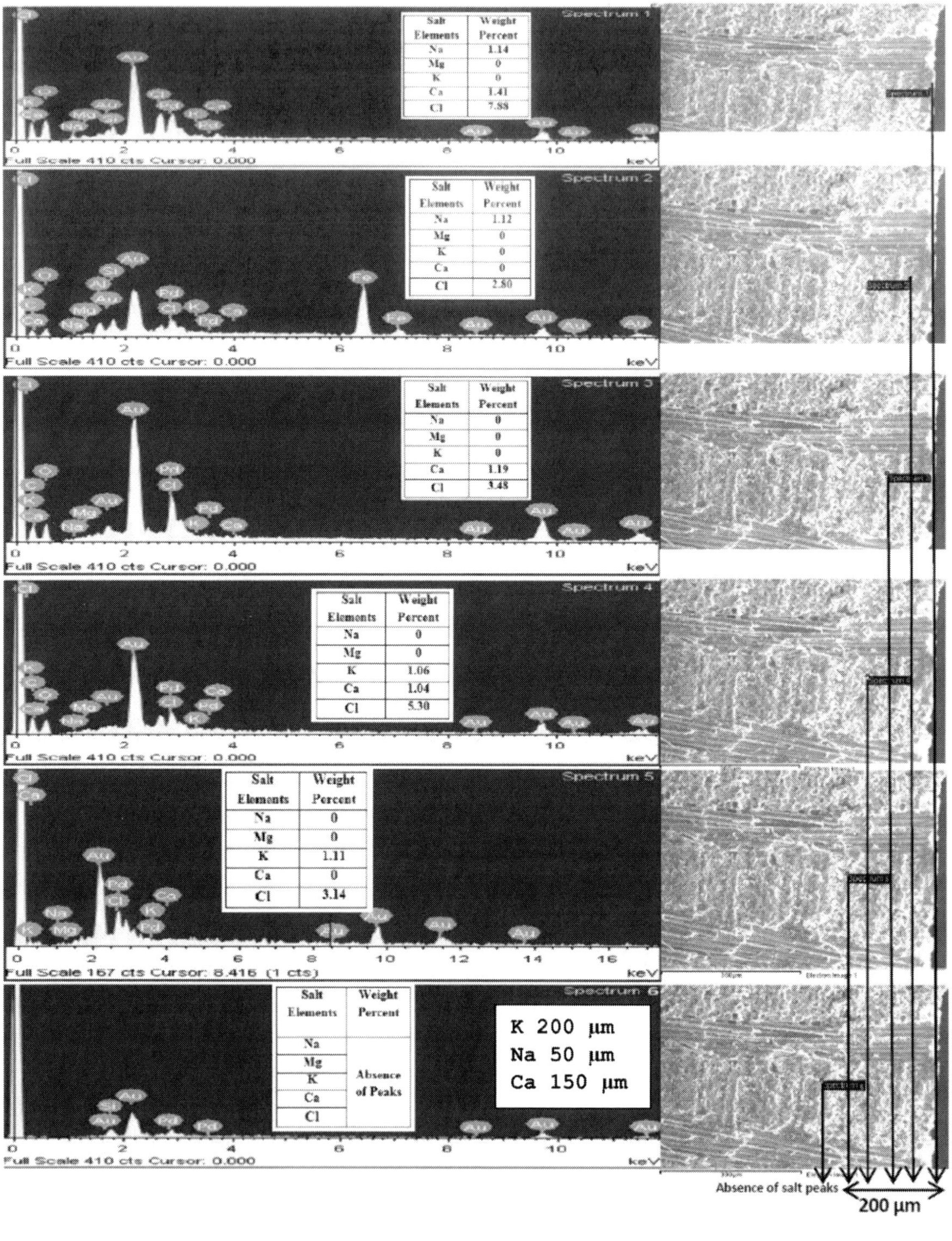

Figure 9. EDS Spectra indicating depth of penetration of salt components after 12 months of immersion period

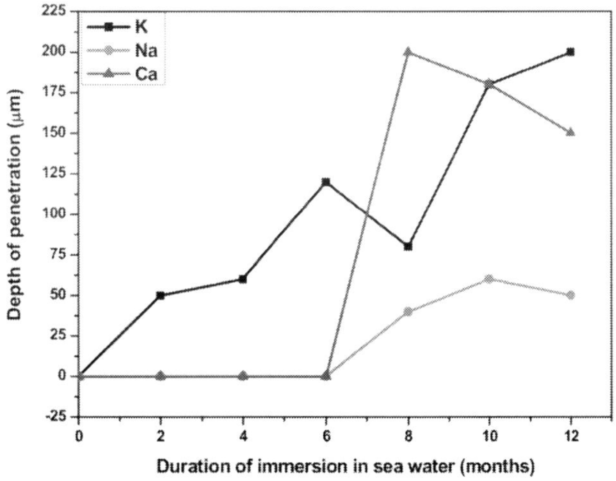

Figure 10. Depth of penetration of salt components with period of sea water immersion

'K' shows higher depths of penetration due to higher ionic mobility compared to the other salt components. After 6 months, 'K' would have interacted with –OH group of epoxy chains (due to having higher reactivity), got arrested, the paired bond inhibiting further penetration of 'K' and encouraging 'Ca' penetration through a free ionic movement. Hence, sudden increase in 'Ca' penetration rate is visible.

Another interesting observation is that despite higher ionic mobility compared to 'Ca', 'Na' penetrates in to the composite body to a lower extent compared to 'Ca'. This may be attributed to the greater affinity of Na^+ ion for this Cl^- ion compared to that of Ca^{++} ion with Cl^- ion. Hence the greater extents of Na-Cl bond pair encounter greater resistance while attempting to penetrate in to the composite body compared to lesser amount of Ca-Cl bond pair.

It is worthwhile to evaluate GFRP composites pertaining to the event of salt penetration. It is obvious that the penetrated salt bodily obstruct moisture ingression by their likely presence in inter laminar space, voids, whiskers and even in micro cracks. While the salts must be obstructing the movement of H_2O molecules along the fibre/matrix interfacial areas followed by diffusion from the interface in to the resin, the direct diffusion of H_2O molecules in to the matrix may not be affected. This is established through experimentation of the authors when GFRP composite samples are immersed in plane distilled water [21]. In the case of plane water the moisture uptake after the lapse of a period of 4 months is seem to be 1.64 %, whereas in the present case moisture ingression even after a period of 1 year is only 1.48 %. Further the high amounts of 'Na' (19,497 ppm) would get a chance of reacting with the negligible amount of 'Na' (0-1 wt %) increasing the molar ratio of 'Na' in the glass fibre and enhancing the process of a bond formation, thereby increasing the strength/stiffness. On the other hand, both 'Ca' (602 ppm) and 'K' (743 ppm) in the sea water may not get a chance for bond formation with the E-glass fibre. Also, the salt penetrations could have resulted in swelling of the matrix bringing about an intimate contact between the matrix and fibre, improving the fibre-matrix adhesion. Simultaneously, the E-glass fibre may get degraded due to leaching of alkali oxides as a consequence of prolonged periods of immersion. The detrimental effects of moisture ingression could have been countered to certain extent by the bond formation on account of penetration of 'Na' in to the composite body and the fibre-matrix adhesion brought about by the swelling of the matrix as a consequence of salt penetration. On the other hand the competing phenomena of plasticisation and leaching of the fibre due to prolonged immersion could have added to the deterioration of the mechanical performance of the GFRP composite.

3.2.6 Mode of Failures

The SEM fractographs (Figure 11) of the sea water immersed samples revealed some chief mode failures. The mode of failures can be any combinations of the following: pot holing, fibre pull-out, matrix crazing, matrix cracking, cusp formation in polymer, fibre breaking, river line mark in polymer showing debonding etc.

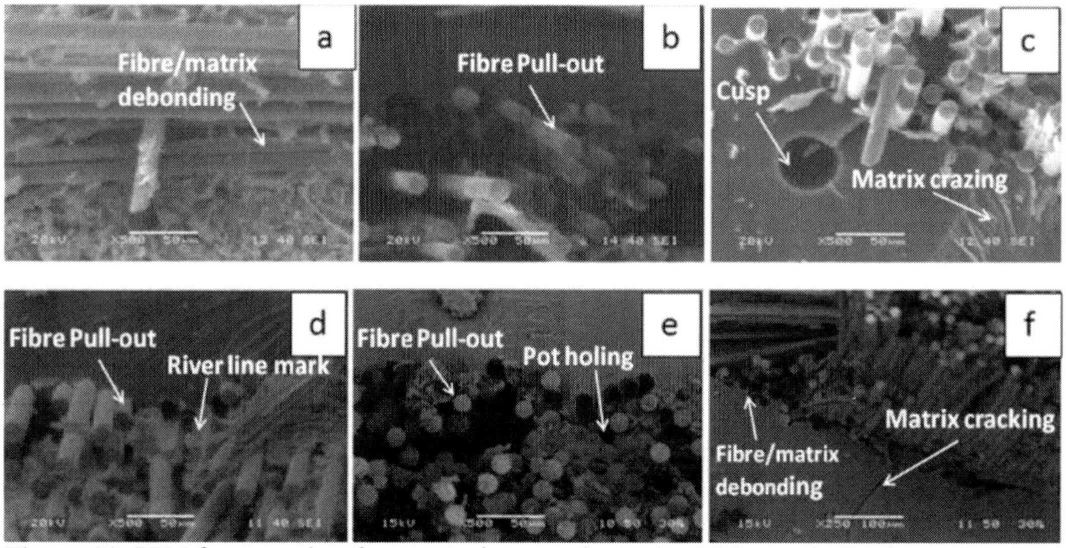

Figure 11. SEM fractographs of sea water immersed samples (a) 2 months (b) 4 months (c) 6 months (d) 8 months (e) 10 months (f) 12 months

4 CONCLUSIONS

The following conclusions can be drawn from this work:

(i) Sea water ingression in to the composite body increases with period of immersion, despite the fact that the presence of penetrated salts bodily obstruct moisture ingression.

(ii) Mechanical properties such as ILSS, stress and strain at rupture, elastic modulus exhibit a general trend of deterioration with the increase of immersion period.

(iii) Different salts penetrate to different depths in the composite body, 'K' showing the highest depth of penetration and 'Na' the lowest depth of penetration after a period of 1 year.

(iv) The Tg values of the sea water immersed samples decreased in general, as a consequence of moisture ingression.

(v) As revealed by SEM fractographs, the mode of failures include pot holing, fibre pull-out, matrix crazing, matrix cracking, cusp formation in polymer, fibre breaking, river line mark in polymer showing de -bonding etc.

REFERENCES

[1] Gellert E P and Turley D M 1999 *Composite Part A* **30** 1259
[2] Ray B C 2005 *J. Reinf. Plast. Compos.* **24** 713
[3] Kootsooks A and Mouritz A P 2004 *Compos. Sci. Technol.* **64** 1503
[4] Smith C S 1990 *Elsevier Appl. Sci. London*
[5] Ellyin F and Rohrbacher C 2000 *J. Reinf. Plast. Compos.* **19** 1405
[6] Aktas A and Uzun I 2008 *Compos. Struct.* **85** 59
[7] Wood C and Bradley W L 1997 *Compos. Sci. Technol.* **57** 1033
[8] Ray B C 2006 *J. Colloid Interface Sci.* **298** 111
[9] Chakraverty A P, Mohanty U K and Biswal B B 2012 *Emer. Mat. Res.* **1** 263
[10] Rutowska M, Krasowska K, Heimwoska A, Steinka E and Janik H 2002 *Polym. Degrad. Stab.* **76** 233
[11] Mourad A I, Abdel-Magid B M, EI-Maaddawy T and Grami M E 2010 *Appl. Compos. Mater.* **17** 557
[12] Assbee K H G and Wyatt R 2007 *Proc. Royal Soc.* 312
[13] Zhou J M and Lucas J P 1995 *Compos. Sci. Technol.* **53** 57
[14] ASTM D 2344/D 2344M-13 2013 Standard Test Method for short-beam strength of polymer matrix composite materials and their laminates, *ASTM International*, West Conshohocken, PA
[15] Bond I, Hucker M, weaver P, Bleay S and Haq S 2002 *Compos. Sci. Technol.* **62** 1051
[16] Maxwell A S, Broughton W R, Dean G and Sims G D 2005 *NPL Report, DEPC MPR 016* ISSN-1744-0270
[17] Luo S, Leisen J and Wong C P 2002 *J. Appl. Polym. Sci.* **85** 1
[18] Wei B 2011 *Corrosion Sci.* **53** 426
[19] Diamant Y, Marom G and Broutman L J 1981 *J. Appl. Polym. Sci.* **26** 3015
[20] Ray P K, Bhusan A, Bera T, Mohanty U, Vadhera S and Ray B C 2003 *Proc. Emerging Trends in Structural Mechanics and Composites, Rourkela* 322
[21] Chakraverty A P, Mohanty U K, Biswal B B and Mishra S C 2014 *Proc. NMD-ATM, Pune*

IOP Conf. Series: Materials Science and Engineering **75** (2015) 012030 doi:10.1088/1757-899X/75/1/012030

Studies on plasma processing of blue dust

S K Samal*[1], Sindhoora L P[1], S C Mishra[1] and B Mishra[2]
[1]National Institute of Technology, Rourkela, Odisha, India.
[2]Dalmia Institute of Scientific & Industrial Research, Rajagangpur, Odisha, India.
E-mail:sumantnitr@gmail.com

Abstract. Plasma smelting was carried out using blue dust and petroleum coke mixtures for five different compositions. By altering percentage of reductant and type of plasma forming gas, recovery rate and degree of metallization were calculated in order to examine the extent of reduction of blue dust. The products were characterized by XRD and optical microscopy techniques. The results of these investigations exhibited that highest degree of metallization and recovery rate of about 98% and 86% respectively, were achieved for nitrogen plasma smelted products.

Keywords: Blue dust, waste utilization, plasma smelting, argon and nitrogen, recovery, degree of metallization.

1. Introduction

Currently iron and steel making is dominated by the route comprising a blast furnace and a basic oxygen converter; meanwhile Ferro alloys are formed in submerged-arc furnaces. Production on reliable source of high quality raw materials such as sinter, pellets, and coke are the principal economic constraints of the process. In addition, the inflexibility of the process capacity, high capital and operational costs and energy lost in between stages still a prominent aspect for reduced economy [1]. Blue dust is a high grade soft hematite ore fines containing more than 90 % Fe_2O_3 enormously available today. For transportation problem and environmental hazardous factor concerned, these iron ore fines are dumped at mine sites. Since its particle size lies in the micron range, blue dust cannot be used directly in the blast furnace for iron making [2, 3]. Utilization of these iron oxide fines and applicability of the same for the blast furnace feed is an approach for production of value added product being wasted [4]. Keeping above facts in sight since last two decades thermal plasma has claimed to be an emerging solution to a numerous processes due its some unique features (high enthalpy, high temperature and high reactivity) and hence implemented in various sectors. Plasma finds significant industrial applications as processes like melting, smelting, smelting and reduction, refining, spark plasma sintering, surface modification and surface coating [5,6,]. Application of plasma technology seems to be an emerging alternative route for iron and steel making and for many of its advantages over any other processes [7-9]. Carbothermal reduction and smelting of metal oxides were experimented effectively in the past research works to get metal as final product with recovery more than 80% [10-13]. Carbothermal reduction of hematite can be possible in three ways [2] as given below;

$Fe_2O_3{}^1 \rightarrow Fe_3O_4{}^2 \rightarrow FeO^3 \rightarrow Fe$

(Single chain reaction)

$Fe_2O_3{}^1 \rightarrow FeO^3 \rightarrow Fe$

(A combination of single and double reaction)

$Fe_2O_3{}^1 \rightarrow Fe$

Content from this work may be used under the terms of the Creative Commons Attribution 3.0 licence. Any further distribution of this work must maintain attribution to the author(s) and the title of the work, journal citation and DOI.
Published under licence by IOP Publishing Ltd

(A triple reaction)
In this piece of research work, an attempt has been made to study the effect of reductant and type of plasma forming gas on reduction of blue dust and characterization of final product.

2. Experimental details

2.1. Materials
Blue dust of average particle size about 100-150μ whose origin is Koida range, Odisha was collected by DISIR, Rajagangpur, Odisha for our study. The major constituents of blue dust was obtained by wet chemical analysis, is given in the Table 1. The loss on ignition was calculated by heating the raw blue dust at 1000^0C for 1hr. The XRD pattern of raw blue dust obtained is given in fig.1.

Table 1. Chemical analysis of Blue dust

Oxides	In wt. %
Fe_2O3	96.87
SiO_2	0.45
Al_2O_3	0.21
TiO_2	Trace
MgO	Trace
LOI	1.48

2.2. Experimentation
The experimental set up used to treat the blue dust consists up a 35 KW DC transferred arc plasma setup, power supply unit, gas feeding system, power and gas control unit etc. On the top, plasma torch is attached in the downward direction. The plasma furnace contains a hollow cylindrical graphite crucible with 145mm outer diameter, wall thickness 15mm and 300mm high that serves as the anode. Hollow graphite rod of 400mm long with 5mm central hole and 35mm outer diameter serves as the cathode. Hollow structure of cathode has designed to have provisions for gas flow. The material to be processed was placed in the anode crucible bed and the arc was initiated by shorting the cathode and the crucible bottom wall (graphite plate). The arc length was increased by raising the cathode up suitably within the crucible to heat the charge placed in the crucible. For charge, material nodules of 8-12mm (diameter) were prepared by taking five different compositions of mixture of blue dust and petroleum coke (5%, 10%, 12%, 15% and 20% of coke). Then nodules were dried at 110^0 C for one hour for moisture removal. Before initiating the arc, reaction chamber was properly cleaned and plasma forming gas is purged for 2min.Then arcing was done by striking cathode with anode. Plasma arc was generated by passing plasma forming gas (Argon and Nitrogen) through cathode and it gradually spreads in all directions and melted down the charge material. The important operating parameters used in this work are illustrated in table 2.

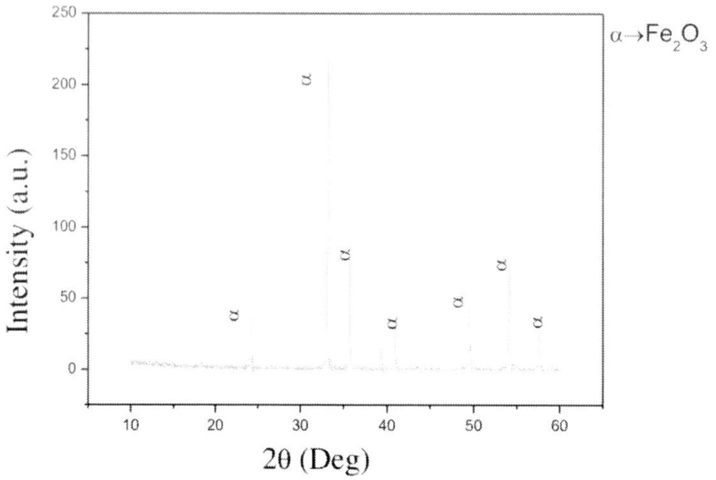

Figure 1. XRD pattern of raw blue dust.

Table 2. Operating parameters for smelting.

Operating Variables	Quantity
Input Voltage	250-300 V
Input Current	50 A
Gas Flow rate	2.5 lpm
Processing time	17 minutes
Charge quantity	300 g

3. Results and discussion

3.1. Recovery

Since carbothermal reduction of blue dust was made in this study, carbon percentage was increased gradually to examine its effect to the extent of reduction. Percentage of recovery was calculated by taking ratio of weight of final product to weight of metal present in the composition before smelting and reduction. Fig.2 represents the plot of variation of percentage of recovery w.r.t. %C in compositions. The curve representing samples smelted in nitrogen plasma lies above that of argon plasma. Maximum recovery of about 86% achieved for sample with 12% coke by nitrogen plasma. Comparatively higher percentage of recovery has been achieved by utilizing nitrogen as plasma gas.

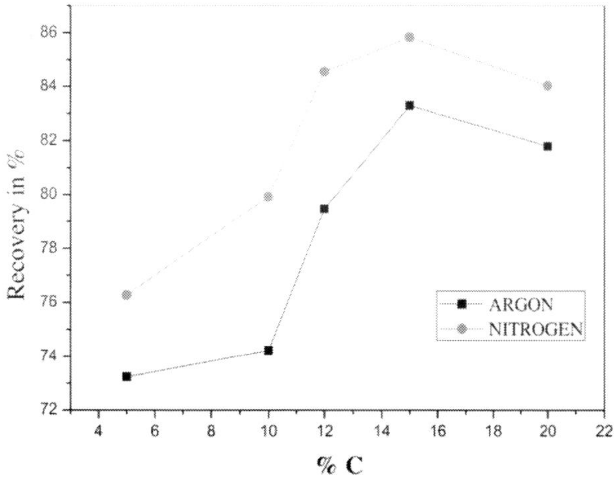

Figure 2.Recovery rate of smelted samples.

3.2. Degree of metallization

Following wet chemical analysis route degree of metallization of all smelted samples were calculated by taking ratio of metallic Fe to total Fe. Highest degree of metallization of about 98% is achieved in case of smelted composition of blue dust and 12% coke.

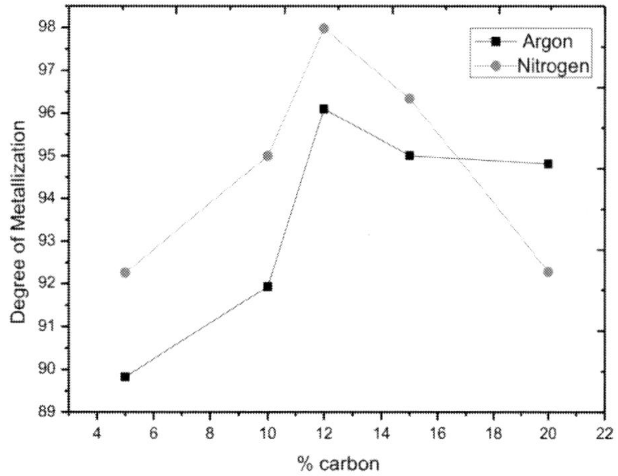

Figure 3. Degree of metallization curve of smelted samples.

Nitrogen plasma smelted samples shows greater DOM values may be due to the fact that, nitrogen being diatomic gas increases the enthalpy of plasma which in turn helps in reduction of blue dust to a greater extent.

IOP Conf. Series: Materials Science and Engineering **75** (2015) 012030 doi:10.1088/1757-899X/75/1/012030

3.3. Microstructural studies

Micrographs of all cold resin mounted polished samples were observed by using Zeusius light emission electron microscope under a constant magnification of 200X as shown in Figure 4 and Figure 5 (a-5%C, b-10%C, c-12%C, d-15%C, e-20%C) . In most samples micro cracks are observed as shown in Figure 4 (a), (b) and Figure 4 (b), (d). It may be due to high cooling rate of smelted samples as in case of plasma processing, temperature falls from thousands of degree to ambient temperature in very short period of time. Elongated flakes are found in Figure 4 (d) and Figure 5(c), whereas in Figure 5(e), elongated flakes with some voids are seen to be present. Length of flakes are in decreasing order in figures 5(c), 4(d) and 5(e) whereas width of flakes are in increasing order in figures 5(c), 5(e) and 4(d) respectively. Hence Figure 5 (c) possesses flakes of highest L/D ratio. In all other cases homogeneous distribution of two phases are observed.

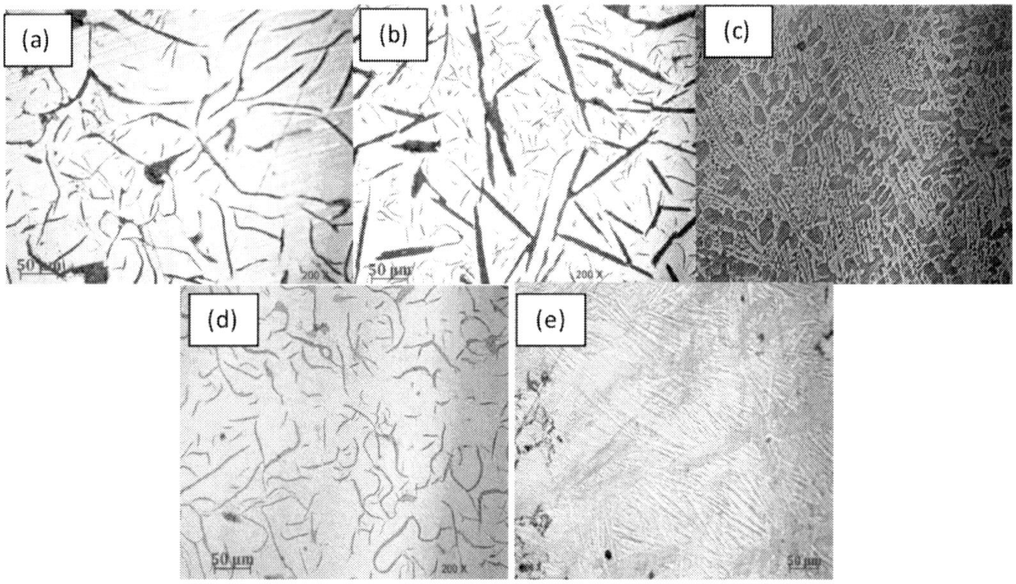

Figure 4. Optical micrographs of nitrogen plasma smelted samples (unetched).

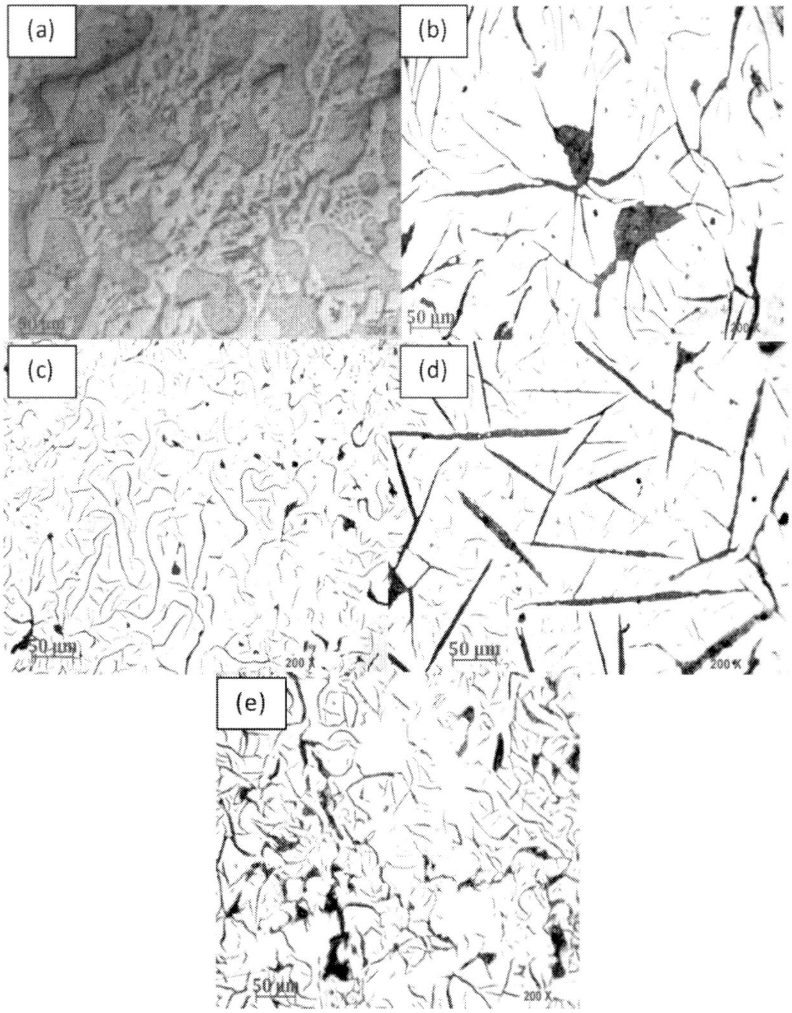

Figure 5. Optical micrographs of argon plasma smelted samples (unetched).

3.4. Phase characterization by XRD

XRD analysis of smelted samples was made to examine presences of phases in each by using Cu-kα (1.54A^0) radiation. X-ray diffraction pattern of all samples are presented in Figure 6.a and Figure 6.b. and in all cases highest intensity peak shows the presence of iron and other peaks shows the presence of cohenite (cementite). Since graphite crucible was used as reaction chamber, decay of graphite bottom plate (more) and crucible walls (lesser) mainly occurs at high temperature and this extra carbon is dissolved into the molten feed and reduces iron oxide to a little more extent. Formation of cementite is due to reaction of carbon with iron (Fe) and wustite (FeO) [14].

$3Fe + C \rightarrow Fe_3C$

$3FeO + 5CO\ (g) \rightarrow Fe_3C + 4CO_2\ (g)$

Again reaction of cementite with wustite can give rise to iron and carbon dioxide.

$2FeO + Fe_3C \rightarrow 5Fe + CO_2$

Hence the brighter phase observed micrographs were confirmed as cohenite (Fe_3C) as clearly visible in Figures 4(c) and 5(a). Figure 4(c) coincides with ledeburite structure as along with cohenite (brighter phase) a dark phase (pearlite) is present. This may be due to rapid erosion of graphite from crucible and cathode rod.

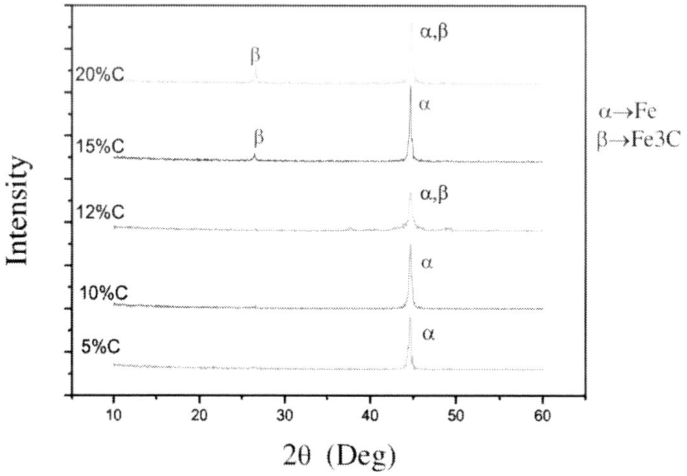

Figure 6.aXRD pattern of all nitrogen plasma smelted samples.

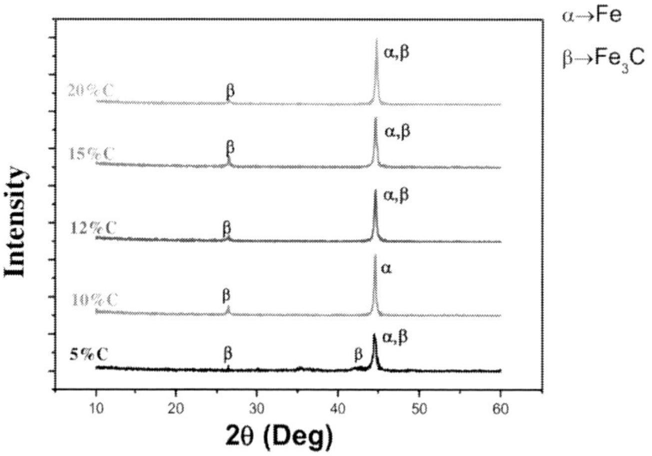

Figure 6.bXRD pattern of all argon plasma smelted samples.

3.5. Hardness

Vickers hardness of smelted samples was performed by using a LECO micro hardness tester LM248AT taking 300gf load and dwell time of 10 second. Three different values of hardness were obtained i.e. 147, 208 and 312(average values), is due to presence of ferrite, pearlite and cementite phases respectively depending on microstructure shown in micrographs. Meanwhile when hardness measured on needle like and elongated phases (cracks and voids) as seen above, sample gets distorted and resulting hardness values varied from 50 to 80 HV.

4. Conclusion

Transferred arc plasma route was used to convert blue dust to industrially useful product like metallic iron. Carbon percentage and type of plasma gas was varied to study its effect on recovery rate and degree of metallization and each case. Nitrogen as plasma forming gas exhibited better rate of reduction as compared to that of argon. Three different phases: ferrite, pearlite and cohenite were identified by optical microscopy, XRD and corresponding hardness values.

References

[1] Ghosh A, Chatterjee A 2012 *Text Book of Iron Making And Steelmaking* **ISBN: 978-81-203-3289-8.**

[2] Sakthivel R, Bhattacharyya D, Eswaraiah C, Das D, Jayasankar K, Mishra B K 2014 Effect of milling on reduction behavior of blue dust *Journalof Alloys and Compounds* **587** 677–680.

[3] Tiwary R K, Narayan S P, Pandey O P 2008 Preparation of strontium hexaferrite magnets from celestite and blue dust by mechanochemical route *J.Mining Metall.* **44** b 91–100.

[4] Chakravarty S, Bhattacharya P, Chatterjee S Sand Singh B N 2000 *Processing of Fines (2)***ISBN: 81 -87053-53-4.**

[5] Mihovsky M 2010 Thermal plasma application in metallurgy *Journal of the University of Chemical Technology and Metallurgy* **45** 1 3-18.

[6] Mishra S C 2012 *Thermal Plasma Application In Metallurgy LAMBART Academic Publication* **ISBN: 9783659235221** Germany.

[7] Jayasankar K, Ray P K, Chaubey A K, Padhi A, Satapathy B Kand PS Mukherjee P S 2012 Production of pig iron from red mud waste fines using thermal plasma technology *International Journal of Minerals, Metallurgy and Materials* **19** 679.

[8] Mimura Kouji, Nanjo Michio 1990Production of pure tantalum by carbon-reductionsmelting and hydrogen plasma-arc melting with refining *Materials Transactions, JIM* **ISSN: 09161821**.

[9] Dembovsky Vladimir 1985 *Plasma Metallurgy, The Principles Materials science monographs* **23.**

[10] Sahoo P K,Galgali R K, Singh S K,Bhattacharjee S, Mishra P K, Mohanty B C1999 Preparation of titania-rich slag by plasma smelting of ilmenite, *Scandinavian Journal of Metallurgy* **ISSN: 03710459** 243-248 Publisher:Munksgaard Int Publ. Ltd.

[11] Nayak B B, Mishra B K, Pradhan S 2010 Smelting of niobium pentoxide in a transferred arc argon plasma and characterization of the smelted product *High Temp. Mater. Process. (USA)* **14** 233-44.

[12] Vijay M, Ananthapadmanabhan P V, Ramachandran K, Hiremath G, Mathai C B,Nalini B, Pillai B C 2013 Carbothermal reduction of sillimanite in a transferred arc thermal plasma reactor, *Int. Journal of Refractory Metals and Hard Materials* **36** 174–178.

[13] J Bessieres J, A Bessieres A, Heizmann I J 1980 Iron oxide reduction kinetics by hydrogen *Int. J. Hydrogen Energy* **5** 585–595.

[14] Srinivasan N Sand L Staffansson L 1990 Theoretical analysis of the fluidized-bed process for the reduction of iron ores *Chemical Engineering Science* **45** 1253-1265.

IOP Conf. Series: Materials Science and Engineering **75** (2015) 012031 doi:10.1088/1757-899X/75/1/012031

Study of mechanical properties, microstructures and corrosion behavior of al 7075 t651 alloy with varying strain rate

A Mukherjee[1], M Ghosh[1*], K Mondal[2], P Venkitanarayanan[3], A P Moon[2], A Varshney[2]

[1] Department of Metallurgy and Materials Engineering, Indian Institute of Engineering Science & Technology, Shibpur, India 711103.

[2] Department of Materials Science and Engineering, Indian Institute of Technology, Kanpur, Uttar Pradesh, India 208015.

[3] Department of Mechanical Engineering, Indian Institute of Technology, Kanpur, Uttar Pradesh, India 208015.

Mukherjee.avijit2012@gmail.com

Abstract

Compression test of Al 7075 T651 was carried out at high strain rates (1138 – 2534 s^{-1}) using Split Hopkinson Pressure Bar and at slow strain rate ($10^{-4}s^{-1}$) in 100KN Universal Testing machine to understand the improvement in mechanical properties and associated changes in microstructures. Cylindrical specimens of 6 mm height and 6 mm diameter were compressed dynamically. The influence of strain rates on mechanical properties, microstructure evolution and corrosion behavior after immersion test in 3.5% NaCl solution was also investigated. Strain rate, withdrawal stress and yield stress were observed to increase with impact velocity in high strain rate tests, while in slow strain rate tests, n value was observed to increase with increasing total strain. Microstructural observations revealed that after high strain rate test, grains of Al matrix were elongated. It was observed that corrosion resistance decreased with increase in impact velocity.

1. Introduction

Al 7075 is a very important alloy for automobile and aeronautics industries [1]. Now a days this material is used quite extensively for automotive body and aircraft structures because it has higher strength and lower weight. Considerable amount of work has been carried out on this material under the high and low strain rates at various loads [2-4]. However, there has been very little work concerning the effects of loads and strain rate on aluminum 7075 and evolution of microstructures after deformation at high and slow strain rates.

In current research, Split Hopkinson Pressure Bar is used [5-7] for high strain rate and uniaxial compression for slow strain rate deformation. By using this technique, it is proved that significant changes occurred in mechanical response under increased load or strain rate.

The objective of this study is to compare the changes in mechanical behavior of aluminium 7075 alloy after deformed in high and slow strain rate at same condition. The stress - strain relation, withdrawal stress, strain-strain rate comparison, n value and microstructure evolution are discussed in terms of test conditions. For corrosion studies immersion test for 30 days in 3.5wt% NaCl solution has been done. The change in corrosion rate has been studied during this test of the deformed samples in high and slow strain rate.

2. Experimental Procedure

Material used is Al7075 T651which has the following composition:

*manojit_ghosh1@rediffmail.com, 9874865163

Content from this work may be used under the terms of the Creative Commons Attribution 3.0 licence. Any further distribution of this work must maintain attribution to the author(s) and the title of the work, journal citation and DOI.
Published under licence by IOP Publishing Ltd

Table 1. Composition of Al7075 alloy in wt%

Composition	Wt%
Zn	6.886
Cu	1.561
Mg	1.605
Si	0.035
Fe	0.83
Cr	0.092
Ti	0.049
Mn	0.023
Li	0.041
Al	Balance

The material received T651 thermo mechanical treatment, consisting of a solution heat treatment for 30 min at 748 K (475°C), a water quenching and a final aging at 394 K (121°C) for 24 h. This material has yield stress (YS) of 502 MPa, ultimate tensile strength (UTS) of 572 MPa. The percentage of failure elongation of this material is up to 11% and young modulus is 72 GPa. Cylindrical specimen was used for the test. The materials were supplied as plates then it was cuts into cylindrical shape which has 6 mm diameter and 6 mm height approximately.
The purpose of SHPB test was to obtain high strain rate properties. The operation of SHPB was done in four conditions of load which are 4 kg/mm², 6 kg/mm², 8 kg/mm² and 10 kg/mm². Four experiments have been performed at each condition to satisfy the condition of repeatability.

2.1. High strain rate test

Specimens were loaded between the incident and transmission bars for the SHPB tests. The pressure was set at 4 kg/mm² then the striker bar launcher was pressurized with air compressor. The specimen was compressed by the impulse from the incident bar generated due to gas gun ignition. This collision created a compression pulse, or stress wave, that propagated through the bar toward the specimen. This wave was called incident wave which was recorded by the incident strain gages. Once the wave reached the specimen, it splited into two smaller waves. The transmitted wave traveled through the specimen and into the transmission bar where the energy was recorded by the transmission strain gauge. The second wave was reflected away from the specimen and traveled back down the incident bar. Both strain gages measured the strain duration and amplification in the bars. The reflected pulse recorded by the incident strain gauge was used to calculate strain. The portion of the compression pulse that continues through the specimen was recorded by transmission gauge and used to calculate stress. Data from the strain gauges was routed through amplifiers and an oscilloscope to where it was stored. Then collected the compressed sample and measured it dimension for future work. This procedure was also done in other conditions of pressure. At the end of the experiment we got all data of strain, strain rate, stress and stress strain curves

2.2. Slow strain rate test

Slow strain rate deformation was carried out in a 100 KN BISS universal testing machine. Specimens were subjected to compressive load during deformation. Specimen dimensions during this test were same as those during the high strain rate deformation test. Slow strain rate compression was carried out at a crosshead speed of 0.5mm/min leading to the strain rate of 10^{-4}/s. Specimens were strained under compressive load to different total strains (ε_T). The extent to which specimens have to be strained, that is the value of the total strain (ε_T), was decided on the basis of the total plastic strain (ε_P) sustained by the specimens at each load during high strain rate tests. During slow strain rate test, specimens were subjected to same values of plastic strain as given in high strain rate test for different loads. To know the amount of total strain to which the specimens had to be compressed for a given plastic strain, a specimen was subjected to total strain of 40%. From the data of this particular test, total strain corresponding to required plastic strains was calculated. Other specimens were than subjected to same total strains to get the desired plastic strains. The specimens were subjected to 14%, 20%, 25% and 30% total strain to get the corresponding plastic strain of 13%, 19%, 24% and 29%, respectively.

2.3. Corrosion Studies

Samples were first polished for corrosion test. After polishing samples were ultrasonically cleaned in acetone to remove any kind of particles, because corrosion is a surface phenomenon therefore any kind of anomaly in the surface (scratches, pits, dirt etc) can play a negative role in corrosion resistance. For immersion test, the circumferential area of the samples was coated with Teflon tape to reduce the area effect from corrosion. The immersion test was carried out in 3.5% NaCl solution.

2.3.1. Immersion test. Immersion test is easier, simpler controlled test method than other test. Immersion tests measures the progress of corrosion damage of the samples in corrosion environment. Here the laboratory tests were carried out the ASTM standard G-103. Two samples of high strain rate test and one sample of slow strain rate were prepared for immersion test. These polished samples were dipped inside different conical flask containing 250 ml of 3.5 wt% NaCl solution (ph-7.5-8.5) for 30 days at room temperature (~ 30°C) and the flasks were sealed with aluminum foil. Teflon tape was used to hang the samples and to avoid any kind of contact corrosion. After completing the test the samples were cleaning with acetone in ultrasonic cleaning machine for 10 to 15 minutes. The cleaning was done preferentially to remove the oxide layer from the surface.

Figure 1. Al 7075 t651 sample in immersion test

After immersion test corrosion rates were estimated by weighing the cleaned specimen in a digital balance (Mettler Toledo XS-205 model).The corrosion rate in meter per year (m/y) was calculated from the weight loss data for the samples obtained after immersion test using this formula:

Corrosion rate = 534 W/ DAT

Where, W = weight loss in mg

D = density in g/cm3 = 2.81 g/cm^3 for aluminum 7075 alloy

A = area of specimen in cm^2

T = total time exposure time in hours

3. Result and Discussion

3.1. High strain rate deformation

3.1.1. Stress, strain & strain rate curve. In Fig. 3(a-d) stress, strain and strain rate curves are plotted for different impact velocities where the strains are 0.13, 0.18, 0.25 & 0.29 and strain rates are 1141

s^{-1}, 1749 s^{-1}, 2153 s^{-1}& 2552 s^{-1}. These figures shows that stress and strain rates are dependent with strain. With strain increasing stress is increased where the strain rate is decreased. But both stresses and strain rates are increased with impact velocity.

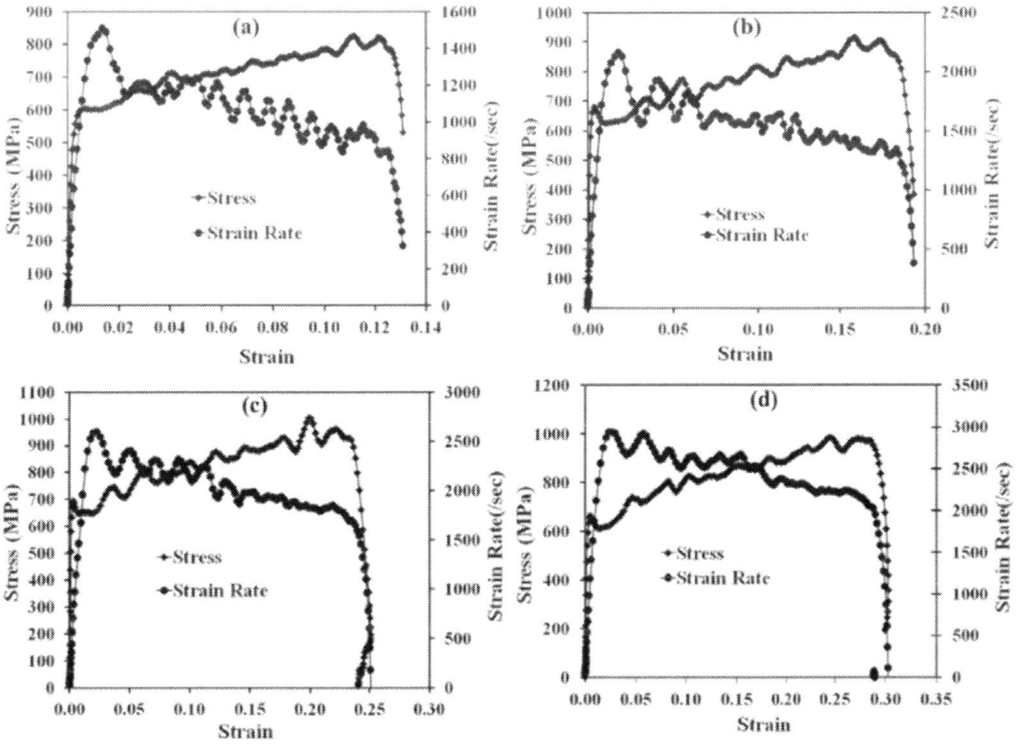

Figure 2(a-d). Stress, strain & strain rate curve at different loads.

3.1.2. Impact velocity vs average strain rate, Withdrawal stresses, Strain at withdrawal stress and yield stress curve. The average values of impact velocity, strain rate, withdrawal stress, strain at withdrawal stress and yield stress are calculated from the raw data which are given below.

Table 2: Average values of strain rate, withdrawal stress, Strain at withdrawal stress and yield stress

Load (kg/mm^2)	*Impact velocity(m/s)*	*Average strain rate (s^{-1})*	*Average withdrawal stress(MPa)*	*Average strain at withdrawal stress*	*Average yield stress (MPa)*
4	1.5	1138	760	0.13	584
6	1.9	1710	850	0.19	636
8	2.2	2164	920	0.24	682
10	2.5	2534	960	0.29	704

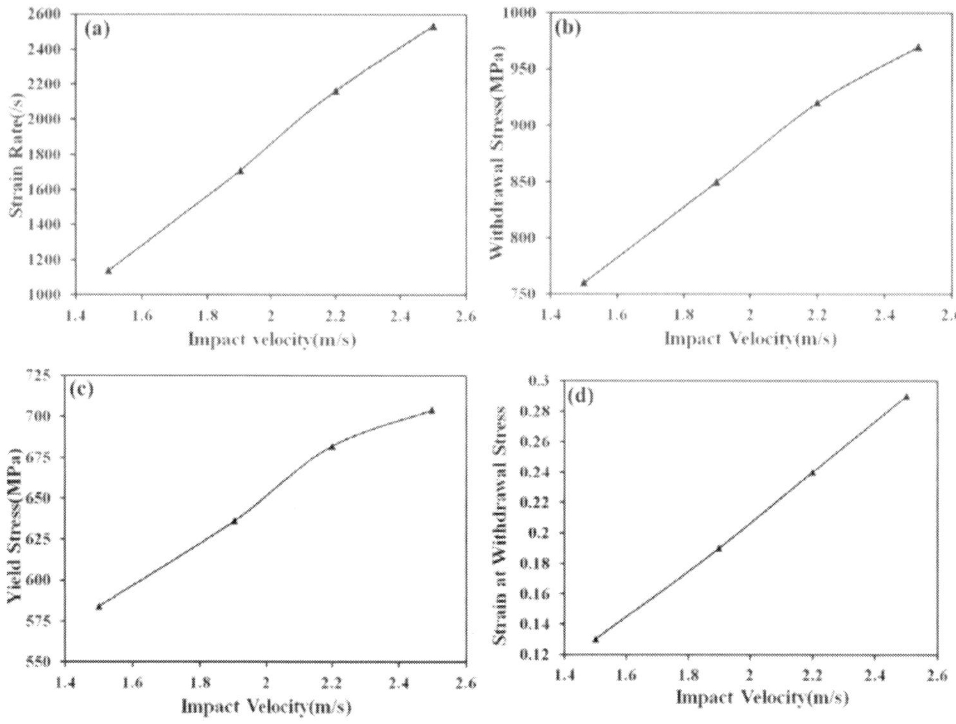

Figure 3(a-d). Impact velocity vs strain rate, withdrawal stress, yield stress and strain at withdrawal stress

From this figure it can be seen that the strain rate, withdrawal stress, yield stress and strain at withdrawal stress are varying proportional to the load. These values are increases with increasing impact velocity.

3.2. Slow strain rate deformation

3.2.1. True stress-strain behavior. The true compressive stress-strain curves for Aluminum 7075 alloy deformed at different strain are shown in fig.5 (a-d). The flow stress as well as the flow curve depends on the strain and strain rate. True stress-strain values were calculated from the engineering stress-strain value. As we know that, true stress, $\sigma_t=\sigma(1+\varepsilon)$ and true strain, $\varepsilon_t=\ln(1+\varepsilon)$ where σ=engineering stress & ε = engineering strain. After calculating value plot the true stress-strain curve which are below:

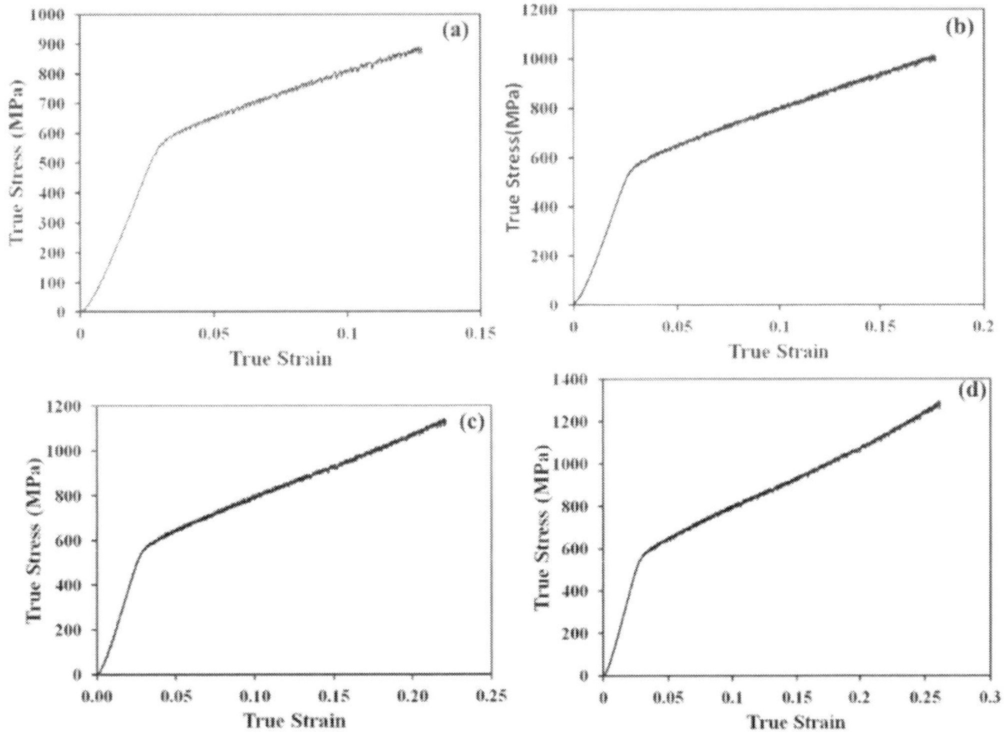

Figure 4(a-d). True stress-strain curves at different strain

3.2.2. n value calculation. The strain hardening exponents (n), determine the rate at which the materials hardens. If value of n increases the effect of strain on material strength and hardness also increases. The most important and widely used application is the evolution of stretch-formability by the n value. This value is calculated by the Holloman equation, $\sigma = k\varepsilon^n$ where, σ = true stress, ε = true strain, k = strength co-efficient. To determine the strain hardening exponents first calculate plastic strain and then taken log value of this equation. So, the equation will be: - $\ln(\sigma) = \ln(k) + n\ \ln(\varepsilon)$. From this equation we calculate the n values which are given below: 0.281, 0.312, 0.343, and 0.383 for various condition of strain.

Table 3. n value at different deformation load

Total strain	n value
0.13	0.281
0.20	0.312
0.25	0.343
0.30	0.383

3.3. Corrosion studies:

3.3.1. Immersion test. The corrosion rate and weight loss are obtained by the immersion test of the sample in 3.5 wt% NaCl for 30 days shows in the chart. It seems that corrosion rate decreases with increasing applied load and weight loss varies in irregular trend. As shown in figure (5 & 6) the corrosion rate is highest for 1.5 impact velocity and lowest for 2.5 impact velocity in both high strain rate & slow strain rate. This variation of corrosion rate shown in the following graph;

Table 4. Change in weight loss and corrosion after immersion test 3.5% NaCl solution

Immersion Samples	*Impact velocity (m/s)*	*Weight loss, Δw (g)*	*Corrosion Rate (m/y)*
High strain Rate test	1.5	5.01 – 2.09	312 – 270
	1.9	4.32 – 2.41	289 – 234
	2.2	4.14 – 2.04	275 – 254
	2.5	3.41 – 1.66	259 – 250
Slow strain Rate test	1.5	2.19	152
	1.9	1.91	141
	2.2	1.87	130
	2.5	1.41	113

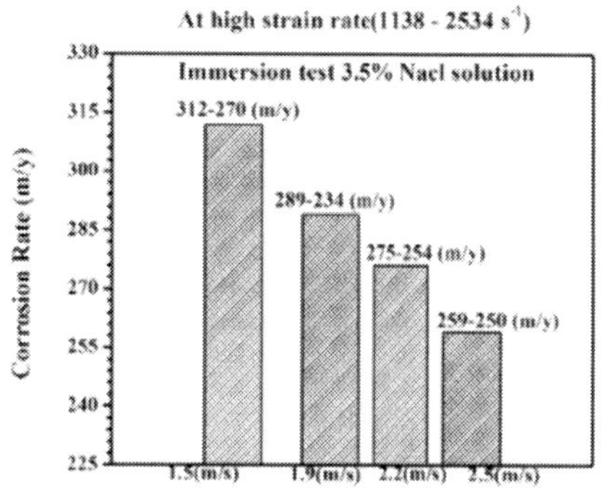

Figure 5. Impact velocity vs corrosion rate in immersion test after deform at high strain rate

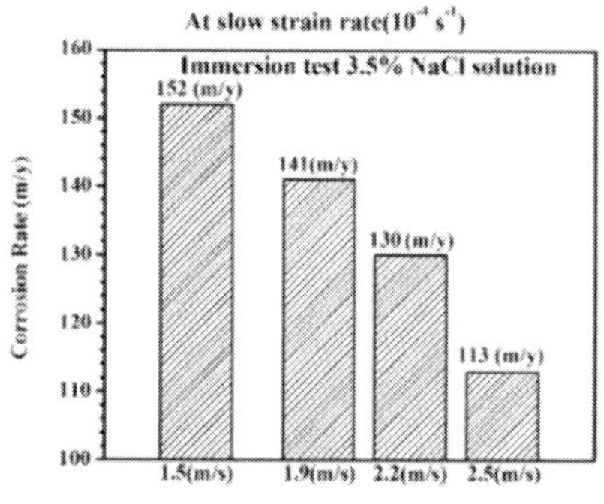

Figure 6. Impact velocity vs corrosion rate in immersion test after deform at slow strain rate

3.4. Microstructure Evolution

3.4.1. High strain rate and slow strain rate. The specimens after impact loading are examined by optical microscopy (OM) and scanning electron microscopy (SEM) in order to determine the effects of strain and strain rate on the evolution of the microstructure. Specimens were sectioned along cross-section or parallel to the applied load for SEM or OM. Specimen were polished in silicon carbide papers up to grit 1000 and also cloth polishing up to 0.1μ and immersed in pure ethanol for 1 min to dissolved water. Then those samples were etched using kroll's reagent (92 ml H_2O + 2 ml HF + 6 ml HNO_3). Fig.8. shows the microstructure of the specimen deformed at high strain rate between 1138 – 2534 s^{-1}. Under this condition, microstructure consist elongated grains in Al matrix. During high strain rate deformation micro voids are developed due to high strain and grow with increase in load. This micro void's coalescences are controlled by the localized plastic flow around them.

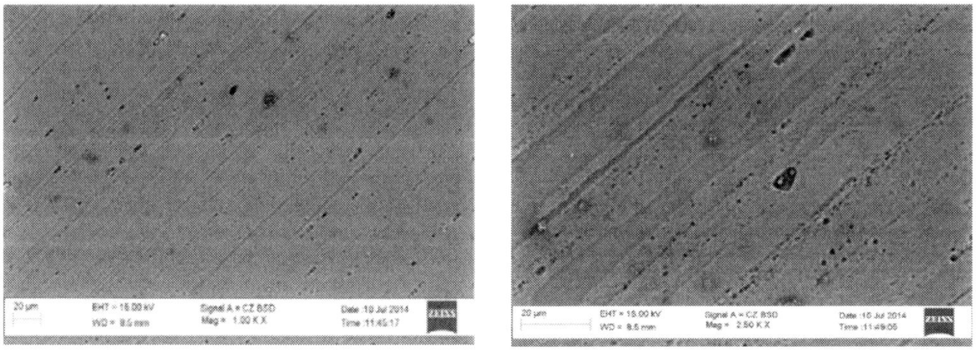

Figure 7. SEM micrographs of the 7075 Al alloy deformed at high strain rates (1138 – 2534 s^{-1})

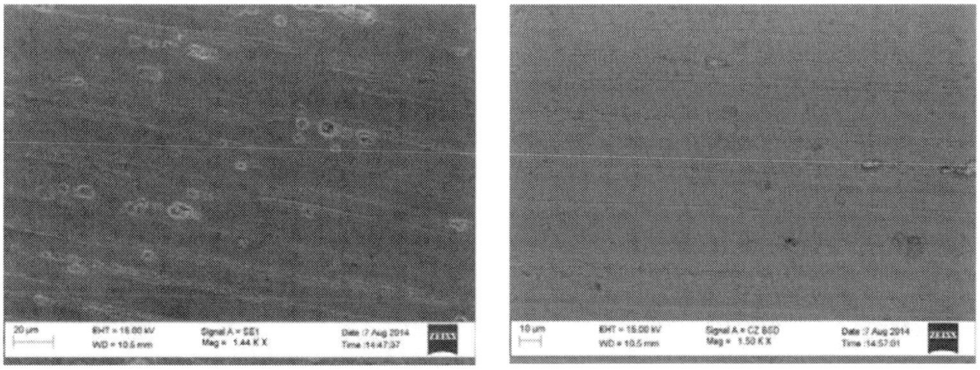

Figure 8. SEM micrographs of the 7075 Al alloy deformed at slow strain rates (10^{-4} s^{-1})

This figure shows the microstructure of the specimen deformed at slow strain rate at 10^{-4} s^{-1}. If the grain morphology is examined, it is found that the grains of the specimens compressed become more and more refined with increase in strain rate. This is due to the increasing dynamic recrystallization.

3.4.3. Corrosion Studies. The surface morphology of the 7075 aluminum alloy sample was examined by SEM immediately after corrosion tests in 3.5 % NaCl solution. SEM images of corroded sample given in Figs.10. show the formation of oxide on the surface of the sample after deformation on high strain rate and slow strain rate. The surface appears a typical localized corrosion. Corrosion products formed on the sample surface are distributed discontinuously. EDX analysis suggests that these corrosion products contain O, Na, Al, Cl, Zn.

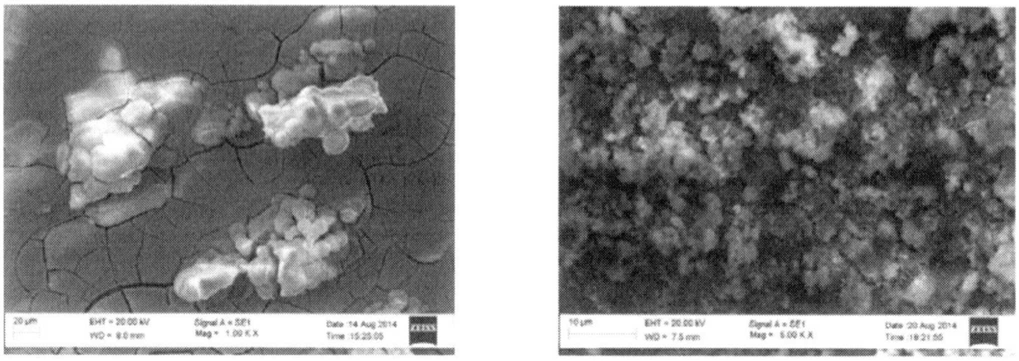

(a) (b)

Fig 10. SEM micrographs of the oxide surface after immersion test in 3.5 wt% NaCl solution for 30 days in Secondary Electron mode: (a) High strain rate. (b) Slow strain rate

4. Conclusion

1. In high strain rate test the mechanical properties like yield stress, withdrawal stress increases with increasing load or impact velocity.
2. Elongated grains of Al along with fine precipitation are observed with microstructures with high strain rates.
2. In slow strain rate n value increases with increasing load.
4. The corrosion rate decreases with increasing load in immersion test in 3.5 wt% NaCl solution.

5. Reference

[1] Smerd R O 2005 *Constitutive Behaviour of Aluminium Alloy Sheet at High Strain Rates* masters degree thesis (Waterloo University, Canada).

[2] Lee W S, Sue W C, Lin C F and Wu C J 2000 *The Strain Rate and Temperature Dependence of the Dynamic Impact Properties of 7075 Aluminum Alloy* vol 100 (Journal of Materials Processing Technology) p 116-122.

[3] Ghosh A K and Gandhi G 1985 *Strength of Metals and Alloys (ICSMA7)* in: H.J. McQueen et al. (Eds.) Pergamon Press Oxford. p.2065.

[4] Zaidi M A and Wert J A 1989 *Aluminum Alloys-Contemporary Research and Applications* in: A K Vesudevan, R D Doherty (Eds.) (Academic Press, NewYork) p.137.

[5] Tao J, Yang K, Xiong H, Wu X, Zhu X and Wen C 2013 *The defect structures and mechanical properties of Cu and Cu-Al alloys processed by split Hopkinson pressure bar* vol A580 (Materials Science & Engineering) p 406-409.

[6] Marais S T, Tait R B, Cloete T J and Nurick G N 2004 *Material testing at high strain rate using split Hopkinson pressure bar* vol1 (Latin American Journal of Solids and Structure) p 319-339.

[7] Sunny G, Yuan F, Prakash V and Lewandowski J 2008 *Design of inserts of Split - Hopkinson Pressure Bar Testing of Low Strain-to-Failure Materials* (Society for Experimental Mechanics) DOI 10. 1007/s11340-008-9145-1.

Synthesis and Characterization of $W_{80}Ni_{10}Mo_{10}$ alloy produced by mechanical alloying

A. Patra[1], S. K. Karak, S. Pal

Department of Metallurgical and Materials Engineering, National Institute of Technology, Rourkela-769008, Orissa, India

Email : patraa@nitrkl.ac.in/anspat.met@gmail.com

Abstract. The present study aims at synthesis and characterization of nanostructured $W_{80}Ni_{10}Mo_{10}$ (wt. %) alloy produced by mechanical alloying (MA). Elemental powders of tungsten (W), nickel (Ni), molybdenum (Mo) were subjected to mechanical attrition in a high energy planetary ball-mill using chrome steel as grinding media and toluene as a process control agent. The crystallite size and lattice strain of the nanostructured powders at different stages of milling (0 h to 20h) was calculated from the X-ray diffraction patterns (XRD). The crystallite size of W in $W_{80}Ni_{10}Mo_{10}$ powder was reduced from 100 μm to 55 nm at 10 h and further reduction to 40 nm at 20 h of milling with increase in lattice strain of 0.25% at 20 h of milling. The lattice parameter of tungsten showed initial expansion upto 0.03% at 10 h of milling and then contraction upto 0.04% at 20 h of milling. The scanning electron microscopy (SEM) also showed mixed morphology of $W_{80}Ni_{10}Mo_{10}$ powders consisting spherical and elongated particles after 20 h of milling. SEM analysis also revealed that particle size reduced from 100 μm to 2 μm with an increase in the milling time from 0 to 20 hours. The dark-field Transmission Electron Microscopy (TEM) observations revealed that the crystallite size of W in milled $W_{80}Ni_{10}Mo_{10}$ alloy is in good agreement with calculated crystallite size from XRD.

1. Introduction

Tungsten is a refractory metal that possesses high melting point (3410°C) and excellent mechanical strength at elevated temperature, highest density of all engineering materials (19.3 g/ml) and tensile elastic modulus of 411 GPa. Therefore it is desired for applications in electrical, electronic, nuclear and space vehicle equipment [1]. On the other hand tungsten suffers from relatively poor fabricability and high ductile-brittle transition temperature. In order to improve the properties and to increase the spectrum of applicability, much effort has been directed towards the development of tungsten alloys in recent years. The work is essentially centered on the goals such as to improve fabricability, particularly at elevated temperatures and to lower the ductile brittle transition temperature of W based alloys. As a result of increasing demand for higher and better mechanical properties, amorphous and nanocrystalline tungsten alloys have attracted enormous attention in recent years [2, 3]. Mechanical properties of nanostructured materials are known to be improved by the refinement of microstructures by mechanical alloying which is a convenient solid state synthesis alternative to melt spinning and similar rapid quenching techniques to develop crystalline and amorphous alloys [4]. Among the various solid state methods, high-energy ball milling has gained quite popularity in recent years

[1] patraa@nitrkl.ac.in/anspat.met@gmail.com (A. Patra)

Content from this work may be used under the terms of the Creative Commons Attribution 3.0 licence. Any further distribution of this work must maintain attribution to the author(s) and the title of the work, journal citation and DOI.
Published under licence by IOP Publishing Ltd

because of its simplicity, ease scale up production, high productivity, relatively inexpensive equipment, and applicability to a wide variety of materials [5].

Present investigation aims at synthesis, characterization and optimization of the process parameters of the current $W_{80}Ni_{10}Mo_{10}$ alloy produced by mechanical alloying (MA) route.

2. Materials and methods

A planetary ball mill (Fritsch Pulverisette P5) was used to mill elemental W, Ni, Mo powder at a mill speed of 300 r.p.m using chrome steel as grinding media and ball to powder weight ratio of 10:1. Milled samples were taken out after 1, 5, 10, 15, 20 h for characterization purpose. The details of the selected alloy and milling parameters are presented in the table 1.

Table 1. Details of alloy composition and milling parameters

Composition (wt.%)	Grinding Medium	Ball to powder weight ratio	Mill Speed (rpm)	Milling Duration (h)	Milling Medium
$W_{80}Ni_{10}Mo_{10}$	Chrome Steel	10:1	300	20	Toluene

A high resolution X- ray diffractometer (Make: Rigaku Japan, Model: Ultima IV) was used to record the X-ray diffraction pattern (XRD) of the mechanically alloyed powders at different stages of milling using Cu-k_{α} radiation (λ=1.541874 Å). The record was matched with the JCPDS data bank to track the evolution of phases during mechanical alloying [6]. The crystal size and lattice strain was calculated by determining the peak position and broadening of peak from the X- ray diffraction pattern [7]. The lattice parameter was calculated from the X-ray diffraction pattern after stripping of $K_{\alpha 2}$ of XRD pattern using precise lattice parameter calculation method [7]. The d values of the high angle diffraction peaks were extrapolated against the function $\cos^2(\theta)/\sin(\theta)$, $\cos^2(\theta)$ [6], to yield the precise measure of d at $\cos(\theta)$ tending to 0. It was found that $\cos^2(\theta)$ function gave best possible fit. The morphology and particle size of the mechanically alloyed (MA) powders at different stages of milling was observed under a scanning electron microscope (SEM) (Make: JEOL, Model: JSM-6084LV). The elemental compositional distribution of milled $W_{80}Ni_{10}Mo_{10}$ powder was analyzed by the energy dispersive X-ray (EDX) analysis attached with SEM. Crystallite size of nanostructured W in mechanically alloyed $W_{80}Ni_{10}Mo_{10}$ powder at 20 hr was detected by transmission electron microscope (TEM) (Make : JEOL, Japan, Model- JEM 2100). Selected area diffraction (SAD) patterns were obtained to identify the crystal structure using appropriate aperture and tilt.

3. Result and Discussion

3.1. X-ray Diffraction (XRD) analysis of milled powders

The XRD patterns of as milled $W_{80}Ni_{10}Mo_{10}$ powder at different milling times (1, 5, 10, 15, 20 h) are displayed in figure 1. The patterns show continuous broadening and reduction in the intensity of the peaks, thus evidencing refinement of crystallite size of W with increase in milling time. It is evident from figure 1 that Ni and Mo undergo into solid solution in the W matrix with increase in milling time. Reduction in crystallite size and plastic strain buildup is attributed to the increase in full-width at half-maximum with increasing milling time [8]. Peak broadening analysis was done with the help of Scherrer equation after elimination of contributions from the strain and instrumental error [6].

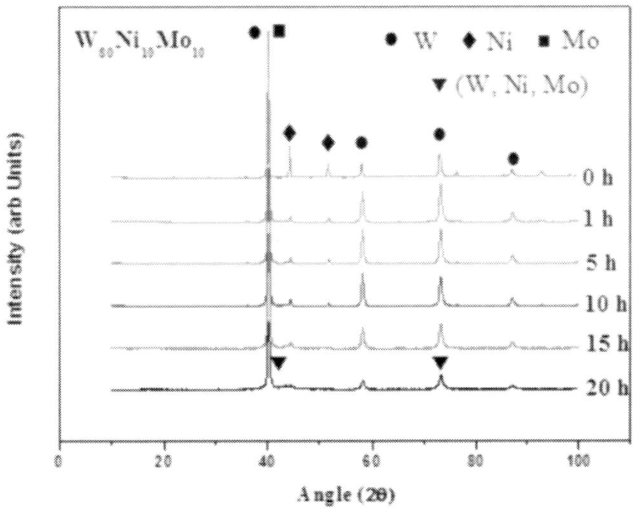

Figure 1. XRD pattern of $W_{80}Ni_{10}Mo_{10}$ alloy milled for 20 h.

The variation of crystallite size and lattice strain is shown in figure 2. The crystallite size is 40 nm at 20 h of milling for $W_{80}Ni_{10}Mo_{10}$ composition. Recent result shows that crystallite size of W based alloys can be brought down to 30 nm at 20 h of milling using tungsten carbide as grinding media [1]. The higher percentage of size reduction in case of tungsten carbide grinding media is attributed to higher hardness of tungsten carbide than chrome steel [1].

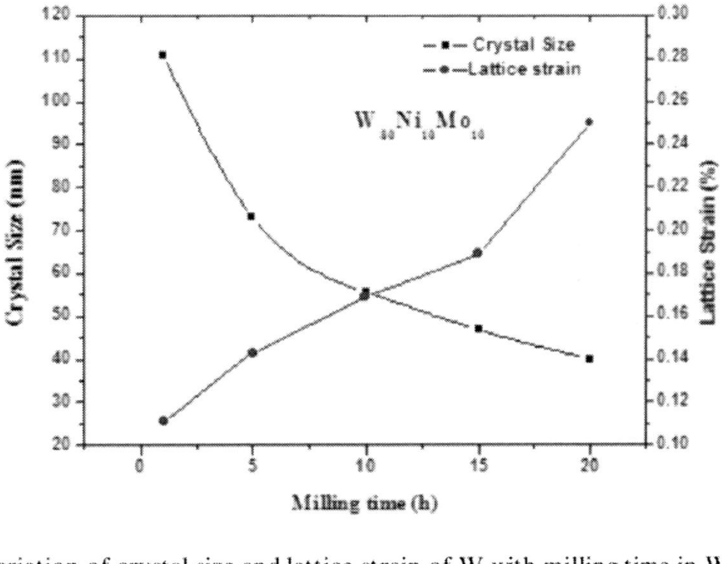

Figure 2. Variation of crystal size and lattice strain of W with milling time in $W_{80}Ni_{10}Mo_{10}$.

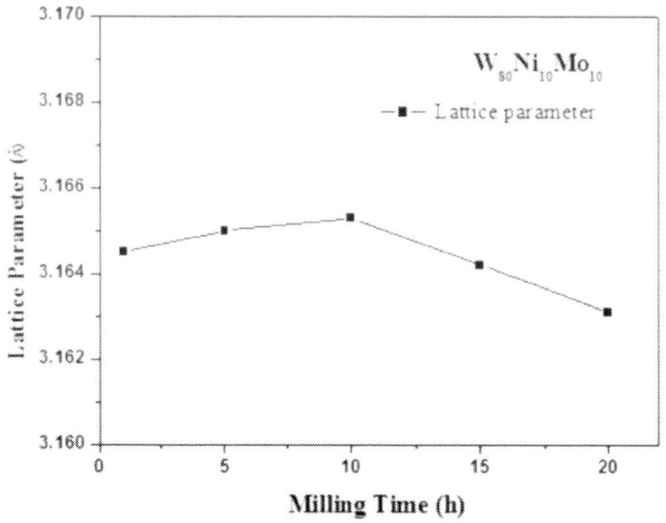

Figure 3. Variation of Lattice parameter of W in $W_{80}Ni_{10}Mo_{10}$ with milling time.

The variation of lattice parameter of W with milling time in $W_{80}Ni_{10}Mo_{10}$ alloy during mechanical alloying is shown in figure 3. It is evident that initially the lattice parameter of W expands upto 0.03% at 10 h and then contracts upto 0.04% at 20 h of milling. This is due to the fact that the initial expansion of lattice parameter of W is attributed to the negative hydrostatic pressure exerted by the formation of nano-crystallites during mechanical alloying. Pabi. *et. al* has recently reported that a significant expansion of Nb lattice during ball milling of pure Nb [9]. On the other hand, the alloying elements Mo and Ni undergoes to form the substitutional solid solution with W, as their atomic radius is marginally lower than that of atomic radius of W which leads to contraction of the lattice after 10 hours of milling [10]. Initially upto 10 hours of milling the reduction in crystallite size is predominant whereas from 10 to 20 hours of milling formation of solid solution (by alloying of Mo and Ni with W) dominates over crystallite size reduction.

3.2. Scanning Electron Microscope (SEM) analysis of milled powders

The SEM micrographs in figure 4 illustrate the change in particle morphology and size from ~100 μm to ~2 μm at different milling time. The W, Ni and Mo particles consist of spherical shape at 0 h which changes to elongated shape after 20 h of milling. The elongated nanoparticles at 20 h of milling are not conducive for flowability due to large interparticle friction therefore responsible for poor green densification property.

Figure 4. SEM images of powder morphology of $W_{80}Ni_{10}Mo_{10}$ alloy after different milling times: (a) 0 h, (b) 1 h, (c) 5 h, (d) 10 h, (e) 15 h, and (f) 20 h.

The EDX spectrum of as milled $W_{80}Ni_{10}Mo_{10}$ powder at 20 h of mechanical alloying is reported in figure 5 which confirms the presence of predominant element W and also minor elements Mo and Ni.

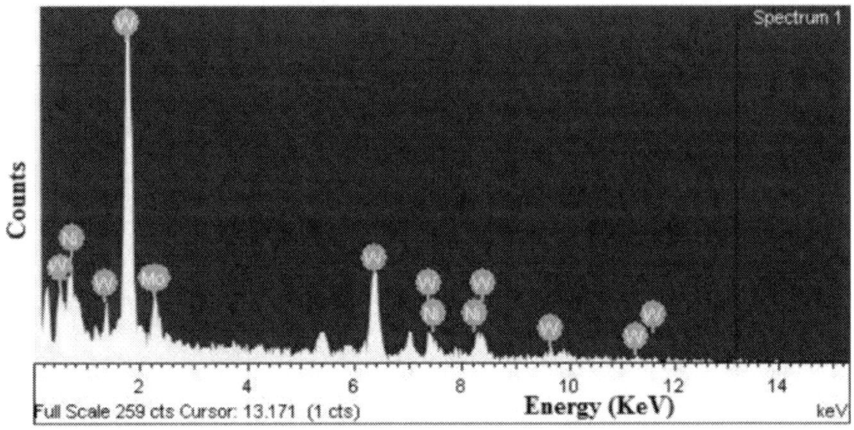

Figure 5. EDX spectrum of $W_{80}Ni_{10}Mo_{10}$ powder milled for 20 h.

3.3. Transmission Electron Microscope (TEM) analysis of milled powder at 20 h

Bright field TEM image along with corresponding SAD pattern of $W_{80}Ni_{10}Mo_{10}$ milled powder at 20 h is shown in figure 6 (a) and (b) respectively. It reveals that the formation nano-crystalline phase at 20 h of milling with size varying from 30 to 40 nm and it is in reasonable good agreement with the result obtained from XRD analysis (figure 2). The SAD pattern consists of continuous rings which denote the presence of fine polycrystallites. Figure 6 (b) reveals the presence of (110), (200), (222) planes of W after the indexing of SAD pattern. The calculated interplaner spacing (*d*) values of BCC-W are 0.22

nm for (110), 0.15 nm for(200), and 0.09 nm for (222) respectively. The results of the interplaner spacing (*d*) values are matching with the standard values of BCC-W from JCPDS data files.

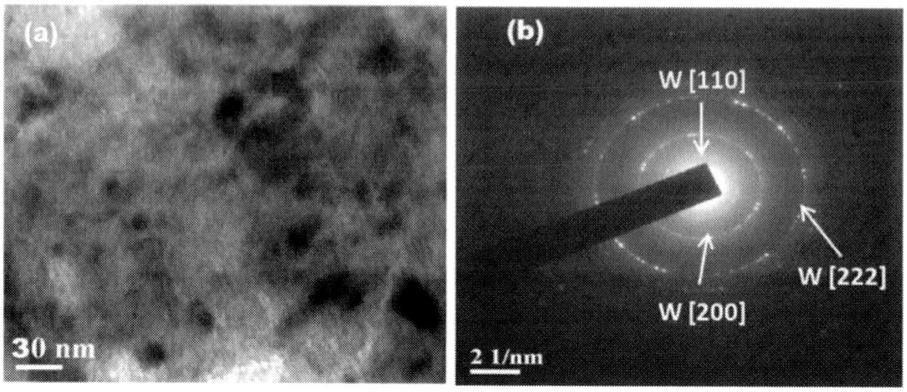

Figure 6. TEM image of $W_{80}Ni_{10}Mo_{10}$ powder milled at 20 h: (a) Bright field TEM image and (b) corresponding SAD pattern.

Conclusion

From the details synthesis and structural characterizations of the present alloy, the following conclusions can be drawn:

- Mechanical alloying is an effective route for synthesis of $W_{80}Ni_{10}Mo_{10}$ alloy powder.
- Crystallite size of tungsten in $W_{80}Ni_{10}Mo_{10}$ gradually decreases with increasing milling time and it records 40 nm at 20 h of milling have been calculated by using Scherrer formula from the XRD patterns.
- The lattice parameter of tungsten initially expands upto 0.03% at 10 h and then contracts upto 0.04% at 20 h of milling.
- The apparent particle size of powders gradually reduces from 100 μm (manually blended) to 2 μm (20 h) by SEM analysis of milled powders.
- The presence of nanocrystalline BCC-W phase with 30-40 nm in size at 20 h of milling is confirmed from the bright field TEM image and corresponding SAD pattern.

References

[1] Patra A 2012 *Microscopy and Analysis.* **26 (5)** 14-15
[2] Glieter H 2001 *Act. Mat.* **48** 1
[3] Koch C 1997 *Nanostruct. Mater.* **9** 13
[4] Suryanarayana C 1998 *Mechanical Alloying* ASM Handbook. **7** 80
[5] Suryanarayana C 2001 *Prog. In. Mat. Sci.* **46** 1-184.
[6] Cullity B 1978 *Elements of X-ray diffraction* **2** (Addison Wesley: Philippines) 86-88
[7] Williamson G *et al.* 1953 *Acta. Met.* **1** 22
[8] Nuthalapati M *et al* 2014 *Metall. Mater. Trans. A* **45A** 3748-3754
[9] Chattopadhayay P *et al.* 2001 Polymorphic transformation and lattice expansion in nanocrystalline niobium revealed by positron annihilation at grain boundary *Appl. Surf. Sci.* **182** 308-312
[10] Singh V 2002 *Physical Metallurgy* **1** (Standard Publisher Distributor: New Delhi) 63- 64

Synthesis of nano-structured duplex and ferritic stainless steel powders by planetary milling: An experimental and simulation study

Shalabh Gupta, R. Shashanka, D. Chaira*

Department of Metallurgical and Materials Engineering, National Institute of Technology Rourkela, 769008, India

*E-mail: chaira.debasis@gmail.com, Phone: +91(661)2462561

Abstract. Nano-structured duplex and ferritic stainless steel powders were prepared by planetary milling of elemental Fe, Ni and Cr powder compositions (duplex: 69Fe-18Cr-13Ni and ferritic:82Fe-17Cr-1Ni) in a dual drive planetary mill for 10 hours. The feasibility for solid solution formation of Cr and Ni in Fe matrix were studied. The samples were collected at regular time intervals and characterized for their morphological and phase analysis using X-Ray diffraction (XRD) and scanning electron microscopy (SEM). It has been observed that as the milling time increases crystallite size decreases and lattice strain increases. Thermodynamic aspects of milled stainless steel powders were performed using Miedema model. The theoretical values ΔH,ΔS and ΔG of and were calculated and were found to be -14.32kJ/mol, 6.93J/mol and -16.39 kJ/mol for duplex stainless steel and -2.81kJmol, 4.35Jmol and -4.11 kJ/mol for ferritic stainless steel respectively.

Keywords: Stainless steel; planetary milling; Miedema model; thermodynamic parameters

1. Introduction

Powder metallurgy route for preparation of stainless steel has been the cause of much scientific attention due to its capability of producing components with a wide range of mechanical properties which are near-net shaped by making use of various alloy selections and processing options. Although powder metallurgy route can be used for different grades of stainless steel powder production but among them duplex and ferritic stainless steels are being extensively studied because of versatility of their applications. Conventionally, duplex stainless steels are used in orthopaedic applications such as medical implants or as surgical cutting tools and applications where high toughness levels are in compatibility with cryogenic applications and creep resistance [1]. Many processing methods have been adopted in order to improve the structure and properties of stainless steel. Methods like hydrostatic extrusion, equi-channel angular pressing, high pressure torsion and mechanical alloying are being used for the structural refinement of metals and alloys [2]. Among all methods mechanical alloying proves to be one of the best powder metallurgy routes to refine the particle size to nano level. Therefore, specially designed dual drive planetary mill (DDPM) was used to prepare nano-structured stainless steel. This high-energy milling is one of the most promising from industrial view point because it involves the use of comparatively less inexpensive equipment for the synthesis of bulk nano-structured powder production; in fact. This has many common aspects with mixing, fine grinding

Content from this work may be used under the terms of the Creative Commons Attribution 3.0 licence. Any further distribution of this work must maintain attribution to the author(s) and the title of the work, journal citation and DOI.
Published under licence by IOP Publishing Ltd

and commination of materials. Alloy formation takes place by diffusion during planetary milling of elemental powder particles which leads to atomic level alloying often accompanied by nano sized structured meta-stable phase transformation [3–5]. The high density of dislocations and grain boundaries created in powders and micro segregation of solute at these defects may lead to the formation of extended solid solutions [6].
Thermodynamic parameters are very important to predict the solid solution formation during planetary milling. Very few literatures are available so far regarding calculation of thermodynamic parameters for the binary systems such as Ti–Al, Ti–Ni, Al–Ni and Al–Ni–Ti using Miedema model [7, 8]. In the present study, we have made an attempt to evaluate the various thermodynamic parameters of ternary system (Fe-Ni-Cr) during planetary milling. Here, our main intention is to predict various thermodynamic parameters during synthesis of nano-structured stainless steel by planetary milling.

2. Experimental

The elemental powder compositions of Fe–18Cr–13Ni (duplex) and Fe–17Cr–1Ni (ferrite) were selected and milled in a specially designed DDPM for 10 h under toluene atmosphere to prevent oxidation. The milling media of the DDPM consist of 600 g stainless steel balls (10 mm diameter) and ball to powder weight ratio of 6:1 was maintained during milling. The angular velocity of the supporting main shaft and jars were kept at 275 and 725 rpm respectively to attain 75 % critical speed. Milled stainless steel powders were characterized by using X-ray diffraction (XRD) in a Philips PANalytical diffractometer using filtered Cu Kα-radiation (λ = 0.1542 nm). Williamson–Hall method was used to calculate crystal size and lattice strain of the milled powders. Morphological analysis of the milled powders was performed by scanning electron microscopy (SEM) using JEOL JSM-6480LV model. Thermodynamic parameters (ΔS, ΔH and ΔG) of stainless steel powders for both compositions were calculated by simulation method using Miedema model.

3. Results and discussions

3.1 Thermodynamic analysis using Miedema model

Considering the formation of disordered A (B, C) solid solution from a mixture of pure A, B and C elements.
Here, ΔH_{ms} *and* ΔS_s and are the enthalpy and entropy of mixing respectively. T is the temperature at which a solid solution is formed.
ΔS_s can be calculated with the assumption of configurational entropy of mixing:

$$\Delta S_s = -R(x_a ln x_a + x_b ln x_b + x_c ln x_c) \quad (1)$$

Where, x_a, x_b *and* x_c are mole fractions for A, B and C in the mixture.
According to Miedema semi-empirical model, the entropy of a solid solution formation consists of three terms for a ternary system:

$$\Delta S = -R(x_a ln x_a + x_b ln x_b + x_c ln x_c) \quad (2)$$

Total enthalpy for formation of solid solution is

$$\Delta H_{Total} = \Delta H_{chemical} + \Delta H_{elastic} + \Delta H_{structural} \quad (3)$$

Where, $\Delta H_{chemical}$, $\Delta H_{elastic}$ and $\Delta H_{structural}$ are the chemical, elastic and structural contributions due to atomic mixing, mismatch in size and the difference in valance electrons and crystal structure of solute and solvent atoms, respectively. The complete form of $\Delta H_{chemical}$ is as follows:

$$\Delta H_{chemical} = \frac{2f(C^s)(X_A V_A^{\frac{2}{3}} + X_B V_B^{2/3}) * [-P(€)^2 + Q(\Delta n_{ws})^{2/3}]}{(n_{ws}^A)^{-1/3} + (n_{ws}^B)^{-1/3}} \tag{4}$$

Where X_a and X_b are the mole fractions of elements A and B; €, V and n_{ws} are work function, molar volume and electron density of constituents, respectively. P and Q are empirical constants and $f(C^s)$ is the concentration function that for solid solution is given by

$$f(C^s) = C_A^s \cdot C_B^s \tag{5}$$

$$C_A^s = \left(\frac{X_A V_A^{\frac{2}{3}}}{X_A V_A^{\frac{2}{3}} + X_B V_B^{\frac{2}{3}}} \right) \tag{6}$$

$$C_B^s = \left(\frac{X_B V_B^{\frac{2}{3}}}{X_A V_A^{\frac{2}{3}} + X_B V_B^{\frac{2}{3}}} \right) \tag{7}$$

$$\Delta H_{elastic} = x_b \Delta E_{A\ in\ B} + x_a \Delta E_{B\ in\ A} \tag{8}$$

Where $\Delta E_{A\ in\ B}$ and $\Delta E_{B\ in\ A}$ are the elastic energy caused by dissolving A in B and B in A, respectively and can be estimated by

$$\Delta Ea\ in\ b = \frac{2K_a \cdot G_b \cdot (\Delta V)^2}{3K_a V_b + 4G_b V_a} \tag{9}$$

$$\Delta Eb\ in\ a = \frac{2K_b \cdot G_a \cdot (\Delta V)^2}{3K_b V_a + 4G_a V_b} \tag{10}$$

Where $\Delta E_{A\ in\ B}$ and $\Delta E_{B\ in\ A}$ are the elastic energy caused by 'A' dissolving in 'B' and 'B' dissolving in 'A' respectively. Where K and G are bulk and shear modulus respectively.

$\Delta S_{structural}$ was neglected as the value is negligible compared to other two enthalpies.

Table 1 shows the list of various parameters used for Fe, Cr and Ni.

Table 1. Values of different parameters for Fe, Ni and Cr

	Fe	Cr	Ni
Molar Volume ($V^{2/3}$), (cm^2)	3.7	3.7	3.5
Electron density ($n_{ws}^{1/3}$), $(d.u.)^{1/3}$	1.77	1.73	1.75
Work function (€), (V)	4.93	4.65	5.2
K (GPa)	170	160	180
G (GPa)	82	115	76

Gibb's free energy for formation of solid solution was calculated by using the following equation:

$$\Delta G = \Delta H_{Total} - T\Delta S \quad (11)$$

Here temperature 'T' for formation of solid solution is assumed 300K.

Table 2 shows the value of different thermodynamic parameters calculated using Miedema model. It is observed from the table that ΔG for duplex stainless steel is higher than ferritic stainless steel. The reason is due to diffusion of higher amount of alloying element in duplex steel than ferritic steel.

Table 2. Values of thermodynamic parameters calculated using Miedema model

	ΔH (kJ/mol)	ΔS (J/mol)	ΔG (kJ/mol)
Duplex	-14.32	6.93	-16.39
Ferrite	-2.81	4.35	-4.12

3.2 X-Ray diffraction study

Fig. 1 (a) and 1 (b) represents the XRD spectra of duplex and ferritic stainless steel powders milled at 0, 2, 5 and 10 h respectively in DDPM using ball to powder weight ratio of 6:1. The sharp crystalline peaks of elemental Fe, Cr and Ni start broadening continuously with milling and gradually move into the Fe lattice. It is evident from both the graphs that intensity of individual elements (Fe, Cr, Ni) decreases gradually with milling. After 10h of milling the duplex stainless steel show strong austenite peaks along with weak ferrite peaks as shown in Fig. 1 (a). In case of ferritic stainless steel, low intense austenite peaks are present along with strong ferrite peaks. Increase in milling time increases the phase transformation from α-Fe to γ-Fe and after 10h both austenite and ferrite peaks resolve in to (111) and (110) planes as shown in Fig. 1 (b). This phase transformation is due to the higher degree of defects and refinement of particles in nano-crystalline size.

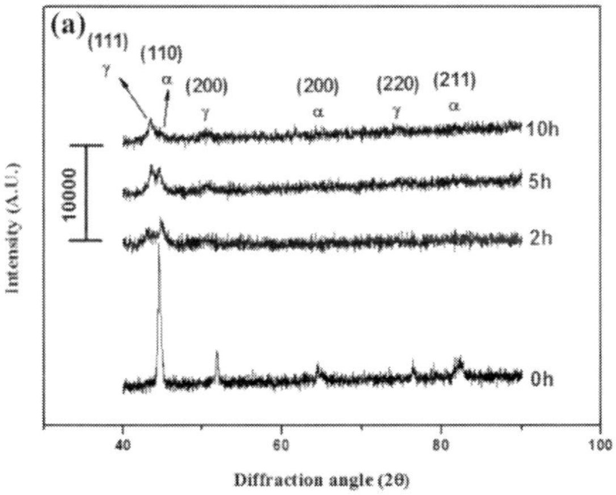

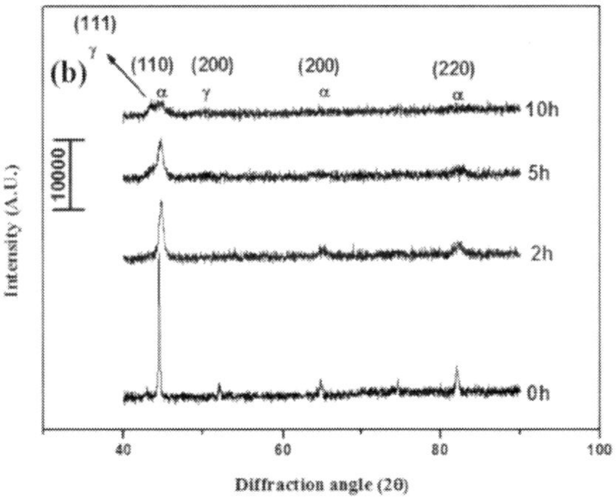

Fig. 1 XRD spectra of 0 to 10h milled **(a)** Austenitic stainless steel **(b)** Ferritic stainless steel powders

The XRD spectra show broadening of XRD peaks with the milling time due to instrumental errors, decrease in crystalline size and lattice strain [9]. Crystal size and lattice strain of both the stainless steels were calculated using Williamson-Hall method [10] by de-convoluting the size and strain by calculating XRD peak width as a function of 2θ. Williamson-Hall equation is as follows,

$$\beta cos\theta = \frac{0.94\lambda}{D} + 4\eta \sin\theta \tag{8}$$

Where, D is crystallite size, β is full width half maxima (FWHM) and η is lattice strain.

During milling crystallite size decreases and strain increases with milling time due to the formation of more defects during milling. Refinement of crystallite size reaches a saturation level at higher milling time and further refinement becomes quite difficult. Due to more and more interaction of powder-ball, powder-powder and powder-jar surface, the lattice strain increases continuously even after 10h of milling. The crystallite size of duplex and ferritic stainless steel decreases from 43 to 6 nm and from 52 to 7nm respectively. Similarly, the lattice strain increases from 0.06 to 1.06 % in case of duplex and from 0.32 to 1.03 % in case of ferritic stainless steel respectively.

3.3 Scanning electron microscopy

Fig. 2 (a) and 2 (b) depicts the SEM micrographs of 0h and 10h milled duplex stainless steel powders. Similarly, Fig. 2 (c) and 2 (d) represent SEM micrographs of ferritic stainless steel milled for 0h and 10h respectively. From the figures it is confirmed that the elemental powder particles are large and irregular in shape before milling but as milling starts particles start to form flat flakes and cold welded due to ductile nature of Fe. Finally particles are work hardened and fragmented in to smaller particles as shown the figure. This refinement of particle size reaches a saturation and further refinement of particles become more difficult after 10h.

Fig. 2 SEM images of austenite stainless steel powder milled for **(a)** 0h **(b)** 10h; and ferritic stainless steel powder milled for **(c)** 0h **(d)** 10h respectively

Energy dispersive spectroscopy (EDS) analysis was carried out to study the elemental composition of stainless steel powders after 10 h of milling. Fig. 3 (a) and 3 (b) depicts the EDS spectra of duplex and ferritic stainless steel powders after 10 h of milling. From the spectra it is evident that there is no much variation in initial elemental composition and final alloy composition.

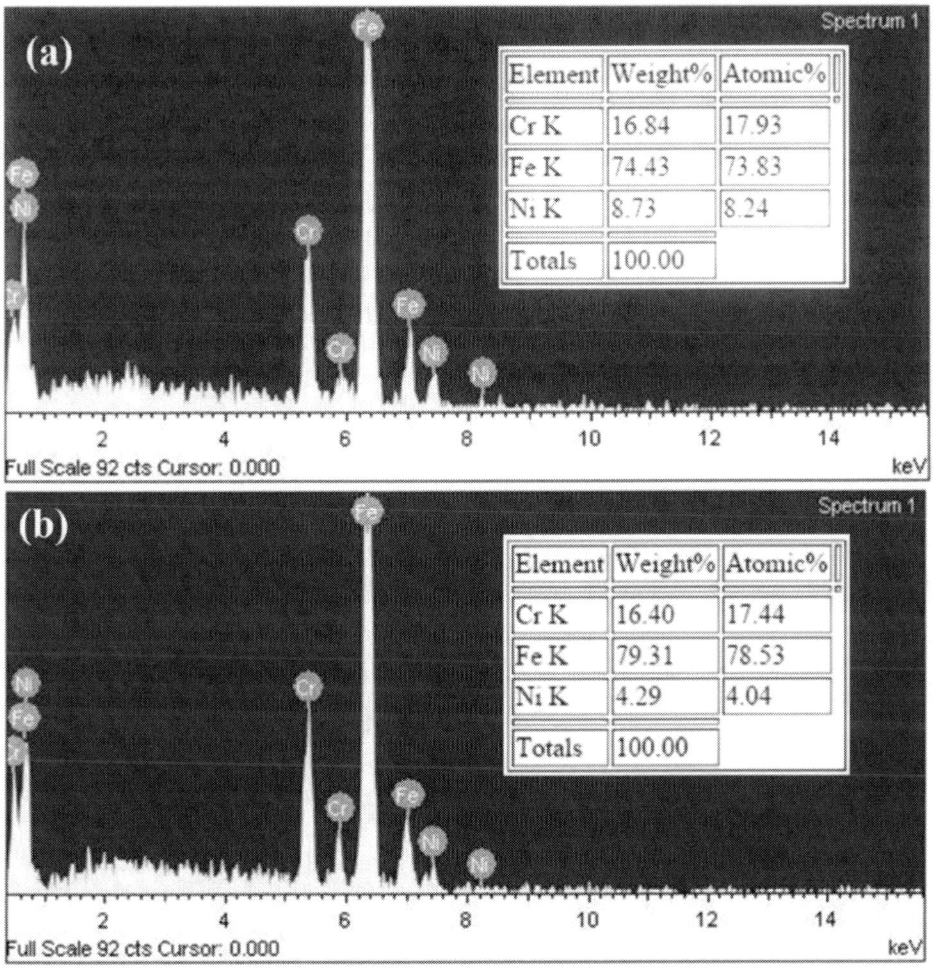

Fig. 3 EDS spectra of **(a)** duplex and **(b)** ferritic stainless steel after 10 h of milling

4. Conclusions

The following conclusions can be made from the present investigation:

1. The austenitic composition was showing ferritic phase till 2 h of milling and becomes austenite after 5 h in DDPM.
2. Austenite phase becomes more dominant at higher milling time
3. The powder particles were of micron range and of random size before milling and as milling starts there will be slight increase in the particle size due to the repeated welding of ductile iron during milling. As milling continues, particles become strain hardened and spherical. After 10 h of milling the crystallite size reduced to 6 and 7 nm in case of austenitic and ferritic stainless steel respectively. The lattice strain goes on increasing in both cases.
4. Thermodynamic parameters (ΔH, ΔS and ΔG) were calculated by using Miedema model and the values found to be -14.31kJ/mol, 6.93J/mol and -16.39kJ/mol for duplex and -2.81kJ/mol, 4.35J/mol, 4.11kJ/mol for ferritic stainless steel respectively.

References

[1] S. Pasebani, I. Charit, Y.Q. Wu, D.P. Butt, J.I. Cole, Mechanical alloying of lanthana bearing Nanostructured ferritic steels, Acta Mater. 61 (2013) 5605–5617.

[2] K.J. Kurzydłowski,Microstructural refinement and properties of metals processed by severe plastic deformation, Bull. Pol. Acad. Sci. Tech. Sci. 52 (2004) 275.

[3] T. Tanaka, K.N. Ishihara, P.H. Shingu, Formation of meta stable phases of Ni–C, Metall. Trans. A. 23 (1992) 2421.

[4] M.J. Tracy, J.R. Groza, Nanophase structure in Nb rich-Nb3Al alloy by mechanical alloying, Nanostruct. Mater. 1 (1992) 369.

[5] M.A. Meyers, A. Mishra, D.J. Benson, Mechanical properties of nano crystalline materials, Mater. Sci. 51 (2006) 427–556.

[6] C. Suryanarayana, Mechanical alloying and milling, Prog. Mater Sci. 46 (2001) 1 – 184.

[7] S.D. Kaloshkin, V.V. Tcherdyntsev, I.A. Tomilin, Yu.V. Baldokhin, E.V. Shelekhov, Phase transformations in Fe–Ni system at mechanical alloying and consequent annealing of elemental powder mixtures, Physica B 299 (2001) 236–241.

[8] M.H. Enayati, M.R. Bafandeh, Phase transitions in nanostructured Fe–Cr–Ni alloys prepared by mechanical alloying, J. Alloys Compd. 454 (2008) 228–232.

[9] R. Shashanka, D. Chaira, Development of nano-structured duplex and ferritic stainless steel by pulverisette planetary milling followed by pressureless sintering, Mater Charact. 99 (2015) 220-229.

[10] R. Shashanka, D. Chaira, Phase transformation and microstructure study of nano-structured austenitic and ferritic stainless steel powder prepared by planetary milling, Powder Technol. 259 (2014) 125–136.

IOP Conf. Series: Materials Science and Engineering **75** (2015) 012034 doi:10.1088/1757-899X/75/1/012034

Wear characteristics of Al-AlN composites produced in-situ by nitrogenation

S Pradhan[1], S K Jena[1], S C Patnaik[1], P K Swain[2] and J Majhi[1]
[1]Department of Metallurgical and Materials Engineering, Indira Gandhi Institute of Technology, Sarang, Dhenkanal-759146, Odisha, India
[2]Department of Chemistry, Indira Gandhi Institute of Technology, Sarang, Dhenkanal-759146, Odisha, India

E-mail:pravatvision2020@gmail.com (P. K. Swain)

Abstract: In the present scenario most of the parts used in automobile and aerospace industries are made of composites. Aluminium metal has a very high strength to weight ratio. Al-AlN metal matrix composites have excellent mechanical properties like good wear resistance, high hardness and high strength to weight ratio which are obtained from low density of aluminium and high hardness and wear resistance of AlN. Therefore, use of Al-AlN MMC leads to advantages in automobile industries as it reduces fuel consumption and gives better structural strength. The composites can be produced ex-situ by introducing AlN as reinforcement into the metal matrix and in-situ by nitrogenation process. In the present process Al-AlN composites were fabricated by treating molten aluminium with a mixture of CaO and NH_4Cl and held at temperature range at 750-930 °C and then cast into a metal mould. Addition of CaO to NH_4Cl is found to have a more pronounced effect as a nitrogenation agent in comparison with NH_4Cl in the molten aluminium in the temperature range of 700-1000°C. Castings of aluminium and Al-AlN composites were prepared by mixing different ratios of aluminium and NH_4Cl, at different casting temperatures and holding time. Wear resistance and hardness of the samples were determined, and microstructure studies were carried out. The composite formed using higher amount of NH_4Cl had higher hardness and wear resistance.

1. Introduction

Aluminium and aluminium alloy based composites are gaining importance due to their specific properties like excellent strength, high ductility, low density and better wear resistance and find applications in automotive and aerospace industries. The in-situ process is a simple, economical method of synthesizing MMCs and is a one-step process of establishing ceramic particles into the metal matrix. AlN due to its non-reacting behaviour with aluminium makes it a suitable reinforcement for forming aluminium matrix composites. $CaCO_3$ is used as a catalyst for efficient nitridation. In the present investigation NH_4Cl+CaO are taken as nitriding agents for the formation of AlN in aluminium. And a comparative study between aluminium and Al-AlN composite is done for wear resistance and hardness characteristics [1, 3].

1.1. Aluminium Nitride

Aluminium nitride has high thermal conductivity. It is stable at high temperatures in inert atmospheres and melts at 2800°C. In a vacuum, AlN decomposes at ~1800°C. In the air, surface oxidation occurs above 700°C, and even at room temperature, surface oxide layers of 5-10 nm have been detected. This oxide layer protects the material up to 1370°C. Above this temperature bulk oxidation occurs. Aluminium nitride is stable in hydrogen and carbon dioxide atmospheres up to 980°C. The material dissolves slowly in mineral acids through grain boundary attack, and in strong alkalis through attack

Content from this work may be used under the terms of the Creative Commons Attribution 3.0 licence. Any further distribution of this work must maintain attribution to the author(s) and the title of the work, journal citation and DOI.
Published under licence by IOP Publishing Ltd

on the aluminium nitride grains. The material hydrolyzes slowly in water. Aluminium nitride is resistant to attack from most molten salts, including chlorides and cryolite.

1.2. Al-AlN Composites

Al-AlN composites are fabricated by milling and hot pressing of composite powders [1]. The physical properties such as thermal expansion, thermal conductivity, electrical conductivity, hardness, Young's modulus, and fracture strength of these composites change with the AlN content [2]. It has been observed that Al-AlN MMCs exhibit better properties than those of Al-SiC and Al-Al_2O_3 composites.

Al-AlN composites have been made by forming aluminium nitride by in situ reactive nitrogen gas injection into molten aluminium alloys experimentally over the temperature range from 700 to 1500°C [4,5]. It has been seen that aluminium nitride can be melt formed in the presence of Mg and Si, with nitrogen and/or ammonia as the reactive gases at temperature above 1100°C. In this, magnesium serves as a catalyst [6]. Magnesium nitride is first formed in the vapour phase by the reaction of vaporized magnesium and nitrogen gas, followed by incorporation of magnesium nitride particles into the molten aluminium. Then by in-situ substitution reaction, aluminium nitride forms between magnesium nitride and aluminium [7,8]. Up to 17 wt. % aluminium nitride in an aluminium alloy has been formed with an average reinforcement size of 3 μm [9]. The potential for this process permits economical liquid phase processing of aluminium nitride-aluminium metal matrix composite with nitrogen gas injection for structural, thermal and wear applications [12,13].

Al-AlN composites have been synthesized using NH_4Cl and NH_4Cl+CaO powders as sources of nitrogen in molten aluminium in the temperature range of 700-1000°C [14]. Active nitrogen dissociated from ammonia at high temperature is used to form AlN. The addition of CaO to NH_4Cl is found to have a more pronounced effect as a nitrogenation agent in comparison with NH_4Cl [10,11].

2. Chemical Reactions

At high temperatures, ammonium chloride dissociates into ammonia and hydrogen chloride as per the following equation [1]:

$NH_4Cl \rightarrow NH_3 + HCl \qquad \Delta G^0 = +91.2$ KJ/mol (1)

Powders of CaO and NH_4Cl when mixed in the ratio of 1:1 by weight and added to aluminium melt at temperatures of 700-900 °C, the following reactions occur:

$2NH_4Cl + 2CaO \rightarrow 2NH_3 + CaCl_2 + Ca(OH)_2 \qquad \Delta G^0 = -33.3$ KJ/mol (2)

$2Al + 2NH_3 \rightarrow 2AlN + 3H_2 \qquad \Delta G^0 = -270.6$ KJ/mol (3)

3. Experimental Procedure

Commercially pure aluminium is used for the experiment. The aluminium was melted in a graphite crucible in a pit type resistance furnace. After attaining a steady temperature, ammonium chloride powder and calcium oxide powder, of 1:1 ratio, wrapped in thin aluminium foil was added into the molten aluminium. The melt was manually stirred by a steel rod until the gas evolution stopped. After removing the dross the melt was poured into a metal mold. The MMC thus prepared was then cut into specific size for further testing.

3.1. Sample -1, Al-AlN Composite

For preparing sample-1, 400 g of aluminium was melted in a graphite crucible and held at 750 °C [15]. Then 96g of ammonium chloride powder and 96g of calcium oxide powder wrapped in thin aluminium foil was added into the molten aluminium. The mixture of the powders was in 1:1 ratio. The melt was held for 30 minutes in furnace and manually stirred by a steel rod. Then the melt was poured into a cylindrical metal mold of 20mm diameter.

3.2. Sample -2, Al-AlN Composite
For preparing sample-2, 400 g of aluminium was taken for melting and held at 900 C. For higher amount of nitrogenation in sample-2, 120g of ammonium chloride powder and 120g of calcium oxide powder wrapped in thin aluminium foil was dipped into 400 g of molten aluminium held at 900 C [15]. Then the temperature was maintained at 930 C and held for 2 hours 30 minutes. During this period the melt was manually stirred periodically to prevent porosity and then it was cast in a horizontal metal mould.

3.3. Hardness Test
Hardness of the samples were determined by using the Brinell hardness tester with the application of 750 kgf load for 15 seconds. A 5 mm diameter steel ball was used as an indentor. Hardness of each sample was measured at least five times.

3.4. Wear Resistance
Wear resistance by weight loss method.
In this test the wear rate is calculated by measuring initial and final weight of specimen before and after the test respectively and their difference gives the amount of material loss due to wear. It is also known as gravimetric analysis.

Let Initial weight of specimen = W_1 gm
Final weight of specimen = W_2 gm
Weight loss due to wear = $W_1 - W_2$ gm
Wear rate of specimen = $(W_1-W_2) / t$
Where t= time of sliding

The metal matrix composite (Al-AlN) specimens were polished and prepared as a pin by machining. The specimen was weighed and the initial weight was noted. Then the sample pin was mounted on the wear testing machine. The machine used to measure the abrasive wear is a DUCOM pin on disk tribometer, which consists of a pin on disc, loading panel and controller. The sample machined as a pin is attached to the pin holder mechanism of the machine. The wear disk is a EN31 Steel disk, hardened to 60HRC. The machine is connected to controller to control and measure rotational speed, applied load, sliding time. The pin was weighed before and after each test and the weight loss was used as the measure of wear. The wear test was conducted by varying the load and keeping the speed and time constant. Then graphs between weight loss vs load were plotted [6, 7].

4. Results & Discussion
4.1. Hardness
Brinell Hardness
(Load= 750 kg, D= 5 mm, Time= 15 secs.)

Table 1. Hardness of aluminium and Al-AlN composites.

Brinell Hardness	Al	Al-AlN (24-24) (Sample-1)	Al-AlN (30-30) (Sample-2)
Test 1 (kgf/mm^2)	54.56	56.55	57.64
Test 2 (kgf/mm^2)	54.60	56.33	58.98
Test 3 (kgf/mm^2)	54.51	56.50	58.60
Test 4 (kgf/mm^2)	54.55	56.49	58.12
Test 5 (kgf/mm^2)	54.54	56.19	58.35
Mean Hardness (kgf/mm^2)	54.55	56.41	58.34

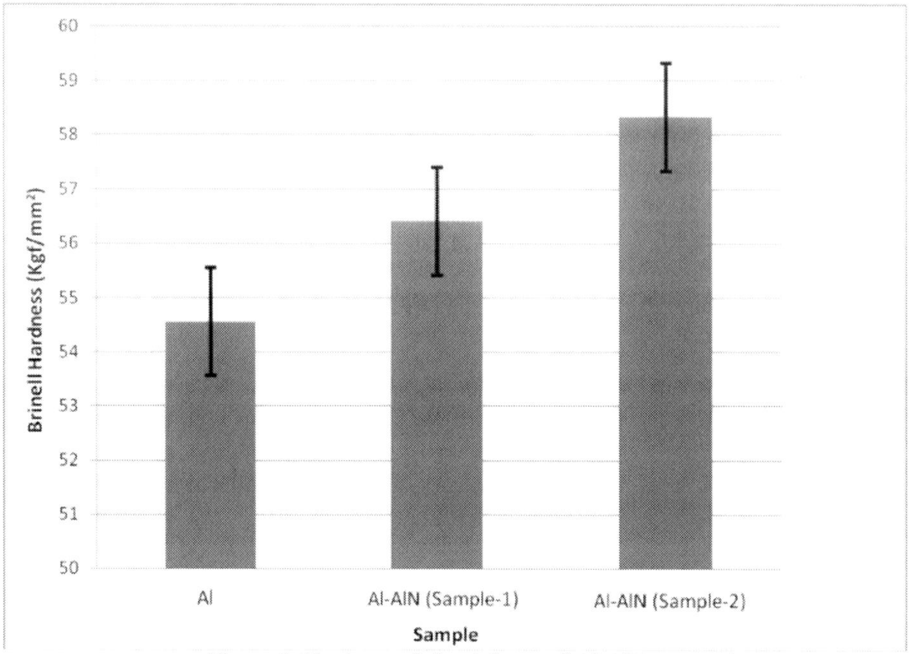

Figure 1. Hardness of aluminium and Al-AlN composites.

Figure 1 shows that Al-AlN composite, sample 1, has higher hardness than aluminium and the hardness of sample-2 which has higher amount of AlN is higher than that of sample 1 indicating that increase in the amount of AlN in the composite increases the hardness of the composite.

4.2. Wear Resistance

4.2.1. Wear of Al and Al-AlN composites at different loads and at 500 rpm speed and 10 minutes.

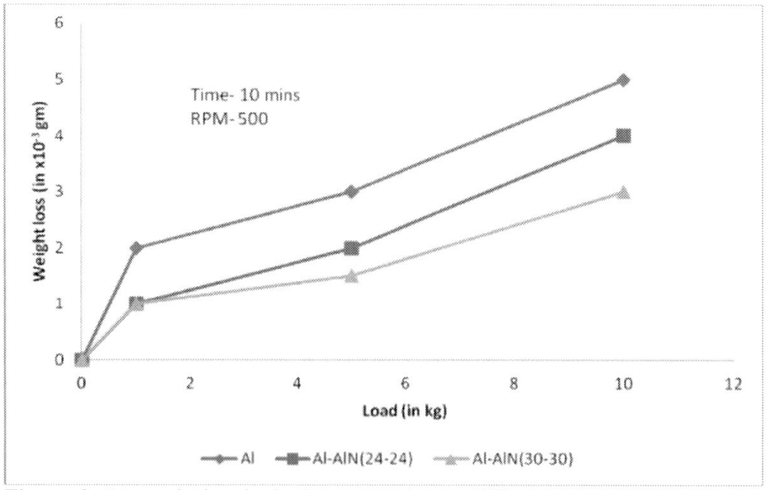

Figure 2. Wear of Al and Al-AlN composites at different loads and at 500 rpm speed and 10 minutes.

Figure 2 shows that wear increases with increasing load in aluminium as well as Al-AlN composites. But Al-AlN (24-24), sample-1, and Al-AlN (30-30), sample-2, composites have greater wear resistance as compared to pure aluminium. Among the composites Al-AlN (30-30), sample-2, composite has superior wear resistance.

4.2.2. Wear of Al and Al-AlN composites at 500 rpm speed and 20 minutes.

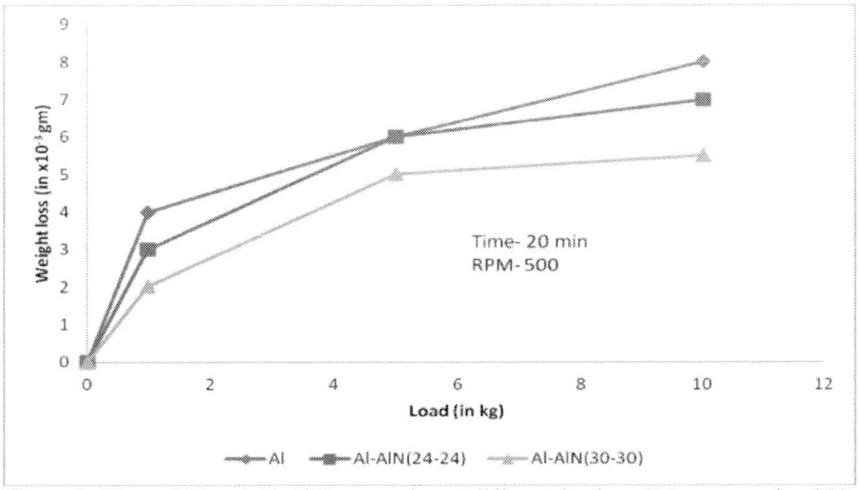

Figure 3. Wear of Al and Al-AlN composites at different loads at 500 rpm speed and 20 minutes.

Figure 3 shows that at 500 rpm speed (Sliding velocity of 2.1 m/s) and 20 minutes the weight loss of all samples have further increased compared to that at 10 minutes. In all samples the weight loss increases with increasing load. The wear resistance of Al-AlN (24-24) composite, sample-1, is slightly better than that of pure aluminium but the Al-AlN (30 -30) composite, sample-2, shows much higher wear resistance compared to both aluminium and Al-AlN (24-24) composite, sample-1.

4.2.3. Wear of Al and Al-AlN composites at different loads at 1600 rpm speed and 10 minutes.

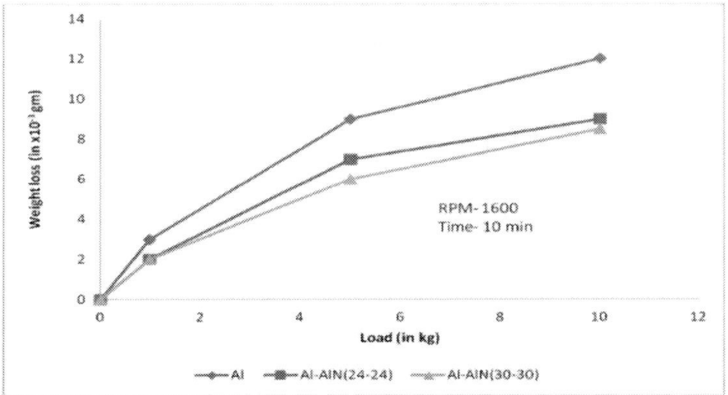

Figure 4. Wear of Al and Al-AlN composites at different loads at 1600 rpm speed and 10 minutes.

Figure 4 shows that at 1600 rpm speed (Sliding velocity of 6.7 m/s) and 10 minutes the weight loss in all samples increases with increasing load and at all loads both Al-AlN (24-24), sample-1, and Al-AlN (30 -30), sample-2, composites have higher wear resistance than pure aluminium. At lower loads the difference in weight loss between aluminium and the composites is not much but at higher loads the difference increases indicating that the composites are more effective in improving the wear resistance at higher loads. Al-AlN (30 -30) composite, sample-2, shows higher wear resistance compared to both aluminium and Al-AlN (24-24) composite, sample-1.

4.2.4. Wear of Al and Al-AlN composites at different loads at 1600 rpm speed and 20 minutes.

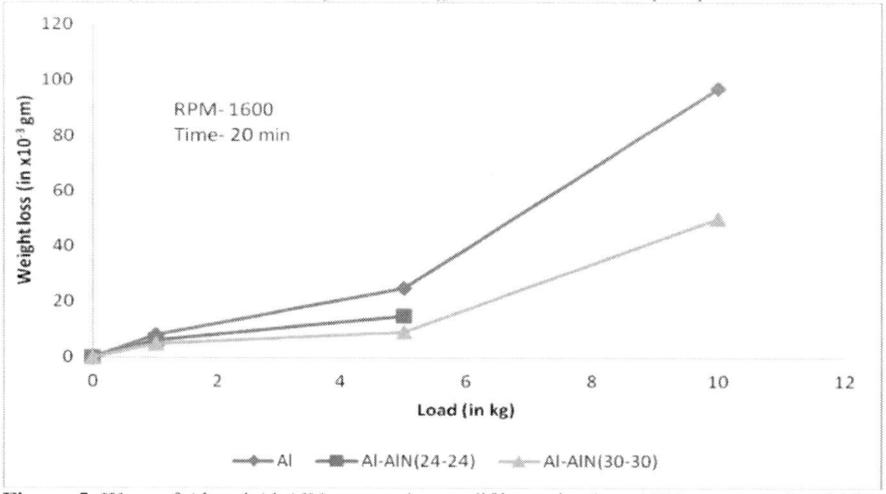

Figure 5. Wear of Al and Al-AlN composites at different loads at 1600 rpm speed and 20 minutes.

Figure 5 shows that at 1600 rpm speed (Sliding velocity of 6.7 m/s) and 20 minutes the weight loss in all samples is much higher than that seen at lesser rpm and at lower times. The weight loss in all samples increases with increasing load. The wear in all samples drastically increases when the load is increased to 10kg as compared to the wear at 5 kg load. The Al-AlN (24-24) composite, sample-1, breaks when loaded to 10kg. Al-AlN (30 -30) composite, sample-2, shows higher wear resistance compared to both aluminium and Al-AlN (24-24) composite, sample-1, at all loads.

5. Conclusions

Al-AlN Metal Matrix Composites could be fabricated in-situ by treating molten aluminium with a mixture of CaO and NH_4Cl and then cast. The composites produced had better mechanical properties like higher hardness and higher wear resistance than aluminium metal. Hardness of the Al-AlN composites increases with addition of more NH_4Cl + CaO, which indicates hardness increases with the increase in the amount of AlN present in the composite. A maximum hardness of 58.34 kgf/mm^2 was achieved in Al-AlN (30-30) composite, sample-2, which was produced by treatment of NH_4Cl + CaO with Al at temp 930^0 C and 2 hours 30 minutes holding period. Increasing the melt temperature, holding period after treatment and periodically stirring lead to better mechanical properties of the composites. Wear resistance was found to increase with higher amount of NH_4Cl + CaO added as compared to Al.

Acknowledgement

The authors thank Director, IGIT Sarang for providing necessary facilities.

References

[1] Fale S, Likhite A and Bhatt J 2011 *Trans. of Indian Inst. of Metals* **64** 111-15
[2] Lindroos V and Talvitie M 1995 *J. Mater. Process. Technol.* **53** 273
[3] Bandyopadhyay N, Ghosh S and Basumallick A 2007*Mater. Manuf. Processes* **22** 679
[4] Hashim J, Looney L and Hashmi M, 2001 *J.Mater. Process.Technol.***119** 324
[5] Tsunekawa Y, Suzuki H and Genma Y 2001 *Mater. Des.* **22** 467
[6] Roy D, Ghosh S, Basumallick A and Basu B 2007 *J. Alloys Compd.* **436** 107
[7] Lee K, Sim H and Kwon H, 2006 *J. Mater. Sci.* **41** 6347
[8] Rosenband V and Gany A 2004 *J. Mater.Process. Technol.* **147** 197
[9] Narula C, Czubarow P and Seyferth D 1998 *J. Mater. Sci.* **33** 1389
[10] Hou Q, Mutharasan R and Koczak M 1995 *Mater. Sci. Eng. A* **195** 121
[11] Chermant J and Mordik B, 1997 *J. Eur. Ceram. Soc.* **17** 1867
[12] Chung S, Yu W and Lin C 1999 *J. Mater. Res.* **14** 1928
[13] Rosenband V and Gany A 2007 *Exp. Therm Fluid Sci.* **31** 461
[14] Far hat Z, Ding Y, Northwood D and Alpas A 1996 *Mater. Sci. Eng. A* **206** 302
[15] Pradhan S, Jena S K, Sahu A, Panda S, Bhuyan P K and Pattnaik A, 2012 *B. Tech Project report on Development of Al-AlN metal matrix composite by nitrogenation*

AUTHOR INDEX